AF361939

Photovoltaics and Electrification in Agriculture

Photovoltaics and Electrification in Agriculture

Editors

Miguel-Ángel Muñoz-García
Luis Hernández-Callejo

MDPI • Basel • Beijing • Wuhan • Barcelona • Belgrade • Manchester • Tokyo • Cluj • Tianjin

Editors
Miguel-Ángel Muñoz-García
Ingeniería Agroforestal
Universidad Politécnica de
Madrid
Madrid
Spain

Luis Hernández-Callejo
Ingeniería Agrícola y Forestal
Universidad de Valladolid
Soria
Spain

Editorial Office
MDPI
St. Alban-Anlage 66
4052 Basel, Switzerland

This is a reprint of articles from the Special Issue published online in the open access journal *Agronomy* (ISSN 2073-4395) (available at: www.mdpi.com/journal/agronomy/special_issues/Photovoltaics-Agriculture).

For citation purposes, cite each article independently as indicated on the article page online and as indicated below:

LastName, A.A.; LastName, B.B.; LastName, C.C. Article Title. *Journal Name* **Year**, *Volume Number*, Page Range.

ISBN 978-3-0365-3290-5 (Hbk)
ISBN 978-3-0365-3289-9 (PDF)

Contents

About the Editors

Miguel-Ángel Muñoz-García

Prof. Dr. Muñoz-García is professor at the ETSI Agronomic, Food and Biosystems at UPM (ETSIAAB), PhD and Telecommunications Engineer. Research career spanning more than 20 years, including private companies (head of R&D projects, in SEPSA (ALBATROS), and in Thyssenkrupp), research centers (CIEMAT) and public universities (UC3M and UPM). Devoted to Photovoltaics, Electronics for Agriculture and Sensors/Control systems at LPF-TAGRALIA Research Group. Holder of two patents in operation (and another under study). Stays at the Beijing Institute of Technology (BIT) in China and at the CEA-INES (Atomic Energy Commission-National Solar Energy Institute) in France. Deputy Director at ETSIAAB for Communication and IT Resources (since 2021) and previously for Communication and Media (2018 to 2021), for Infrastructure and Economic Affairs (2013-2016) and Department Secretary (2011-2013). Committee Member of Technical Committee for Standardization: CTN 068 agricultural and forestry machinery on behalf of UNE for ISO/TC23/SC19/WG, Scientific Committees of the European Photovoltaic Solar Energy Conference and Exhibition, and of the Iberian Congress of Agroengineering.

Luis Hernández-Callejo

Prof. Dr. Luis Hernández-Callejo is Electrical Engineer at Universidad Nacional de Educación a Distancia (UNED, Spain), Computer Engineer at UNED and PhD at Universidad de Valladolid (Spain). Professor and researcher at the Universidad de Valladolid. His areas of interest are: renewable energy, microgrids, photovoltaic energy, wind energy, smart cities, artificial intelligence. Prof. Dr. Hernández Callejo is an editor in numerous scientific journals, and is a guest editor in many special issues. He has directed four Doctoral Theses, and at the moment, he is directing six Doctoral Theses. He is a professor in wind energy, solar energy, microgrids, and he collaborates with many universities in Spain and in the rest of the world.

Preface to "Photovoltaics and Electrification in Agriculture"

Dear Colleagues,

According to the United Nations sustainable objectives, agriculture is the single largest employer in the world, sustaining the livelihoods of 40 percent of the world's population, many of whom continue to live in poverty. Agriculture uses a large amount of water, which in turn requires a lot of energy to be transformed to the point where it can actually be used. Such energy is usually electric, which implies a great economic cost and also greenhouse gas emissions, since it is usually of non-renewable origin.

However, photovoltaic energy is a great opportunity to reduce both costs and emissions, even more so with the drop in prices that has occurred in recent years, reaching prices per watt of less than €0.5. In this new situation, numerous opportunities for the use of photovoltaic energy appear in agricultural applications.

This Special Issue is focused on applications, uses, and research related to photovoltaic solar energy and agriculture, both in energy generation in rural areas for agricultural uses, and in its use, problems, and opportunities. Novel works related to new discoveries to known problems will be accepted, as well as analyses of opportunities and improvements in photovoltaic systems and elements used for agriculture.

Miguel-Ángel Muñoz-García, Luis Hernández-Callejo
Editors

agronomy 

Editorial

Photovoltaics and Electrification in Agriculture

Miguel A. Muñoz-García [1,*] and Luis Hernández-Callejo [2]

1 Department of Agroforestry Engineering, Escuela Técnica Superior de Ingeniería Agronómica, Alimentaria y de Biosistemas, Universidad Politécnica de Madrid, Av. Puerta de Hierro 2, 28040 Madrid, Spain
2 Departamento de Ingeniería Agrícola y Forestal, Universidad de Valladolid, 42004 Soria, Spain; luis.hernandez.callejo@uva.es
* Correspondence: miguelangel.munoz@upm.es; Tel.: +34-9-1067-0968

1. Introduction: The Importance of Electricity for Agriculture

The editorial introduces a Special Issue entitled "Photovoltaics and Electrification in Agriculture". Agriculture requires not only tillage and fertilization but also water supply and, in some cases, heating and cooling. These needs go hand in hand with the use of energy, which, increasingly, is electrical energy. An option that has dropped a lot in price in recent years is photovoltaic energy. This type of energy has experienced an explosion in terms of its expansion worldwide and has been revealed as a viable solution to rapidly increase the electrical power of non-fossil origin. However, the use of panels must compete with the use of the soil for cultivation, and in many cases, it could displace the use of the soil for cultivation, something that would not be desirable either from a production point of view or from an ecological point of view. For this, a new concept of soil sharing for crops and energy production is being developed in what is called "agrovoltaics". This shared production model is analyzed in this document. In addition, the electrification of agriculture allows the introduction of elements, such as sensors, the IoT, and intelligent control. The internet connection opens the doors to technologies such as those based on data, digital control, and what is called precision agriculture, both for cultivation in greenhouses and for regular cultivation. This would not be possible without an electrical energy source that allows powering the inter-connected elements, photovoltaics being the best candidate again. However, above all, we must not forget the issue of CO_2 emissions due to the use of energy in agriculture. In this sense, photovoltaic energy can reduce the carbon footprint and provide one of the cheapest energy sources available. All these topics are analyzed in this Special Issue, focusing on photovoltaics and its uses and impact on agriculture.

Energy is inherent in ancient and modern agriculture, in one form or another. It is necessary for tillage, for pumping, and for the transport and transformation of the products. Modern agriculture manages to produce enough food for the growing world population (even though it is not correctly distributed, and this generates famines in many areas of the planet). This is possible thanks not only to the use of new agriculture techniques but also to the help of the energy necessary to carry out these techniques. In most cases, the origin of this energy could be electrical, even more and more in transport, with electric vehicles, and without any doubt, for pumping, heating, cooling and in the powering of the control systems associated to the new agriculture based on expert systems.

The origin of the energy must be renewable to guarantee the achievement of neutrality of greenhouse gas emissions and therefore, probably the most important source will be photovoltaic energy. According to Eurostat [1] areas such as EU, agriculture poses around 3.3% of the final energy consumption. From it, 55% comes from fossil fuel sources, despite the global consumption decreased by 8.1% in the past two decades. Besides this, it must be pointed out that globally, electricity is only around 20% of the energy consumed, but it has been increasing in the past decades, when it was less than 10% fifty years ago, according to IEA [2].

Electricity also makes it possible to technify crops in so-called precision agriculture. In this way, agriculture optimizes the resources used, reduces the carbon footprint and the environmental impact, and manages to produce more, better, and more efficiently. Precision agriculture is the evolution of agriculture towards sustainability. All of this requires efficient technology and energy with the least possible impact, such as photovoltaic energy.

2. Special Issue Overview: General Topics

2.1. The Use of Photovoltaics in Greenhouses

The new context of climate change, in turn, generates new forms of production where the environmental parameters in which it takes place must be controlled more and better. The paradigm of this type of system is greenhouse cultivation. In this type of cultivation, not only the interior temperature is controlled, or parameters such as relative humidity, the amount of CO_2 and the nutrients applied in the case of hydroponic cultivation but also the amount of radiation that reaches the plants. In countries with high insolation, shading is used either by means of meshes or by "liming" the roof. Photovoltaic solar energy as an energy source in greenhouses can also be used as a shading element for certain crops. However, the analysis of the impact of the solar panels on the crop is a topic being studied nowadays and a topic of great interest for the future development of solar energy applied to greenhouses.

The use of alternative systems, such as semi-transparent photovoltaic panels, can add value to greenhouse cultivation. This type of panel lets a part of the radiation pass through, as it does not include a reflective layer on the back. Consequently, the panel may lose a small percentage of efficiency, which is offset by the benefits to the crop. In the work of Aira et al. [3], the influence of radiation loss on production and crop quality is analyzed. In their conclusions, the shaded crop adapted to the new conditions, partially compensating for the decrease of radiation.

The application of agrovoltaic technology in arid areas, such as the southwestern United States, using organic solar panels, was carried out by Waller et al. [4] in a study exploring the influence of certain photovoltaic materials that do not completely block so-called photo active radiation (PAR), applied to hydroponic cultivation under greenhouse, something that makes it a very interesting case and with a great future for its high efficiency not only in terms of energy but also in the application of water and nutrients. Their conclusions show that, as in other cases of similar studies, the crop adapts to the decrease in radiation so that finally, the amount of harvest by weight is similar.

In greenhouse cultivation, fixed-plane surfaces are available with a certain orientation. The estimation of radiation on such plane, depending on the location, must be carefully analyzed to determine the impact and viability on the crop and on the photovoltaic production itself, something that has been considered in the work of Díez et al. [5], applying new methodology, including anisotropic models of the sky. The model is able to calculate the incidence angle of the radiation at any time besides the irradiation of a considered placement, with increased accuracy. As one of the conclusions, it allows predicting the hourly radiation reaching the crop and then the expected production.

2.2. Viability of Photovoltaics Coexisting with Traditional Agriculture

An agrovoltaic system has certain benefits in terms of the temperature of both the panels and the crop itself. In the study by Othman et al. [6], it is analyzed how the crop itself, with its evapotranspiration, reduces the temperature of the solar panels, which, as is known, is the main responsible for their loss of efficiency. Likewise, the placement of panels reduces the evapotranspiration itself and therefore, the hydric stress of the crop, especially in areas with high solar radiation. All these topics are currently the object of analysis and research, and each new study contributes to the improvement of the knowledge in this area.

The areas in countries with desert climates in Africa and the Middle East present peculiarities that make the analysis of the viability of photovoltaic systems on crops interesting. The case of Africa is especially interesting since, on the one hand, climate change can affect this continent much more intensely and, on the other, population growth is significantly higher than on the rest of the planet. Therefore, optimizing the use of resources is extremely necessary in order to achieve the necessary social and political stability for a region where the only option many young people find is to emigrate. The Srijana et al. study [7] shows how traditional farming systems based on the use of diesel for irrigation and those that only use rainwater are at a clear disadvantage compared to agrovoltaic systems that use solar energy for irrigation. Moreover, the concept of land equivalent ratio (LER) is calculated with positive results that demonstrate how the combination of photovoltaics and agriculture results in better use of land.

In the Moreda et al. study [8], an exhaustive theoretical study was carried out, including nine types of rotating crops on a surface of 24 ha on the hypothesis of crops in the southern area of Spain, one of the most representative areas of Mediterranean agriculture. Two systems for obtaining irrigation water were compared: Surface water and extracted from wells. The conservative analysis of the profitability of the shared use of the land for solar panels and rotary cutters was carried out in periods of four years. The conclusion is moderately optimistic, however, the rise in fossil electricity prices, as well as the emission reduction targets, make the results more than reasonable.

Not only the regions that until now were classified as arid, but also those semi-arid or even those that were less warm, are affected by the increase in temperatures due to climate change as well as the decrease in average rainfall. This forces other regions like the rest of the Mediterranean areas to adapt their crops to this new situation.

The economic feasibility of agrovoltaic systems is an issue depending not only on the type of crop and photovoltaic system but also on the area of the analysis. Pascaris et al. [9] analyzed the agrovolvaic systems from a human point of view, interviewing north-American farmers and detecting in their conclusions how farmers are concerned not only with the present situation but also the long-term issues like productivity, market potential, just compensation, and system flexibility, in order to decide to implement or not the new energy systems over their crops. Although the lifetime of the systems has been extended to 25 or 30 years, and the reliability has been increased, there are other considerations related to the market that can affect the profitability and therefore, the viability of these systems. However, this is an issue that affects not only agrovoltaic systems but everything that concerns the agricultural market in general, generating increasing uncertainties that do not have to do with technology itself but also with the geopolitical situation and global market tensions.

We can also find studies in areas with high humidity, such as central Europe. In the study by Wexelex et al. [10], trials were carried out for two different years with the cultivation of celery under solar panels in irrigated cultivation. Although the years were not conducive to the analysis, it can be observed that in any case, the agrovoltaic system does not pose a significant problem for the crop, which would allow concluding that the use of the soil can be shared with the consequent improvement in efficiency of land use.

2.3. Precision Agriculture and Photovoltaics

The electrification of the tasks associated with agriculture encompasses not only the more traditional methods, such as tillage or pumping, but also those related to the so-called precision agriculture.

There are many activities that benefit from the application of intelligent systems, such as selective spraying and fertilization, which can be applied with robots that include an internet connection (internet of things IoT), and which benefit from autonomy that provides them with a photovoltaic power system like the one proposed by Chand et al. in their study [11]. The document concludes that this field still has a long way to go and that there are still few experiences, which is why it is an area of great interest for research and the application of results to the automation of cultivation tasks and the improvement of precision in agriculture.

In agricultural applications, an important aspect is the remoteness from the connection points when implementing the Internet of Things (IoT), which influences the energy consumption of transmission systems and forces to study strategies to optimize the use of data and power. This aspect is addressed by Swain et al. in their work [12], where they study a long-range system applying Matlab algorithms. In their conclusions, they determine the improvement that occurs by including hybrid systems in terms of range and then addressing real-time systems. All this is applicable to the new generation of agriculture 4.0.

2.4. Research in Electrification Applied to Rural Areas

Not only photovoltaics is covered in this Special Issue, but also cases of research in electricity production and transport applied to rural areas. This is the case of Pindado et al. in the work about the importance of a precise selection of the protections as fuses [13] in the context of a rapid growth of renewable connected plants with a distributed scheme. In this kind of topology, the fault of one of the plants can affect the others and, with the increasing number of plants, the adequate selection of the fuses to operate when necessary is of vital importance.

Rural electrification is not yet complete in many areas, such as areas in Palestine, limiting local development that is fundamentally agronomic. The contribution of photovoltaic energy to solve the problem of rural electrification is crucial due to lower prices, modularity, and increased reliability, something that is reflected in the Ibrik study [14], something also reinforced by the low impact on greenhouse gas emissions of this solution (which we remember is not null, since it exists in the manufacture, installation, and dismantling of the plant). In addition, electrical micro-grids are a boost both to local development in rural areas and to the elimination of the need for new power lines.

3. Conclusions

The electrification of the tasks associated with agriculture is an unstoppable and necessary process. The use of photovoltaic energy will be one of the energy sources with the greatest impact on this process. In the context of the decarbonization of all sectors, including agricultural and livestock production and the rural world, photovoltaic energy is undoubtedly one of the mechanisms with the easiest implementation.

In summary, the abovementioned works demonstrate in their analysis and conclusions that:

- Photovoltaic energy is the most competitive electrical energy option for the agricultural sector at the present time due to the drastic drop in component prices.
- The cultivation can be developed under photovoltaic panels coexisting in the so-called "agrovoltaics" with an increase in profitability or land use.
- Precision agriculture or agriculture 4.0, based on the Internet of Things (IoT), benefits from the use of photovoltaic solar energy for its purposes.
- Greenhouses are an ideal agricultural production system for the integration of photovoltaic panels of different technologies, such as organic, semi-transparent, or amorphous silicon panels.

Funding: This research received no external funding.

Informed Consent Statement: Not applicable.

Conflicts of Interest: The authors declare no conflict of interest.

References

1. EuroStat. Share of Energy Consumption by Agriculture in Final Energy Consumption, EU, 2009 and 2019. 2021. Available online: https://ec.europa.eu/eurostat/statistics-explained/index.php?title=Agri-environmental_indicator_-_energy_use (accessed on 29 October 2021).
2. IEA. *Global Share of Total Final Consumption by Source*; IEA: Paris, France, 2018; Available online: https://www.iea.org/data-and-statistics/charts/global-share-of-total-final-consumption-by-source-2018 (accessed on 29 October 2021).

3.   Aira, J.-R.; Gallardo-Saavedra, S.; Eugenio-Gozalbo, M.; Alonso-Gómez, V.; Muñoz-García, M.; Hernández-Callejo, L. Analysis of the Viability of a Photovoltaic Greenhouse with Semi-Transparent Amorphous Silicon (a-Si) Glass. *Agronomy* **2021**, *11*, 1097. [CrossRef]
4.   Waller, R.; Kacira, M.; Magadley, E.; Teitel, M.; Yehia, I. Semi-Transparent Organic Photovoltaics Applied as Greenhouse Shade for Spring and Summer Tomato Production in Arid Climate. *Agronomy* **2021**, *11*, 1152. [CrossRef]
5.   Diez, F.; Martínez-Rodríguez, A.; Navas-Gracia, L.; Chico-Santamarta, L.; Correa-Guimaraes, A.; Andara, R. Estimation of the Hourly Global Solar Irradiation on the Tilted and Oriented Plane of Photovoltaic Solar Panels Applied to Greenhouse Production. *Agronomy* **2021**, *11*, 495. [CrossRef]
6.   Othman, N.F.; Yaacob, M.E.; Su, A.S.M.; Jaafar, J.N.; Hizam, H.; Shahidan, M.F.; Jamaluddin, A.H.; Chen, G.; Jalaludin, A. Modeling of Stochastic Temperature and Heat Stress Directly Underneath Agrivoltaic Conditions with Orthosiphon Stamineus Crop Cultivation. *Agronomy* **2020**, *10*, 1472. [CrossRef]
7.   Bhandari, S.N.; Schlüter, S.; Kuckshinrichs, W.; Schlör, H.; Adamou, R.; Bhandari, R. Economic Feasibility of Agrivoltaic Systems in Food-Energy Nexus Context: Modelling and a Case Study in Niger. *Agronomy* **2021**, *11*, 1906. [CrossRef]
8.   Moreda, G.; Muñoz-García, M.; Alonso-García, M.; Hernández-Callejo, L. Techno-Economic Viability of Agro-Photovoltaic Irrigated Arable Lands in the EU-Med Region: A Case-Study in Southwestern Spain. *Agronomy* **2021**, *11*, 593. [CrossRef]
9.   Pascaris, A.S.; Schelly, C.; Pearce, J.M. A First Investigation of Agriculture Sector Perspectives on the Opportunities and Barriers for Agrivoltaics. *Agronomy* **2020**, *10*, 1885. [CrossRef]
10.   Weselek, A.; Bauerle, A.; Zikeli, S.; Lewandowski, I.; Högy, P. Effects on Crop Development, Yields and Chemical Composition of Celeriac (Apium Graveolens L. Var. Rapaceum) Cultivated Underneath an Agrivoltaic System. *Agronomy* **2021**, *11*, 733. [CrossRef]
11.   Chand, A.; Prasad, K.; Mar, E.; Dakai, S.; Mamun, K.; Islam, F.; Mehta, U.; Kumar, N. Design and Analysis of Photovoltaic Powered Battery-Operated Computer Vision-Based Multi-Purpose Smart Farming Robot. *Agronomy* **2021**, *11*, 530. [CrossRef]
12.   Swain, M.; Zimon, D.; Singh, R.; Hashmi, M.F.; Rashid, M.; Hakak, S. LoRa-LBO: An Experimental Analysis of LoRa Link Budget Optimization in Custom Build IoT Test Bed for Agriculture 4.0. *Agronomy* **2021**, *11*, 820. [CrossRef]
13.   Pindado, S.; Alcala-Gonzalez, D.; Alfonso-Corcuera, D.; del Toro, E.G.; Más-López, M. Improving the Power Supply Performance in Rural Smart Grids with Photovoltaic DG by Optimizing Fuse Selection. *Agronomy* **2021**, *11*, 622. [CrossRef]
14.   Ibrik, I. Micro-Grid Solar Photovoltaic Systems for Rural Development and Sustainable Agriculture in Palestine. *Agronomy* **2020**, *10*, 1474. [CrossRef]

 *agronomy*

Article

Semi-Transparent Organic Photovoltaics Applied as Greenhouse Shade for Spring and Summer Tomato Production in Arid Climate

Rebekah Waller [1], Murat Kacira [1,*], Esther Magadley [2], Meir Teitel [3] and Ibrahim Yehia [2]

1   Department of Biosystems Engineering, The University of Arizona, Tucson, AZ 85721, USA; rebekahewaller@email.arizona.edu
2   Triangle Research and Development Center, P.O. Box 2167, Kfar-Qari 30075, Israel; esther.magadley@gmail.com (E.M.); yehia1002@gmail.com (I.Y.)
3   Institute of Agricultural Engineering, Agricultural Research Organization, The Volcani Center, HaMaccabim Road 68, P.O. Box 15159, Rishon LeZion 7528809, Israel; grteitel@volcani.agri.gov.il
*   Correspondence: mkacira@arizona.edu

Abstract: Recognizing the growing interest in the application of organic photovoltaics (OPVs) with greenhouse crop production systems, in this study we used flexible, roll-to-roll printed, semi-transparent OPV arrays as a roof shade for a greenhouse hydroponic tomato production system during a spring and summer production season in the arid southwestern U.S. The wavelength-selective OPV arrays were installed in a contiguous area on a section of the greenhouse roof, decreasing the transmittance of all solar radiation wavelengths and photosynthetically active radiation (PAR) wavelengths (400–700 nm) to the OPV-shaded area by approximately 40% and 37%, respectively. Microclimate conditions and tomato crop growth and yield parameters were measured in both the OPV-shaded ('OPV') and non-OPV-shaded ('Control') sections of the greenhouse. The OPV shade stabilized the canopy temperature during midday periods with the highest solar radiation intensities, performing the function of a conventional shading method. Although delayed fruit development and ripening in the OPV section resulted in lower total yields compared to the Control section (24.6 kg m^{-2} and 27.7 kg m^{-2}, respectively), after the fourth (of 10 total) harvests, the average weekly yield, fruit number, and fruit mass were not significantly different between the treatment (OPV-shaded) and control group. Light use efficiency (LUE), defined as the ratio of total fruit yield to accumulated PAR received by the plant canopy, was nearly twice as high as the Control section, with 21.4 g of fruit per mole of PAR for plants in the OPV-covered section compared to 10.1 g in the Control section. Overall, this study demonstrated that the use of semi-transparent OPVs as a seasonal shade element for greenhouse production in a high-light region is feasible. However, a higher transmission of PAR and greater OPV device efficiency and durability could make OPV shades more economically viable, providing a desirable solution for co-located greenhouse crop production and renewable energy generation in hot and high-light intensity regions.

Keywords: organic photovoltaics; greenhouses; tomato; shading; arid region

Citation: Waller, R.; Kacira, M.; Magadley, E.; Teitel, M.; Yehia, I. Semi-Transparent Organic Photovoltaics Applied as Greenhouse Shade for Spring and Summer Tomato Production in Arid Climate. *Agronomy* **2021**, *11*, 1152. https://doi.org/10.3390/agronomy11061152

Academic Editors: Miguel-Ángel Muñoz-García and Luis Hernández-Callejo

Received: 18 May 2021
Accepted: 31 May 2021
Published: 4 June 2021

Publisher's Note: MDPI stays neutral with regard to jurisdictional claims in published maps and institutional affiliations.

1. Introduction

Greenhouse-integrated photovoltaic (PV) technologies are increasingly seen as a promising solution for sustainable greenhouse agriculture. Higher annual yields and lower water consumption compared to conventional farming make greenhouse production particularly attractive for space-limited and water-limited regions. However, the climate control systems and other electrical components involved in greenhouse operations consume large amounts of energy [1,2]. PV systems that are structurally integrated with the greenhouse enable the co-production of renewable energy and crops on the same land footprint, which is advantageous from a resource-use efficiency perspective [3].

Different types of PV technologies have been tested for integration with greenhouse systems. One design strategy for PV integration with greenhouses involves the positioning of conventional opaque PVs (conv-PVs) along the greenhouse structure with gaps between them, thereby allowing a fraction of sunlight to enter the greenhouse. The challenge in this is ascertaining the correct percentage of roof coverage, balancing PV electricity generation with adequate lighting conditions inside the greenhouse [4,5]. In order to provide more control of the lighting conditions in the greenhouse with integrated conv-PVs, some have proposed designs with dynamically-controlled PV blind systems that can respond to outdoor conditions and sunlight variations [6–8].

Thin-film semi-transparent or transparent PV technologies (STPV), despite showing relatively lower efficiencies compared to conv-PV, are gaining attention for their application to greenhouse structures. In contrast to conv-PV, STPV absorbs only a portion of incident light for electricity generation, leaving unabsorbed photons available for plant photosynthesis [9,10]. Recent advances in fabrication methods have enabled significant progress in thin-film STPV development [11]. There are many types of technologies within this emerging class of PV (e.g., copper indium gallium selenide, cadmium telluride, amorphous silicon, perovskite, dye-sensitized solar cells (DSSCs), organic (OPV), and hybrid cells), but all share the advantages of relatively low manufacturing, transportation, and installation costs, and overall preferable life cycle characteristics compared to conv-PVs that are deposited on thick rigid substrates [12–14]. For greenhouse applications, STPVs could potentially be deployed in larger coverage areas than would otherwise be acceptable with conv-PVs in terms of shading due to their transparency. From an electricity generation standpoint, this design advantage could compensate for relatively lower efficiency of STPVs compared to conv-PVs.

Of the STPV technologies, OPV has a number of material properties that make it uniquely promising technology for integration with greenhouses, including solution processability, spectral tuneability, and mechanical flexibility. Additionally, with regard to production, the solution processability of OPVs means that OPV device fabrication can be implemented with roll-to-roll printing technologies, thus enabling large-scale, continuous production at a fraction of the capital cost and energy input of conv-PV systems, and still lower than other STPV alternatives [15,16]. Environmentally, OPVs are considered preferable even to other STPV technologies, given the fact that OPVs' material components are largely sourced from abundant earth materials and are recyclable/recoverable [13].

The spectral properties of OPVs are important to consider for integration into any transparent or translucent surfaces (e.g., windows, screens, greenhouse covers) in order to determine the effect on transmitted light. A key figure of merit for any STPV technologies is the light utilization efficiency (LUE), equal to the product of the power conversion efficiency (PCE) and the average visible transmission (AVT), which is the weighted transmission spectrum of the STPVs against the photopic response of the human eye [17]. For greenhouse applications, the plant response to light transmitted through the OPV material is of greater consequence than the human eye response. This issue was addressed by Emmott et al. [18] in a modeling study of the techno-economic potential for OPV-integrated greenhouses, in which a crop growth factor (G) was used instead of AVT in an optical model that quantified the effect of various spectrally-selective OPV materials on greenhouse crops. This metric G considered the relative photosynthetic efficiency of the average crop plant at different light wavelengths, based on the work of McCree [19]. Based on this analysis, the authors demonstrated that attaining higher efficiencies and high transparency contact materials were the most important factors in determining OPV suitability for a greenhouse, rather than the absorbance or lack thereof of the OPV material in the photosynthetically active radiation (PAR) range, which includes light wavelengths between 400 to 700 nm.

In hot regions, excessive solar radiation during the summer season requires the application of shading methods (e.g., whitewashing, external shade netting, internal shade screens, transparent spray agents reflecting NIR light, etc.) in combination with the greenhouse ventilation/cooling systems to lower cooling loads and maintain desired growth conditions for the crop. Depending on the method, solar radiation transmittance can be decreased by 30–50% compared to non-shaded greenhouses [20]. For these locations, greenhouse-integrated OPVs could serve the dual purpose of electricity generation and shading. This potential role for greenhouse-integrated OPVs in high solar insolation regions was addressed by Okada et al. [5], in which lettuce yield and electricity production in an OPV-integrated greenhouse system were simulated in a hot, arid climate (Arizona). It was determined that OPV roof coverage ratios of 50% and 100% during the summer could extend the growing season due to the shading provided, and 49% OPV coverage was sufficient to meet the energy demands of the off-grid greenhouse modeled in the study, with the achievement of the desired lettuce crop yields. Likewise, in an energy-balance modeling analysis of an OPV-integrated greenhouse design conducted by Ravishankar et al. [21], assuming an OPV cell efficiency of approximately 10%, with 85% roof coverage of a glass greenhouse (219 m^2) covered by photoactive OPV area, it was shown that the OPV system was able to produce more than enough electricity to cover the energy requirements of a greenhouse tomato production system in an arid climate such as Arizona. Building on this work, Hollingsworth et al. [22] examined the life-cycle environmental and economic impacts of OPV-integrated greenhouses compared to non-PV-powered and conventional-PV-powered greenhouses and concluded that OPV-integrated greenhouses could outperform the alternatives if there were not significant reductions in crop yields due to OPV-related shading effects.

Evidently, the potential scope of the use of OPVs for greenhouse applications is in large part dependent on understanding the effects of the OPV light modification on the greenhouse microclimate and crop. The response of greenhouse crops to the lighting conditions resulting from the integration of STPV technologies with the greenhouse structure is an increasingly researched topic [23]. A number of these studies have experimented with tomato plants, a major greenhouse vegetable crop worldwide, which is often shaded in high-light regions to mitigate heat stress and fruit quality issues caused by high radiation intensities and air temperatures [24].

Li et al. [25] evaluated resource use efficiency, greenhouse microclimate, and crop yields in a greenhouse covered with glass laminated with luminescent solar concentrator-based PV cells. The PV system evaluated exceeded the energy demands of the 103 m^2 research greenhouse facility located in Tucson, Arizona, equipped with wet-pad and fan cooling system and a natural-gas-based heating system, growing cherry tomatoes with 25.3% more light use efficiency (LUE) in the PV-covered greenhouse compared to the control greenhouse that was covered with double-layer acrylic.

Hassanien et al. [10] investigated the effect of shading by mono-crystalline silicon semi-transparent panels mounted on top of a polycarbonate-covered greenhouse, occupying 20% of the roof area, on the growth of container tomatoes. It was determined that there were no significant differences in the growth of the tomato plants in the PV-integrated greenhouse compared to the unshaded greenhouse.

Ntinas et al. [26] tested a hydroponic tomato cultivation system for medium-sized and cherry varieties in two glass greenhouses, with one greenhouse outfitted with a DSSC cover and the other greenhouse serving as a control. A variety of plant physiological and productivity parameters were measured to determine the effects of the filtered lighting resulting from the integrated DSSC cover. Illuminance was reduced by 20% in the DSSC greenhouse compared to the control. Although the results showed that plants grown in the DSSC greenhouse were found to have relatively lower yields overall, it was found that the shading in the DSSC greenhouse during the summer season was beneficial for the qualitative characteristics of the tomato fruits.

Friman-Peretz et al. [27] compared the microclimate and tomato crop response between a polytunnel with OPVs installed on the roof, contributing 23% shading to the growing space, and a control greenhouse over two seasons in 2018 and 2019 in a Mediterranean climate. Air temperature and humidity levels inside the OPV greenhouse were found not to be significantly different compared to the control. Radiation distribution inside the OPV greenhouse was less homogenous than the control due to the gaps between the OPV strips deployed on the greenhouse roof. The cumulative yield and average fruit mass were higher in the OPV greenhouse in 2018 and not significantly different in 2019 when a 25% shade cloth was installed on the control greenhouse.

Until now, greenhouse studies evaluating the effects of integrated STPVs, including OPVs, on the tomato crop response and microclimate have experimented with limited/partial shading treatments (less than 25%). As mentioned previously, for greenhouses located in hot and sunny regions, higher shading may be required/desired during the spring and summer production seasons. In order to determine the feasibility of using semi-transparent OPVs as a realistic alternative for conventional greenhouse shading methods, it is critical to test higher shading treatments, which are more representative of typical cultural practices in such regions. Higher shading treatment also involves larger areal coverage by the OPV materials on/in greenhouses—the relatively low efficiency of OPVs compared to silicon PVs necessitates a larger coverage areas, which for this design application would also be desired in order to achieve high shading treatments. The present study addresses the lack of large-scale experimental research into the application of OPVs as a seasonal greenhouse shading method for hot and arid regions. Flexible, semi-transparent, roll-to-roll printed OPV arrays were installed as a roof shade cover on a greenhouse hydroponic tomato production system in a dry, desert region (southern Arizona) characterized by high air temperatures and radiation intensities in the spring and summer. Analyzing the effects of a relatively high OPV shading treatment on the greenhouse microclimate and tomato crop growth and yield, this study provides insights and recommendations for the design and application of greenhouse-integrated OPV shading elements.

2. Materials and Methods

2.1. Study Greenhouse

The study greenhouse was located at the University of Arizona Controlled Environment Agriculture Center in Tucson, Arizona (latitude: 32°16′ N, longitude: 110°56′ W, altitude: 728 m). Figure 1 shows a 3-D schematic of the greenhouse structure, which was developed in a Rhinoceros CAD environment (Robert McNeel & Associates, Seattle, WA, USA) [28]. The greenhouse had a 9.1 m × 14.6 m footprint with 1.8 m sidewalls and a height of 4.9 m at the roof apex, a gothic-arch roof profile, and a true north-south orientation (Golden Pacific Structures, Cincinnati, OH, USA). The greenhouse cover material was double-layer, air-inflated 8-mm low-density polyethylene (LDPE) plastic with a light transmittance of approximately 75% for all solar radiation wavelengths and 65% in the PAR range. The greenhouse climate control system (EnviroStep, Wadsworth, Arvada, CO, USA) controlled air temperature via evaporative cooling with a wet-pad and two exhaust fans positioned on the northern and southern walls, respectively. Two horizontal airflow (HAF) fans located close to the roof in the northwest and southwest quadrants were continuously operating (day and night) to facilitate air distribution throughout the growing space. External environmental conditions were monitored at a climate station positioned at the northern apex of the greenhouse roof, measuring ambient relative humidity and air temperature (HMP60, Vaisala, Helsinki, FI, USA) and horizontal shortwave irradiance with a pyranometer (SP-510, Apogee Instruments, Logan, UT, USA).

2.2. OPV Device Characterization and Installation on Greenhouse Roof

The OPV devices used in this study were PBTZT-stat-BDTT-8 based full solution coated, flexible, semi-transparent organic photovoltaic cells (manufactured by ARMOR Solar Films GmbH, formerly known as OPVIUS GmbH, Kitzingen, Germany). Each 800 mm × 1000 mm OPV panel was comprised of 4 serially-connected OPV modules, with each module containing of ten 12.5-mm-by-660-mm serially-connected cells, with 2.5-mm gaps ('dead area') between the cells; the active area (i.e., areas covered by the cells) constituted 75.8% of each panel. Eight OPV panels were laminated together and wired in parallel to form an OPV array that measured 6400 mm ×1000 mm × 0.6 mm and weighed approximately 6 kg, with a total active area of 3.4 m^2 per array. The OPV arrays had been deployed in a different design configuration on the greenhouse roof since October 2019 for a previous study characterizing the OPV array electrical performance; 6 OPV arrays were producing power at the time of measurement for the present study.

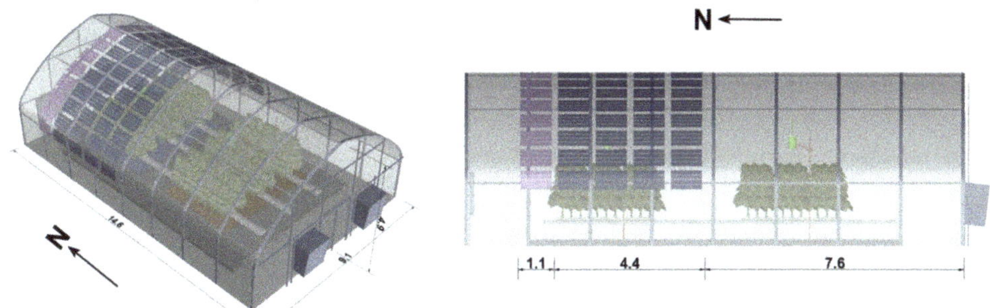

Figure 1. 3-D schematic of the study greenhouse from the southwest view (**left**) with the organic photovoltaics (OPV) shade element installed on the northern section of the greenhouse roof; the view from the west side of the greenhouse (**right**) shows the position of the OPV shade element; two additional OPV arrays (shown in pink) were installed on 11 June 2020. Dimensions are in meters.

The OPV arrays were positioned on the northern section of the greenhouse in a contiguous pattern to fully cover both the east and west greenhouse roof pitches and partially cover the sidewalls (Figure 1). Nylon cord was used to connect the corners of the opposite-facing OPV arrays at the greenhouse roof apex. The opposite end of each array was tied to the greenhouse base near the ground. Additionally, heavy-duty LDPE adhesive tape (GGR Supplies, Miami, FL, USA) was used to adhere the OPV arrays to the polyethylene cover. The OPV shade cover initially comprised eight OPV rolls. Two additional OPV arrays were installed on 11 June 2020 on the northern edge of the OPV shade cover to increase the shaded area, as the solar elevation angle had increased over the course of the study period (see Section 2.2). The total OPV area installed on the greenhouse roof was 51.2 m^2 initially, and then 64 m^2 with the two additional OPV rolls. Thus, the percentage of the OPV shade cover active area was 38.8% initially and then 48.5% with the additional two OPV arrays installed.

Electrical monitoring of the OPV arrays was conducted using an automated current-voltage (I-V) curve measurement system, which was programmed in the Python language. The system operated via serial communication protocol between a laptop (HP, Palo Alto, CA, USA) and a DC programmable electronic load device (8542B, B&K Precision Corporation, Yorba Linda, CA, USA). The OPV arrays were connected to the electronic load, and I-V curves of the OPV arrays were then taken in 10-min intervals during daylight hours (5:00–20:00 h), with the OPV array held at open-circuit between I-V curve measurements. The electronic load device used was capable of monitoring one OPV array at a time, and thus each OPV array was connected to the I-V curve measurement system for a one day of data collection and then subsequently disconnected.

2.3. Shading Effect of the OPV Arrays

Due to the changing solar zenith over the course of the study period, the position of the shading on the greenhouse floor resulting from the OPV shade cover shifted. Figure 2 illustrates the shifting position of the shadows caused by the overhead OPV shade cover on the greenhouse floor during the measurement period; the shadow positions were simulated using the Sun analysis feature in the Rhinoceros CAD program [28], which takes location, date, and time as inputs to calculate the solar angle, and then represents the sun at the calculated angle as a strong directional light cast on the 3-D geometry. The OPV planting area was positioned such that the central sampling region (see Figure 3) would be continuously shadowed by the OPV coverage for the duration of the measurement period. The northern section of the growing area that was shaded by the OPV is hereafter referred to as the 'OPV section'. The non-OPV shaded section is hereafter referred to as the 'Control section'. The Control section was initially unshaded for 98 days of the study, from 11 March 2020 to 10 June 2020, and then a 30% shade net was installed on the greenhouse roof where the control plants were located on 11 June 2020, enabling a comparison between the OPV shade cover and conventional shading method. The shade net resulted in a relatively proportional decrease in the transmittance of all solar radiation wavelengths in the Control section; its spectral properties are explained further in Section 3.3. The period in which the Control section was unshaded and the period in which the shade net was deployed are distinguished in the reported results.

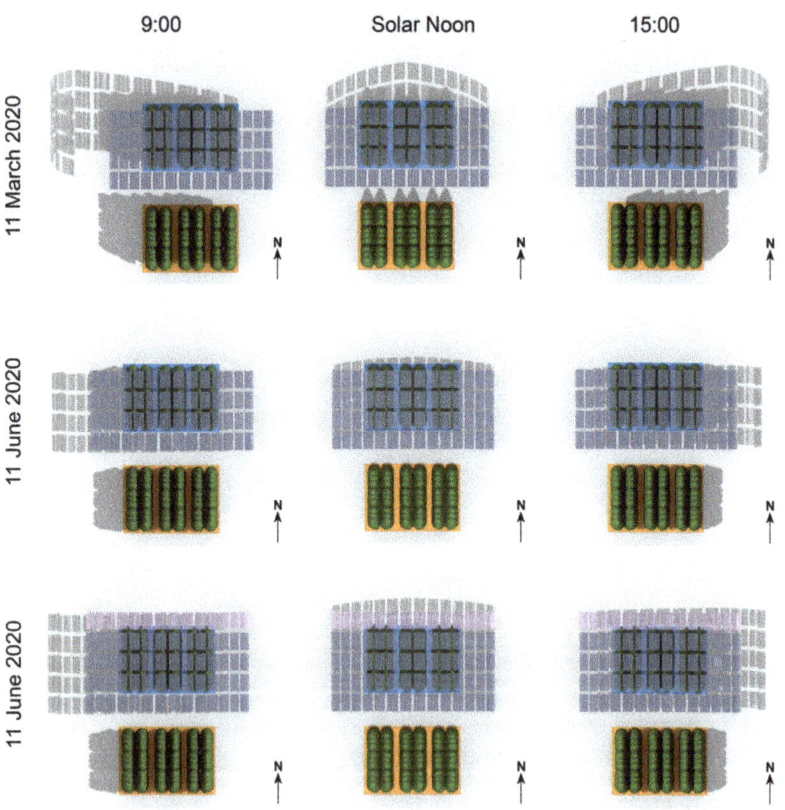

Figure 2. Simulated shadows cast by OPV shade element installed on greenhouse roof on the OPV planting area floor (highlighted in blue), showing the highest amount of southern exposure at the beginning of the measurement period (11 March 2020) and the highest amount of northern exposure (11 June 2020), when two additional OPV arrays were installed to increase coverage.

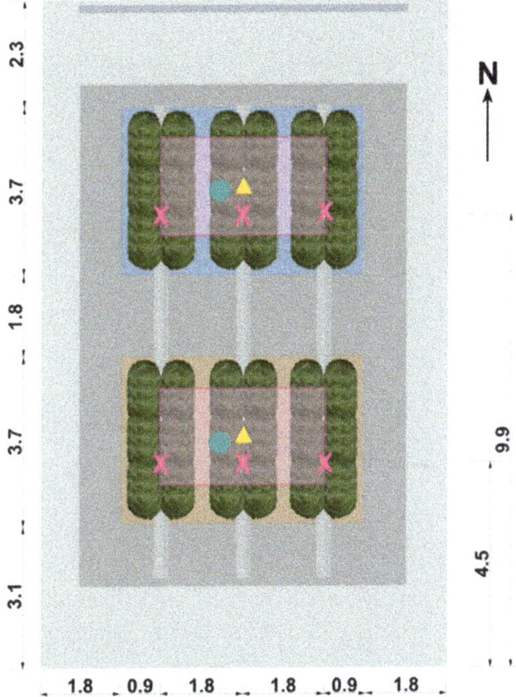

Figure 3. Top view of greenhouse growing area layout, with the OPV and Control planting sections highlighted in blue and orange rectangles (each 17.3 m^2), respectively. The pink rectangles represent the sampling region in each section. The positions of solar radiation sensors (red 'X'), relative humidity/air temperature sensors (blue circle), and infrared radiometers (yellow triangle) are shown. Dimensions are in meters.

2.4. Growing System and Microclimate Monitoring

Figure 3 shows a schematic view of the greenhouse growing area layout and positions of the environmental sensors. One hundred and eight grafted tomato seedlings ('Rebelsky' variety) were transplanted into the growing system on 5 March 2020, and crop data collection began on 11 March 2020. The tomato plants were planted in stonewool cubes fixed to stonewool slabs (GRODAN, Roermond, The Netherlands) in double rows along three gutters with a length of 9.7 m and a gradient of approximately 1%. The distance between the middle of the gutters was 1.8 m. An approximate 1.8-m gap separated the southernmost OPV-shaded plants and the northernmost control plants, with a planting area of 3.7 m × 5.4 m (17.3 m^2) for both the OPV and Control sections (see Figure 2). Each section contained 54 tomato plants, with 9 plants in each of the six rows, resulting in a planting density of 2.70 plants m^{-2} in each section. The growing system was designed and operated as a fully automated recirculating hydroponic system. The OPV and Control sections received the same irrigation schedule and nutrient solution (Hoagland's solution), with EC and pH levels controlled with an automated dosing unit (Intellidose, Autogrow, Auckland, NZ, New Zealand). For the measurement of plant transpiration, lysimeters were positioned to capture irrigation input and drainage from six plants located in the center of each section. The volume, electrical conductivity (EC), and pH of both irrigation input and drainage were measured each morning at around the same time with a handheld EC/pH meter (HI-9814, Hanna Instruments, Smithfield, RI, USA).

All sensor positions in the growing area are shown in Figure 3. Global shortwave irradiance was measured at the canopy-level using 6 shortwave pyranometers (SP-510, Apogee Instruments, Logan, UT, USA) positioned in the east, center, and west planting areas in both the OPV and Control sections. All solar radiation measurements reported for the OPV and Control sections are averaged values from the east, center, and west pyranometers in the two sections. One air temperature/humidity sensor (HMP60, Vaisala, Helsinki, FI, USA) was located centrally at the canopy-level in each section. Canopy temperature was continuously monitored from 14 May 2020 onward using two infrared radiometer sensors (SI-111-SS, Apogee Instruments, Logan, UT, USA) positioned in the center of the OPV and Control sections (Figure 3).

The measurement period for the crop spanned 126 days (18 weeks), concluding on 11 July 2020. Crop data were collected weekly for 12 randomly tagged sample plants— 6 plants per section, located in the inner four rows close to the center of each section (Figure 3). The growth parameters measured both vegetative and reproductive behavior in the crop, specifically head growth (i.e., terminal growth), stem diameter, leaf length, distance of flower growth to the bottom of the apical meristem (i.e., 'tip'), number of fruiting trusses, and the number of open and closed flowers. Yield data were also collected weekly beginning 62 days after transplant (DAT) once fruit had begun to ripen. Yield parameters included total fruit mass, the number of fruits, and average fruit mass. The independent two-sample Student t-test with a p-value of 0.05 was used to determine statistical significance in the growth and yield parameters measured between the OPV and Control sections.

Lighting quality was measured using a portable spectroradiometer (PS-300, Apogee Instruments, Logan, UT, USA) both outside the greenhouse and in the same positions as the pyranometers on two clear-sky days (20 and 22 May 2020) in the morning, midday, and afternoon, with additional measurements taken of the Control section once the shade net was installed. The spectroradiometer measurements indicate the intensity of shortwave radiation wavelengths between 300–1000 nm incident on the plant canopy in both the OPV and Control sections. Estimates of PAR, typically represented in μmol m^{-2} s^{-1}, incident on the crop canopy in the OPV and Control sections were calculated. These estimates were based on simultaneous measurements taken with a handheld full-spectrum quantum meter (MQ-501, Apogee Instruments, Logan, UT, USA) and a handheld shortwave pyranometer (MP-200, Apogee Instruments, Logan, UT, USA) taken in multiple locations under the OPV shade and under the polyethylene in the Control section. A generalized ratio for PAR to shortwave radiation was determined: in the OPV section, the ratio was 1.81 μmol m^{-2} s^{-1} of PAR for every 1 W m^{-2} of shortwave radiation; in the Control section, the ratio was 2.11 μmol m^{-2} s^{-1} of PAR for every 1 W m^{-2} of shortwave radiation. These ratios were applied to the continuous canopy-level shortwave radiation measurements to estimate canopy-level PAR, which is presented in the analysis as weekly average values. All continuously monitored conditions inside and outside the greenhouse were sampled every 15 s and averaged every 10 min, with average values recorded on two dataloggers (CR-3000 and CR-1000, Campbell Scientific, Logan, UT, USA).

3. Results and Discussion

3.1. Comparison of Microclimate in OPV and Control Sections

Figure 4 shows the average hourly fluctuations in air temperature, relative humidity, and canopy-level shortwave solar radiation in the OPV and Control sections of the study greenhouse for each month of the growing period. The average values for daytime and nighttime periods, along with outdoor conditions, are summarized in Table 1. The average daytime temperature was quite similar in the OPV and Control sections, differing by less than 0.5 °C on average for the duration of the growing period. Given the relatively short distance between the two planting sections and air circulation provided by the ventilation system and HAF fans, this result was expected. With the air temperature being relatively consistent across the OPV and Control sections, the relative differences in crop performance can be attributed mainly to the deployment of OPV on the greenhouse roof, and other environmental factors related to the OPV shade treatment.

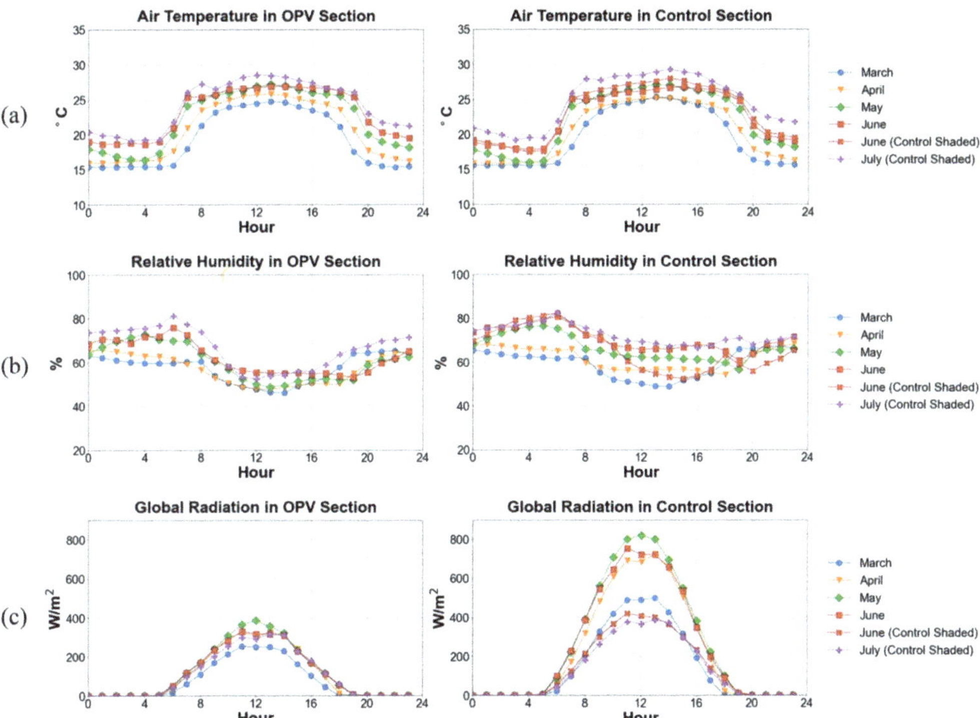

Figure 4. Average hourly values by month for (**a**) air temperature, (**b**) relative humidity, and (**c**) global radiation at the canopy level in the OPV (**left**) and Control (**right**) sections of the greenhouse during the growing period. Global radiation values are averaged from sensors on east, center, and west pyranometers in both OPV and Control sections. A 30% shade net was deployed on the Control section, beginning on 11 June 2020.

The difference in relative humidity (RH) levels between the OPV and Control sections was more pronounced: the OPV section had lower average RH compared to the Control section in both day and night periods for all months, with the magnitude of difference increasing in May and June, corresponding with the highest light-intensity period, until the shade net was installed on the Control section. The daily solar radiation in the OPV section was approximately 35–40% lower than the Control section until the shade net was deployed, at which point the OPV section had 10–15% lower light levels compared to the Control section. It has been established that there is a strong, linearly increasing relationship between greenhouse tomato plant transpiration and solar radiation incident above the crop canopy [29]. The microclimatic differences in relative humidity observed in the OPV and Control sections, which were larger when the Control section was unshaded, can be attributed to the relatively higher transpiration in the Control section canopy, which was also indicated by the direct measurement of transpiration conducted in this study (Table 2), as a result of the relatively higher solar radiation intensities to which it was exposed.

Table 1. Average daytime and nighttime measurements for air temperature, relative humidity, and canopy-level daily solar radiation sum in the OPV-shaded ('OPV') and non-OPV-shaded ('Control') sections of the greenhouse, and outside above the greenhouse. See Materials and Methods section for description of sensor locations.

		Month					
		March	April	May	June	June	July
Control Section Shade Net (Y/N)		N	N	N	N	Y	Y
Day air temperature ($°C$)	OPV	22.8	24.0	25.3	25.7	26.2	26.8
	Control	23.2	23.6	25.2	25.3	26.0	27.0
	Outside	19.2	24.6	30.8	32.2	34.9	35.2
Night air temperature ($°C$)	OPV	15.7	16.7	18.1	19.7	19.5	21.0
	Control	16.0	16.6	18.0	19.3	19.4	21.3
	Outside	12.3	15.0	21.1	24.3	25.1	28.3
Day relative humidity (%)	OPV	52.0	52.0	59.4	64.4	57.6	62.8
	Control	54.1	57.8	65.8	69.4	63.6	71.7
	Outside	38.1	22.4	15.7	17.7	13.4	25.2
Night relative humidity (%)	OPV	62.5	59.4	64.4	65.4	63.2	72.0
	Control	64.8	65.8	69.4	72.1	67.7	73.1
	Outside	64.2	42.1	30.2	28.2	22.5	35.8
Daily solar radiation ($MJ/m^2/day$)	OPV	6.48	9.32	10.7	9.76	9.86	9.14
	Control	13.6	20.4	22.7	21.4	12.5	11.1
	Outside	18.5	26.7	28.7	28.0	30.3	24.6

Table 2. Average monthly values for daily plant transpiration (mL plant^{-1} day^{-1}) in OPV and Control sections.

		Months				
		April	May	June	June	July
Control Section Shade Cloth (Y/N)		N	N	N	Y	Y
Daily Transpiration (mL plant^{-1} day^{-1})	OPV	1587	2264	3316	3101	2381
	Control	1680	2491	4105	3195	2382

Table 2 shows the average monthly values of daily plant transpiration (mL plant^{-1} day^{-1}) during the growing period. For the plants shaded by the OPV cover, the transpiration values were generally lower than the plants in the Control section. The differences in transpiration were higher in May (227 mL plant^{-1} day^{-1} higher in the Control section) and highest in June (789 mL plant^{-1} day^{-1} higher in the Control section); after installation of the shade cover on the Control section, the daily average transpiration of the Control section decreased by 910 mL plant^{-1} day^{-1}, which made the difference between the OPV and Control sections only 94 mL plant^{-1} day^{-1}. In July, the daily average difference in transpiration between the OPV and Control sections was virtually the same. One of the primary functions of shading devices on greenhouses in hot regions is to conserve water, which is spent in crop transpiration as a plant cooling mechanism, as well as in evaporative cooling systems (if used). The unshaded Control plants had 10.2% higher transpiration than the OPV-shaded plants between April and June; once the shade net was installed, the shaded Control plants had 6.9% higher transpiration than the OPV plants in June and July, indicating that water can be saved through greenhouse shading, and in this case, using OPV shade elements.

3.2. Effects of OPV Shading on Canopy Temperature

Figure 5 compares the average hourly canopy temperature of the OPV-shaded plants to that of the Control plants, revealing relatively larger diurnal fluctuations in canopy temperature in the Control section. The OPV shade cover appears to stabilize canopy temperature during the midday period, with the highest solar radiation intensities, whereas the temperature of the unshaded Control canopy increased during this period.

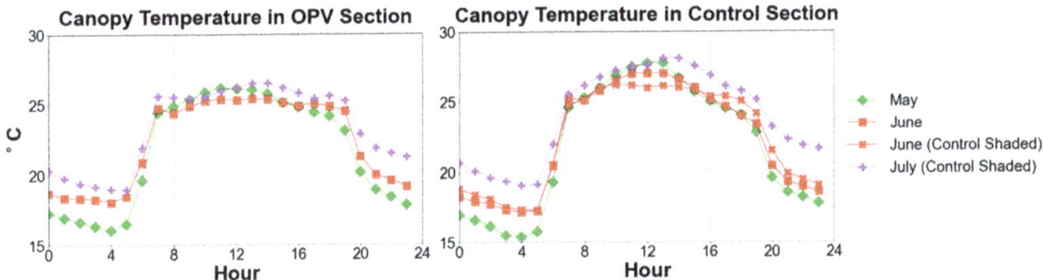

Figure 5. Average hourly values by month for canopy temperature (°C) in the OPV (**left**) and Control (**right**) sections of the greenhouse. A 30% shade net was deployed on the Control section, beginning on 11 June 2020.

Figure 6 presents these results in more detail, showing the canopy temperature, air temperature, and canopy-level solar radiation in the OPV and Control sections of the greenhouse for different environmental conditions in the month of June. For the clear sky day in which no shade net was installed on the Control section (3 June 2020), the canopy temperature in the Control section was 2–3 °C higher than the OPV section during the midday period. Although the air temperature was controlled to relatively consistent levels between the two sections throughout the day, higher solar radiation intensities in the Control section between 10:00–16:00 h increased the canopy temperature over the air temperature level, whereas the OPV-shaded canopy maintained relatively stable temperatures throughout the daylight hours and below the air temperature level. Two days later, when the sky was mostly cloudy, it can be seen that the canopy temperature differences were smaller between the two sections, with sudden increases in canopy temperature coinciding with increased solar radiation levels when the cloud-cover broke. During a clear-sky day later in June when the shade net was deployed over the Control section (23 June 2020), canopy temperatures in that section were more stable throughout the day, although still 1–2 °C higher than the OPV section during the midday period, coinciding with the largest difference in solar irradiance levels between the two sections (approximately 100 W m^{-2}) between 10:00–16:00 h.

Overall, these results, shown in Figure 6, illustrate the effect of solar radiation intensity on plant leaf temperature. Up to a certain point, given sufficient carbon dioxide (CO_2) levels, higher light levels can boost the photosynthetic rate. However, excessive radiation, and especially direct radiation, can also result in overheating of and subsequent tissue damage in the plant, and especially in fruit tissues (which are not as efficient in cooling via transpiration as leaves) [30]. One of the challenges of greenhouse climate control in a desert environment such as southern Arizona is managing the large diurnal outdoor climate fluctuations (e.g., low humidity, high solar radiation intensity, and cooler nights) in order to provide desirable and consistent microclimate conditions for the crop. Shading methods are an integral component of climate control strategies in desert greenhouses during the summer season, assisting ventilation/cooling systems in minimizing microclimate fluctuations [31]. It can be seen in Figure 6 that the OPV coverage served this function as a shade element for greenhouse tomato production during the high-light-intensity period and in the summer season.

3.3. OPV Effects on Lighting Conditions

Figure 7a shows the lighting conditions measured at the canopy-level in both the OPV and Control sections, with and without the shade net deployed on the Control section. The lighting conditions in the Control section had linearly increasing transmittance from the 300–1000 nm range, with relatively lower UV transmittance; the polyethylene film was coated with anti-UV blocking agents for better stability. The shade net in the Control section caused a relatively constant reduction in the transmittance of all wavelengths measured. For the OPV section, there was a sharp increase in transmittance between 380–440 nm. Between 450–650 nm, there was relatively constant transmittance of around 28%, and then another peak between 660–750 nm.

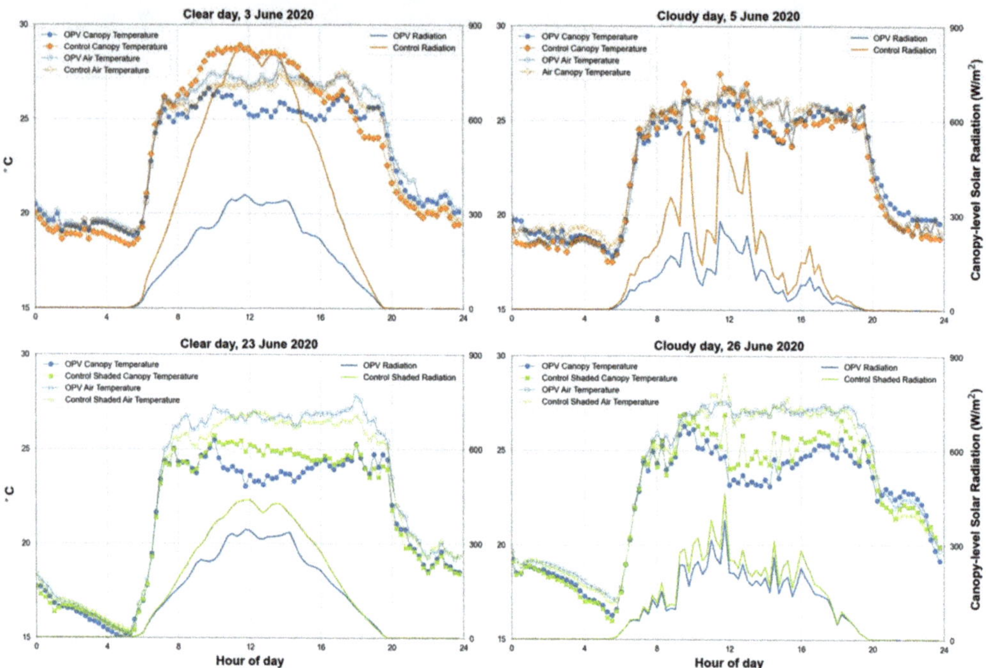

Figure 6. Canopy temperature, air temperature, and canopy-level solar radiation in the OPV and Control sections, comparing conditions in clear (**left**) and cloudy (**right**) sky conditions, and the Control section when it was unshaded (**top**) and when it was shaded by the 30% shade net (**bottom**).

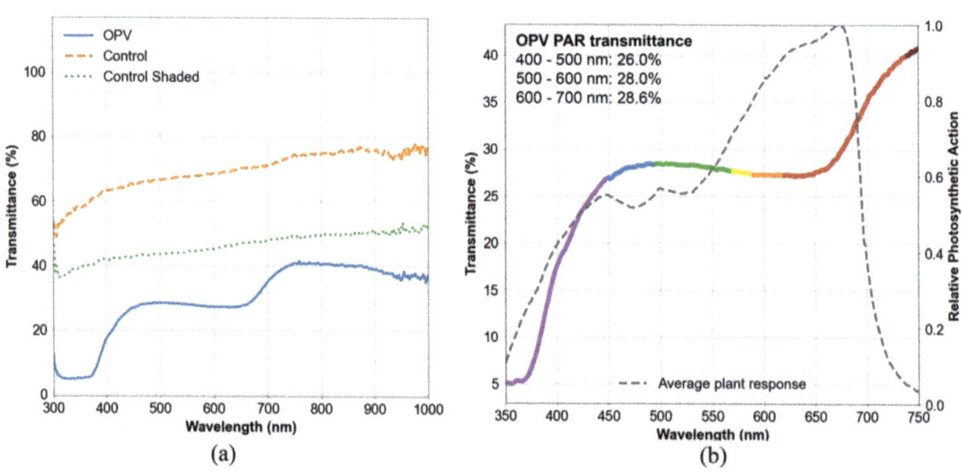

Figure 7. Canopy-level lighting quality in OPV and Control sections: (**a**) comparing OPV and Control sections with and without a shade net installed, measured at midday with a clear sky; (**b**) comparing the light transmittance in the PAR region in the OPV section to the relative photosynthetic action (efficiency), shown in the dashed gray line, which was adapted from McCree [19].

Figure 7b shows the lighting conditions incident on the canopy in the OPV section alongside the averaged relative photosynthetic action of 22 herbaceous crops (i.e., the CO_2 assimilation rate of plants at different ambient light wavelength ranges) [19]. The transmittance of a wavelength-selective semi-transparent OPV shade cover would ideally complement the relative photosynthetic efficiency of the crop in specific wavelength ranges. It can be seen in Figure 7b that the OPV shade cover used in this study was not spectrally optimized for this purpose, although it is worth mentioning that blue photons (between 400–500 nm) are used less efficiently in photosynthesis compared to orange and red photons (600–700 nm) [32]. OPV films that primarily absorb in the near-infrared range (beyond 780 nm) have been developed for greenhouse (and other) applications [33–35], although such technology has not yet been commercialized. However, the goal of spectral optimization should be weighed against other design factors involved in greenhouse-integrated OPV shading strategies. These design factors pertain to the technical features and performance of the OPV (e.g., PCE, transparency, device lifetime), the deployment strategy that is used (e.g., partial vs. full roof coverage), the available solar resources of the target location and in different seasons, and economic considerations (e.g., cost of electricity, cost of OPVs).

For a clear sky, the percentage of solar radiation in the PAR range is approximately 45%; for a fully clouded sky, this percentage increases to approximately 60%, due to the clouds blocking a relatively greater portion of UV and NIR radiation [36]. Although they are mostly transparent, greenhouse cladding materials do not transmit all solar radiation wavelengths equally—although the double-layer polyethylene cover on the greenhouse for this study transmitted ~75% of all solar radiation wavelengths, it can be seen in Figure 7a that the transmittance of PAR radiation was only ~65%; the OPV shade cover decreased transmittance of PAR radiation to the OPV-shaded area to ~28%.

The accumulated daily PAR radiation incident on the crop canopy is known as the daily light integral (DLI) and is measured in $mol_{PAR}\ m^{-2}\ day^{-1}$. Figure 8 presents the estimated weekly averages for the DLI in the OPV and Control sections for the measurement period (see the Materials and Methods section for the calculation procedure). For greenhouse tomato cultivation, the minimum recommended DLI is 15 [37]; a DLI of 30 or above has been suggested as optimal [38], assuming that other environmental conditions (e.g., air temperature, CO_2 availability, root zone conditions) are also maintained at optimal levels. It can be seen in Figure 8 that the DLI values in the OPV section were below the recommended minimum of 15 for 3 weeks, from transplanting until the beginning of April, and then remained above 15 for the duration of the study period. When the Control section was not shaded, the DLI was reduced in the OPV section between 57–65%; the installation of the shade net on the Control section in June narrowed this difference to 31–35%. In terms of targeted DLI levels for tomato cultivation, the OPV shade treatment was excessive when considering the optimal recommended DLI levels.

3.4. Crop Growth and Yield Performance

Figure 9 shows weekly measurements of the vegetative and reproductive growth parameters of the tomato plants located in the OPV and Control sections during the measurement period. The tomato plants grown under the shade of the OPV generally displayed more vegetative growth, specifically, accelerated head growth (i.e., stem elongation) and leaf length on average, and showing a fuller canopy compared to the Control section; this result is attributed to shade avoidance behaviors in the OPV plants, particularly in the initial month after transplant. The differences in head growth and leaf length between the two sections became less pronounced later in the growing season, once the plants had begun the reproductive stage (63 DAT onward). Once the shade net was deployed on the Control section (97 DAT), all growth trends were not found to be significantly different.

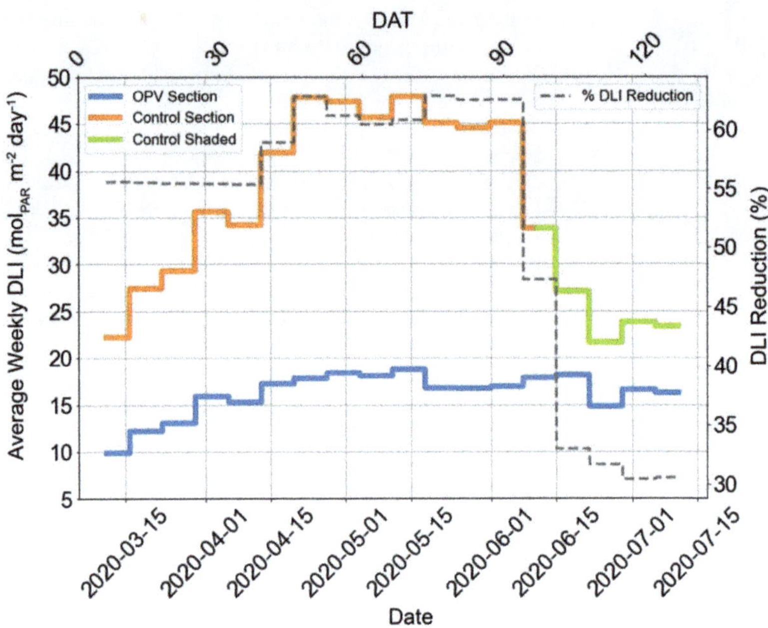

Figure 8. Estimated weekly averages for the daily light integral (DLI, in mol$_{PAR}$ m^{-2} day^{-1}) in OPV and Control sections with corresponding dates and days after transplant (DAT).

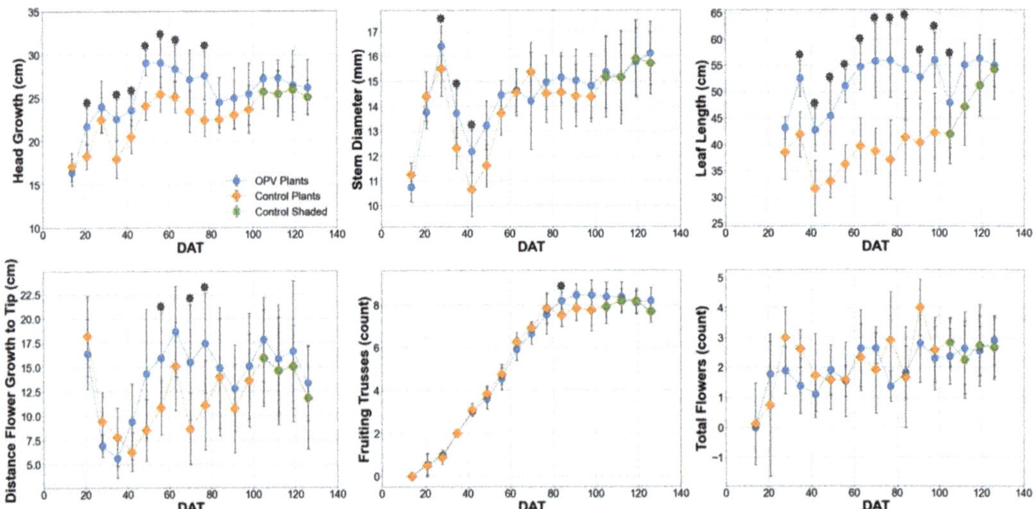

Figure 9. Weekly plant growth in the OPV and Control sections; parameters measured include head growth, head diameter, leaf length, distance of first open flower to plant tip, number of fruiting trusses, and the total number of flowers; days after transplant (DAT); each data point represents the averaged values of 6 sample plants, with bars representing the standard deviation; the ✳ indicates a significant difference based on a two-sample *t*-test statistic, calculated for the OPV and Control sections.

Figure 10 presents the yield productivity in the OPV and Control sections for 10 weekly harvests. The delayed fruit development and ripening in the OPV section is evident in the lower average yields, number of fruits, and average fruit mass compared to the Control section in the first three harvests. Interestingly, the initial three-week period after transplant, in which the DLI levels were estimated to be under the recommended minimum level of 15 (Figure 8), corresponds with the period of lag in the yield productivity of the OPV plants relative to the Control plants. Constrained solar radiation, specifically with reduced DLI, under shade treatments has been shown to limit the growth and development of greenhouse tomatoes [39]. Although the total cumulative yield was somewhat lower in the OPV compared to the Control section (24.6 kg m^{-2} and 27.7 kg m^{-2}, respectively), beginning with the fourth harvest on 83 DAT (28 May 2020), the average yield productivity values in the OPV and Control sections were similar. Between DAT 76–97, reduced yields can be seen in the Control section, whereas the OPV plants showed increased yields, indicating that the OPV plants benefitted from the microclimate conditions created by the OPV cover, mitigating heat stress, which led to yields similar in the control group and even further increases after DAT 80. The deployment of the shade net on the Control section appeared to have a beneficial effect on yield performance in that section.

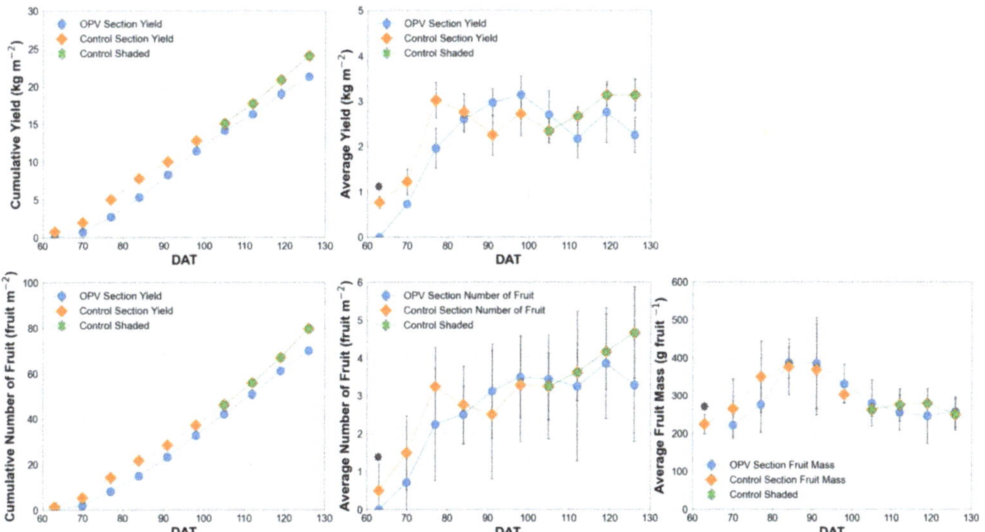

Figure 10. Weekly yield data in OPV and Control Sections for 10 harvests, with the first harvest 62 DAT (7 May 2020) and last harvest 125 DAT (7 July 2020). Each data point represents the averaged value of 6 sample plants located in the inner rows of each section, with bars representing the standard deviation. The * indicates a significant difference between the samples for that week of measurement.

Figure 11 compares the cumulative yield (in kg) in the OPV and Control sections to the cumulative PAR radiation received by the canopy in each section. The linear equations indicate the yield per mole of PAR energy (calculated as g mol$_{PAR}$$^{-1}$) received by the canopy—essentially, the light use efficiency (LUE) of the tomato plants in producing fruit. The LUE in the OPV section was approximately twice as high as that in the Control section, with 21.4 g mol$_{PAR}$$^{-1}$ in the OPV section compared to 10.1 g mol$_{PAR}$$^{-1}$ in the Control section. Once the shade cover was deployed on the Control section, the LUE of the Control section increased to 18.2 g mol$_{PAR}$$^{-1}$, slightly lower than the OPV section. From these results, it can be seen that the plants grown under the OPV shade were able to adapt to the lower light levels reasonably well, with less stress observed under high light intensities, achieving higher weekly yields during the hot and high-light periods, and with comparable yields to those of the plants grown under a conventional shade net when it was deployed on the Control section. These results clearly indicate the potential of using OPV as a shade element, while also being able to generate electrical energy within the same greenhouse footprint.

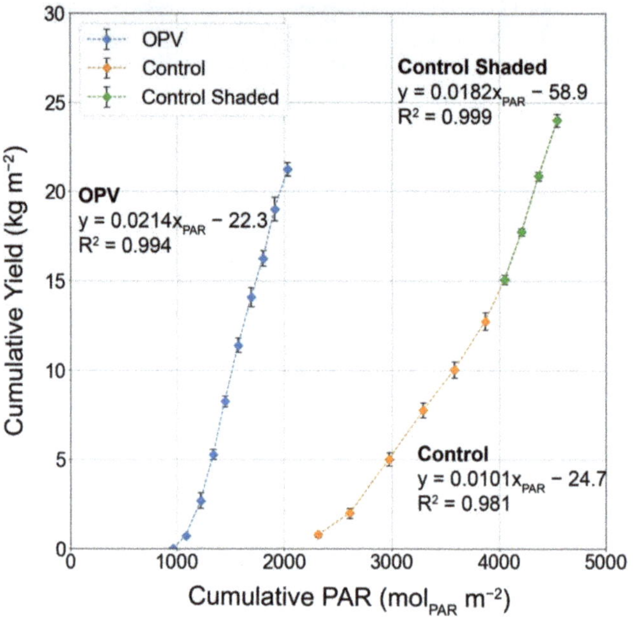

Figure 11. Cumulative yield (in kg m^{-2}) versus cumulative photosynthetically active radiation (PAR, in mol$_{PAR}$ m^{-2}) incident on the canopy in OPV and Control sections. Points represent the averaged values of 6 plants in each section, with bars representing the standard deviation.

3.5. OPV Power Production

Table 3 presents data on the daily power output of six OPV arrays deployed on the east and west pitches of the greenhouse roof. Measurements were taken during a six-day period (23–28 May 2020) with clear skies and an average total daily solar radiation of 8.62 kWh m^{-2} day^{-1}. The electrical performance varied between individual OPV arrays: despite similar solar radiation conditions, the daily power output ranged between 0.056–0.088 kWh m^{-2} day^{-1}. At the time of this measurement, the OPV devices had already experienced a degradation in performance, having been deployed on the greenhouse roof since October 2019 for a previous study that focused on the power generation performance of the OPV arrays. In that study, the average efficiency of the OPV arrays was found to be 1.82%. For the greenhouse (133 m^2 footprint), with the hardware used in this study, the daily energy requirement in the summer season can range between 10–20 kWh day^{-1} (0.075–0.150 kWh m^{-2} day^{-1}) depending primarily on the extent of operation of the wet-pad and fan ventilation/cooling system. Thus, for the daily greenhouse energy requirement to be met during the summer season, an OPV system deployed to fully cover the greenhouse roof would need to achieve 1.1–2.1% efficiencies, which is in the performance range of the OPV devices used in this study. As the OPV industry continues to advance and scale up commercial production, improvement in the quality and stability of large-area OPV products is anticipated. For integrated OPV greenhouse power system design, it should be noted that unless grid-connection is available, batteries or alternative energy storage technologies would need to be included.

Table 3. Power generation performance of six identical OPV arrays deployed on the greenhouse roof, measured on clear-sky days from 23–28 May 2020.

OPV	Location on Greenhouse Roof (East vs. West Pitch)	Date of Measurement	Daily Outdoor Solar Radiation (kWh m^{-2} day^{-1})	OPV Daily Energy Output (kWh m^{-2} day^{-1})
OPV1	East	23 May 2021	8.56	0.056
OPV2	West	25 May 2021	8.56	0.069
OPV4	East	28 May 2021	8.72	0.064
OPV5	West	24 May 2021	8.72	0.080
OPV6	East	27 May 2021	8.64	0.088
OPV7	West	26 May 2021	8.50	0.058

Beyond considerations of OPV performance, however, the design configuration of the OPV arrays used in the present study (i.e., the extensive and symmetrical OPV coverage over the greenhouse roof on both east-facing and west-facing surfaces) resulted in a relatively distributed output of power over the course of the day. This can be seen in Figure 12, which presents the diurnal power generation patterns of the OPV arrays. This installation strategy is advantageous in terms of meeting greenhouse energy requirements throughout the day.

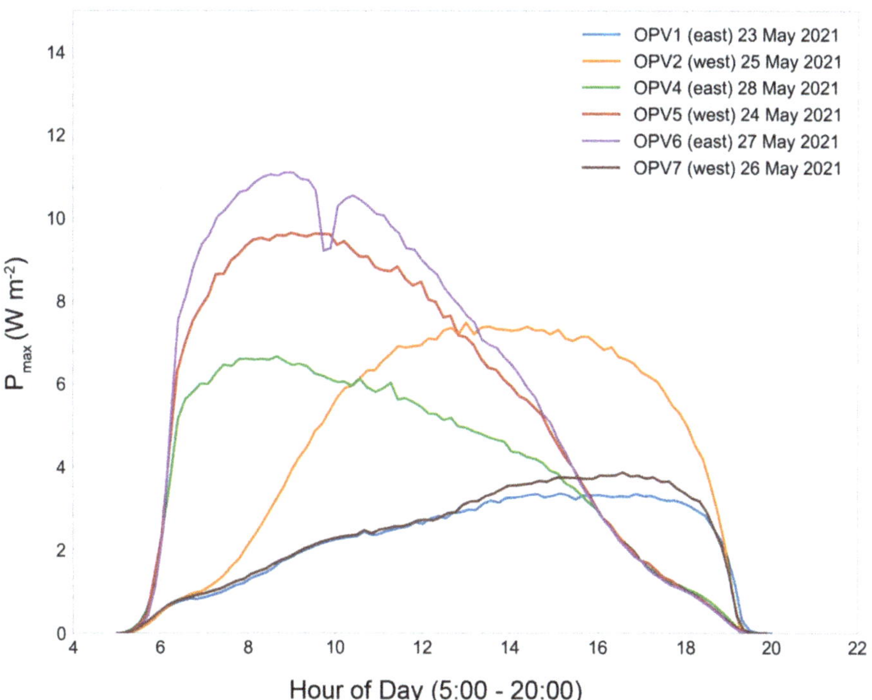

Figure 12. Diurnal power generation patterns (W m^{-2}) of the OPV arrays deployed on the greenhouse roof. The area beneath each curve represents the total daily energy output of each OPV array (in Wh m^{-2}).

4. Conclusions

This study demonstrated the application of commercially manufactured, semi-transparent, flexible, roll-to-roll printed organic photovoltaic (OPV) arrays (3.4 m^2 active area) as a shade element for greenhouse tomato production in a hot, arid climate. The OPV arrays decreased the transmittance of all solar radiation wavelengths in the shaded area by approximately 40% and photosynthetically active radiation (PAR, 400–700 nm) by approximately 37%.

During the hottest months of the measurement period (May–July), the OPV shade provided a suitable climate for tomato crop production, stabilizing canopy temperature during the times of day with the highest solar radiation intensities, performing the function of a conventional shading method. Constrained solar radiation levels in the OPV section for three weeks following transplant in early March, in which the daily light integral (DLI) values were estimated to be lower than the recommended minimum of 15 mol_{PAR} m^{-2} day^{-1}, resulted in more vegetative growth and delayed fruit development and ripening compared to the Control section, indicating shade avoidance behavior, and leading to lower average yields in the first three (of 10 total) weekly harvests. Beginning with the fourth harvest, however, yield, fruit number, and fruit mass in the OPV and Control sections were similar. Trends in yield productivity for the OPV plants showed increased performance during high-light-intensity periods in May and June. The light utilization efficiency (LUE), which is the relationship between cumulative yield and cumulative PAR radiation, measured in g mol_{PAR}^{-1}, was approximately twice as high in the OPV section (21.4 g mol_{PAR}^{-1}) compared to the Control section when it was unshaded (10.1 g mol_{PAR}^{-1}); during the period in which the Control section was shaded, the LUE increased to 18.2 g mol_{PAR}^{-1}.

Although the electrical performance of the OPV arrays used for this study varied, the east-west orientation and extensive coverage of the OPVs over the curved greenhouse roof meant that power production could be relatively evenly distributed throughout the day. It can be concluded that the daily electrical energy requirements for the greenhouse in this study with its installed hardware during the summer production season in this region could be met with full roof coverage by OPVs and with efficiencies between 1.1–2.1%, which is in the performance range of the devices used in this study.

Future investigations of OPV applications to greenhouses should test crop types with different light requirements, seasonal effects, effects of OPVs on plant water and nutrient efficiencies, OPVs with different spectral characteristics (higher light transmittance in the PAR spectrum), the effects of degradation on OPVs' spectral characteristics over time, and different OPV installation strategies. The dynamics of energy production/consumption in integrated OPV greenhouse systems should also be explored in future work.

Ultimately, this study demonstrated a readily available design strategy for OPV applications to greenhouses in high-light regions, where shading methods are required for spring and summer greenhouse crop production.

Author Contributions: Conceptualization, R.W., M.K., E.M., M.T., and I.Y.; methodology, R.W.; software, R.W.; validation, R.W.; formal analysis, R.W.; investigation, R.W. and M.K.; resources, M.K.; data curation, R.W.; writing—original draft preparation, R.W.; writing—review and editing, R.W., M.K., E.M., M.T., and I.Y.; visualization, R.W.; supervision, M.K.; project administration, M.K.; funding acquisition, M.K., M.T., and I.Y. All authors have read and agreed to the published version of the manuscript.

Funding: This research was supported by Research Grant No. US-4885-16 from BARD, United States –Israel Binational Agricultural Research and by the National Science Foundation Grant No. DGE1735173 IndigeFEWSS project.

Data Availability Statement: The data presented in this study are available on request from the corresponding author.

Conflicts of Interest: The authors declare that they have no known competing financial interests or personal relationships that could have appeared to influence the work reported in this paper.

References

1. Barbosa, G.; Gadelha, F.; Kublik, N.; Proctor, A.; Reichelm, L.; Weissinger, E.; Wohlleb, G.; Halden, R. Comparison of Land, Water, and Energy Requirements of Lettuce Grown Using Hydroponic vs. Conventional Agricultural Methods. *Int. J. Environ. Res. Public Health* **2015**, *12*, 6879–6891. [CrossRef]
2. Cook, R.L.; Calvin, L. Greenhouse Tomatoes Change the Dynamics of the North American Fresh Tomato Industry. *U.S. Dept. Agric. Econ. Res. Serv.* **2005**, *2*, 1–65.
3. Stanghellini, C.; Montero, J.I. Resource Use Efficiency in Protected Cultivation: Towards the Greenhouse with Zero Emissions. *Horticulturae* **2012**, *927*. [CrossRef]
4. Marrou, H.; Guilioni, L.; Dufour, L.; Dupraz, C.; Wery, J. Microclimate under Agrivoltaic Systems: Is Crop Growth Rate Affected in the Partial Shade of Solar Panels? *Agric. For. Meteorol.* **2013**, *177*, 117–132. [CrossRef]
5. Okada, K.; Kacira, M.; Teitel, M.; Yehia, I. Crop Production and Energy Generation in a Greenhouse Integrated with Semi-Transparent Organic Photovoltaic Film. *Acta Hortic.* **2018**, *1227*, 231–240. [CrossRef]
6. Li, Z.; Yano, A.; Yoshioka, H. Feasibility Study of a Blind-Type Photovoltaic Roof-Shade System Designed for Simultaneous Production of Crops and Electricity in a Greenhouse. *Appl. Energy* **2020**, *279*, 115853. [CrossRef]
7. Marucci, A.; Zambon, I.; Colantoni, A.; Monarca, D. A Combination of Agricultural and Energy Purposes: Evaluation of a Prototype of Photovoltaic Greenhouse Tunnel. *Renew. Sustain. Energy Rev.* **2018**, *82*, 1178–1186. [CrossRef]
8. Marucci, A.; Monarca, D.; Cecchini, M.; Colantoni, A.; Cappuccini, A. Analysis of Internal Shading Degree to a Prototype of Dynamics Photovoltaic Greenhouse through Simulation Software. *J. Agric. Eng.* **2015**, *46*, 144–150. [CrossRef]
9. Bambara, J.; Athienitis, A.K. Energy and Economic Analysis for the Design of Greenhouses with Semi-Transparent Photovoltaic Cladding. *Renew. Energy* **2019**, *131*, 1274–1287. [CrossRef]
10. Hassanien, R.H.E.; Li, M.; Yin, F. The Integration of Semi-Transparent Photovoltaics on Greenhouse Roof for Energy and Plant Production. *Renew. Energy* **2018**, *121*, 377–388. [CrossRef]
11. Lee, T.D.; Ebong, A.U. A Review of Thin Film Solar Cell Technologies and Challenges. *Renew. Sustain. Energy Rev.* **2017**, *70*, 1286–1297. [CrossRef]
12. La Notte, L.; Giordano, L.; Calabrò, E.; Bedini, R.; Colla, G.; Puglisi, G.; Reale, A. Hybrid and Organic Photovoltaics for Greenhouse Applications. *Appl. Energy* **2020**, *278*, 115582. [CrossRef]
13. Muteri, V.; Cellura, M.; Curto, D.; Franzitta, V.; Longo, S.; Mistretta, M.; Parisi, M.L. Review on Life Cycle Assessment of Solar Photovoltaic Panels. *Energies* **2020**, *13*, 252. [CrossRef]
14. Ramanujam, J.; Bishop, D.M.; Todorov, T.K.; Gunawan, O.; Rath, J.; Nekovei, R.; Artegiani, E.; Romeo, A. Flexible CIGS, CdTe and a-Si:H Based Thin Film Solar Cells: A Review. *Prog. Mater. Sci.* **2020**, *110*, 100619. [CrossRef]
15. Riede, M.; Spoltore, D.; Leo, K. Organic Solar Cells—The Path to Commercial Success. *Adv. Energy Mater.* **2021**, *11*, 2002653. [CrossRef]
16. Søndergaard, R.; Hösel, M.; Angmo, D.; Larsen-Olsen, T.T.; Krebs, F.C. Roll-to-Roll Fabrication of Polymer Solar Cells. *Mater. Today* **2012**, *15*, 36–49. [CrossRef]
17. Traverse, C.J.; Pandey, R.; Barr, M.C.; Lunt, R.R. Emergence of Highly Transparent Photovoltaics for Distributed Applications. *Nat. Energy* **2017**, *2*, 849–860. [CrossRef]
18. Emmott, C.J.M.; Röhr, J.A.; Campoy-Quiles, M.; Kirchartz, T.; Urbina, A.; Ekins-Daukes, N.J.; Nelson, J. Organic Photovoltaic Greenhouses: A Unique Application for Semi-Transparent PV? *Energy Environ. Sci.* **2015**, *8*, 1317–1328. [CrossRef]
19. McCree, K.J. The Action Spectrum, Absorptance and Quantum Yield of Photosynthesis in Crop Plants. *Agric. Meteorol.* **1971**, *9*, 191–216. [CrossRef]
20. Ahemd, H.A.; Al-Faraj, A.A.; Abdel-Ghany, A.M. Shading Greenhouses to Improve the Microclimate, Energy and Water Saving in Hot Regions: A Review. *Sci. Hortic.* **2016**, *201*, 36–45. [CrossRef]
21. Ravishankar, E.; Booth, R.E.; Saravitz, C.; Sederoff, H.; Ade, H.W.; O'Connor, B.T. Achieving Net Zero Energy Greenhouses by Integrating Semitransparent Organic Solar Cells. *Joule* **2020**, *4*, 490–506. [CrossRef]
22. Hollingsworth, J.A.; Ravishankar, E.; O'Connor, B.; Johnson, J.X.; DeCarolis, J.F. Environmental and Economic Impacts of Solar-Powered Integrated Greenhouses. *J. Ind. Ecol.* **2020**, *24*, 234–247. [CrossRef]
23. Aroca-Delgado, R.; Pérez-Alonso, J.; Callejón-Ferre, Á.J.; Velázquez-Martí, B. Compatibility between Crops and Solar Panels: An Overview from Shading Systems. *Sustainability* **2018**, *10*, 743. [CrossRef]
24. Peet, M.M.; Welles, G. Greenhouse tomato production. In *Tomatoes*; Heuvelink, E., Ed.; CAB International: Wallingford, UK, 2005; pp. 257–304.
25. Li, W. Resource Use Efficiency under Photovoltaic Integrated Greenhouse Glazing. Master's Thesis, Department of Biosystems Engineering, The University of Arizona, Tucson, AZ, USA, 2014.
26. Ntinas, G.K.; Kadoglidou, K.; Tsivelika, N.; Krommydas, K.; Kalivas, A.; Ralli, P.; Irakli, M. Performance and Hydroponic Tomato Crop Quality Characteristics in a Novel Greenhouse Using Dye-Sensitized Solar Cell Technology for Covering Material. *Horticulturae* **2019**, *5*, 42. [CrossRef]
27. Friman-Peretz, M.; Ozer, S.; Geoola, F.; Magadley, E.; Yehia, I.; Levi, A.; Brikman, R.; Gantz, S.; Levy, A.; Kacira, M.; et al. Microclimate and Crop Performance in a Tunnel Greenhouse Shaded by Organic Photovoltaic Modules—Comparison with Conventional Shaded and Unshaded Tunnels. *Biosyst. Eng.* **2020**, *197*, 12–31. [CrossRef]

28. *Rhinoceros 3D v7*; Robert McNeel & Associates: Seattle, WA, USA, 2020; Available online: https://www.rhino3d.com/ (accessed on 3 June 2021).
29. Jolliet, O.; Bailey, B.J. The Effect of Climate on Tomato Transpiration in Greenhouses: Measurements and Models Comparison. *Agric. For. Meteorol.* **1992**, *58*, 43–62. [CrossRef]
30. Masabni, J.; Sun, Y.; Niu, G.; del Valle, P. Shade Effect on Growth and Productivity of Tomato and Chili Pepper. *HortTechnology* **2016**, *26*, 344–350. [CrossRef]
31. Giacomelli, G.A.; Kubota, C.; Jensen, M. Design Considerations and Operational Management of Greenhouse for Tomato Production in Semi-Arid Region. *Int. Soc. Hortic. Sci.* **2005**, *691*, 525–532. [CrossRef]
32. Bugbee, B. Toward an Optimal Spectral Quality for Plant Growth and Development: The Importance of Radiation Capture. *Acta Hortic.* **2016**, *1134*, 1–12. [CrossRef]
33. Liu, Y.; Cheng, P.; Li, T.; Wang, R.; Li, Y.; Chang, S.Y.; Zhu, Y.; Cheng, H.W.; Wei, K.H.; Zhan, X.; et al. Unraveling Sunlight by Transparent Organic Semiconductors toward Photovoltaic and Photosynthesis. *ACS Nano* **2019**, *13*, 1071–1077. [CrossRef] [PubMed]
34. Shi, H.; Xia, R.; Zhang, G.; Yip, H.-L.; Cao, Y. Spectral Engineering of Semitransparent Polymer Solar Cells for Greenhouse Applications. *Adv. Energy Mater.* **2019**, *9*, 1803438. [CrossRef]
35. Song, W.; Fanady, B.; Peng, R.; Hong, L.; Wu, L.; Zhang, W.; Yan, T.; Wu, T.; Chen, S.; Ge, Z. Foldable Semitransparent Organic Solar Cells for Photovoltaic and Photosynthesis. *Adv. Energy Mater.* **2020**, *10*, 2000136. [CrossRef]
36. Meek, D.W.; Hatfield, J.L.; Howell, T.A.; Idso, S.B.; Reginato, R.J. A Generalized Relationship between Photosynthetically Active Radiation and Solar Radiation. *Agron. J.* **1984**, *76*, 939–945. [CrossRef]
37. Runkle, E. *DLI "Requirements" Technically Speaking*; Greenhouse Product News Magazine: Sparta, MI, USA, 2019.
38. Dorais, M. The Use of Supplemental Lighting for Vegetable Crop Production: Light Intensity, Crop Response, Nutrition, Crop Management, and Cultural Practices. In Proceedings of the Canadian Greenhouse Conference, Niagara Falls, ON, Canada, 9 October 2003.
39. Kläring, H.P.; Krumbein, A. The Effect of Constraining the Intensity of Solar Radiation on the Photosynthesis, Growth, Yield and Product Quality of Tomato. *J. Agron. Crop Sci.* **2013**, *199*, 351–359. [CrossRef]

agronomy

Article

Analysis of the Viability of a Photovoltaic Greenhouse with Semi-Transparent Amorphous Silicon (a-Si) Glass

José-Ramón Aira [1], Sara Gallardo-Saavedra [2], Marcia Eugenio-Gozalbo [3], Víctor Alonso-Gómez [4], Miguel-Ángel Muñoz-García [5] and Luis Hernández-Callejo [2,*]

[1] Department of Construction and Architectural Technology, Escuela Técnica Superior de Arquitectura, Universidad Politécnica de Madrid, Avda. Juan de Herrera 4, 28040 Madrid, Spain; joseramon.aira@upm.es
[2] Department of Agricultural and Forestry Engineering, Campus Duques de Soria, Universidad de Valladolid, 42004 Soria, Spain; sara.gallardo@uva.es
[3] Department of Didactics of Experimental, Social and Mathematical Sciences, Campus Universitario Duques de Soria, Universidad de Valladolid, 42004 Soria, Spain; marcia.eugenio@uva.es
[4] Department of Applied Physics, Campus Universitario Duques de Soria, Universidad de Valladolid, 42004 Soria, Spain; victor.alonso.gomez@uva.es
[5] Department of Agroforestry Engineering, Escuela Técnica Superior de Ingeniería Agronómica, Alimentaria y de Biosistemas. Universidad Politécnica de Madrid, Av. Puerta de Hierro 2, 28040 Madrid, Spain; miguelangel.munoz@upm.es
* Correspondence: luis.hernandez.callejo@uva.es; Tel.: +34-975129418

Citation: Aira, J.-R.; Gallardo-Saavedra, S.; Eugenio-Gozalbo, M.; Alonso-Gómez, V.; Muñoz-García, M.-Á.; Hernández-Callejo, L. Analysis of the Viability of a Photovoltaic Greenhouse with Semi-Transparent Amorphous Silicon (a-Si) Glass. *Agronomy* 2021, *11*, 1097. https:// doi.org/10.3390/agronomy11061097

Academic Editor: Jung Eek Son

Received: 24 March 2021
Accepted: 27 May 2021
Published: 28 May 2021

Publisher's Note: MDPI stays neutral with regard to jurisdictional claims in published maps and institutional affiliations.

Abstract: For decades, society has been changing towards an energy mix that enhances the use of renewable sources and a more distributed generation of energy. The agricultural sector is included in this trend, which is why several studies are currently being carried out focused on the use of solar energy in greenhouses. This article aims to demonstrate the viability of a greenhouse that integrates, as a novelty, semi-transparent amorphous silicon photovoltaic (PV) glass (a-Si), covering the entire roof surface and the main sides of the greenhouse. The designed prototype is formed by a simple rectangular structure 12 m long and 2.5 m wide, with a monopitch roof, oriented to the southwest, and with a 35° inclination. The greenhouse is divided into two contiguous equal sections, each with an area of 15 m^2, and physically separated by an interior partition transparent wall. The surface enclosure of one of the sections is made of conventional glass, and the one of the other, of PV glass. How the presence of semitransparent PV glass influences the growth of horticultural crops has been studied, finding that it slightly reduces the production of vegetal mass and accelerates the apical growth mechanism of heliophilic plants. However, from a statistical point of view, this influence is negligible, so it is concluded that the studied technology is viable for horticultural production. The energy balance carried out indicates that the energy produced by the PV system is greater than the energy consumed by the greenhouse, which shows that the greenhouse is completely viable and self-sufficient for sites with the adequate solar resource.

Keywords: sustainable greenhouse; semi-transparent photovoltaic panels; amorphous silicon; building-integrated photovoltaics; distributed generation; microgrid

1. Introduction

Currently, most of the energy used in the world comes from fossil fuels [1]. The building sector is responsible for 40% of emissions of greenhouse gases and 38% of the global energy demand [2], mainly consumed for maintaining thermal comfort conditions [3].

The application of renewable energy technologies in buildings can effectively help reduce the consumption of fossil fuels and thus contribute to a more sustainable global energy model [4]. The goal of independence from fossil fuels makes most countries encourage renewable energy generation, thus, having a more diversified energy mix.

Among renewable energy generation technologies, solar photovoltaic (PV) is presented as one of the most interesting, since solar energy is available anywhere in the world [5]. Furthermore, it should be noted that in the last decade the cost of PV modules has fallen by more than 80%, while the cost of fossil fuels, such as gasoline or diesel, which compete with renewable energies in electricity generation, have increased more 250% [6]. This technology has been implemented progressively in materials and constructing structures, giving way to what today is known as integration of PV solar energy in buildings or building-integrated PV technology (BIPV) [7]. Combined with distributed generation (DG), breaking the paradigm of centralized generation (in which energy is generated away from consumption points), a very interesting BIPV concept is achieved, being the one proposed in this article.

The farming community is also aware of this change in the energy mix production, so many researchers are carrying out studies focused on the use of solar energy in agro-industrial buildings, and specifically in greenhouses. Thus, there are several studies focused specifically on the state of the art of solar energy applied in greenhouses [8–10]. Other research analyzes the influence of solar panel orientation and shading on electricity or horticultural production [11–14], or the way solar radiation is distributed indoors and its influence on horticultural production [15].

The overall cumulative radiation inside the greenhouse decreases depending on the coverage rate (PV_R, or ratio of the horizontal surface of the greenhouse that is covered by solar panels placed on the roof) so that the reduction is equal to 0.8% for each 1% increase in PV_R [16]. In a recent study, the most suitable PV_R coverage ratio was analyzed according to the type of crop in 14 greenhouses in southern Europe [17]. The structures with a PV_R of 25% were compatible with the cultivation of all the considered species, including those with a high light demand (tomato, cucumber, sweet pepper), with an estimated insignificant or limited yield reduction (less than 25%). Low light species (such as asparagus) and low light crops can be grown with a maximum PV_R of up to 60%. Limiting the roof coverage with opaque solar panels has promoted numerous investigations in recent years on semitransparent PV cells in BIPV applications, particularly in greenhouses, forming a research line with increasing scientific interest [18–21], even with the implementation of organic PV cells [22–24].

Moreover, some crops require different light intensities depending on their vegetative cycle (during germination, flowering, the fruiting, etc.). Some research in recent years has focused on the use of PV panels that are capable of modifying their inclination to allow more or less light to enter the greenhouse [25–27].

This article aims to demonstrate the technical, economic and environmental feasibility of a greenhouse in which semi-transparent amorphous silicon (a-Si) PV glass panels are integrated on the entire surface of the roof, and of the main sides of the greenhouse (south west and northeast). How the greenhouse performs its horticultural production functions, while it recovers the electrical energy necessary to be able to supply the needs of the microgrid that makes it up, will be analyzed. These needs should be adjusted, at all times, to the climatic conditions of the site, trying to make sure that the demanded energy can be covered with the available PV DG, while trying to maintain horticultural production at optimal levels. As stated, this paper presents two novel points in relation to other related research. On the one hand, amorphous silicon semi-transparent glasses technology is used for power generation, which to our knowledge has never been used in greenhouses and on the other hand, photovoltaic glass is placed on the entire surface of the greenhouse, i.e., PVR of 100%.

2. Materials and Methods

2.1. Prototype Description

The greenhouse prototype is formed by a simple rectangular structure with a length of 12 m and a width of 2.5 m, with a monopitch roof, oriented to the southwest, and with an inclination of 35°, with the objective of receiving the greatest solar radiation. The site is located at the European Center for the Training, Research and Development of Alternative Energies of the Campus Duques de Soria of the University of Valladolid

(UVa). The elements of the structure are made of pine lumber Soria (*Pinus sylvestris* L.) with protective treatment for outdoor use without ground contact, Figure 1. Wood is a material in high demand today in sustainable construction because it acts as a sink for CO_2, keeping it captured in its plant structures throughout the lifetime of the infrastructure [28].

Figure 1. PV greenhouse located on the Duques de Soria University Campus of the UVa.

The greenhouse is divided into two equal and contiguous sections, each with an area of 15 m², physically separated by a transparent interior partition wall. The enclosure surface of the sections are made of conventional glass (not PV active) in one section, and PV amorphous silicon (a-Si) glass in the other. The dimensions, mechanical properties (strength and rigidity), and thermal (thermal transmittance) of conventional and PV glasses are similar, so that they do not interfere with the study. This partition has the purpose of being able to compare the horticultural production obtained in both sections in order to evaluate the viability of PV glasses. The PV glass chosen for the design was 034-BN-12450635-30-1, whose main characteristics are described in Table 1.

Table 1. Properties of PV glass.

Model	034-BN-12450635-30-1
Setting:	Glass-glass (without inert chamber)
Dimensions:	1245 mm × 635 mm × 7.96 mm (3.2 glass + 0.76 PVB encapsulation + 4 glass)
Transparency:	Colorless glass with 30% transparency
PV technology type:	Amorphous silicon
Rated power:	22 W per module//1320 W in total over the entire active area (60 modules)
Special treatments:	No special treatments

Initially the control parameters (temperature, relative humidity of the air and CO_2 concentration) were determined. Sensors were installed for their monitoring, both inside and outside the greenhouse. A triple sensor (which measures these three parameters simultaneously) was placed outside the greenhouse, another triple sensor inside the greenhouse in the conventional glass section, and another triple sensor inside the PV glass section.

Depending on the desired type of cultivation, it is necessary that the control parameters remain within specific value ranges for optimal horticultural production. To modify and thus correct control parameters, the following actuators or control systems were installed: bidirectional fans to cause ventilation or extraction as required, heating to warm up, and nebulizers to produce water steam.

In addition, two more sensors were installed to measure PAR radiation inside the greenhouse, one in the conventional glass section and the other in the PV glass section; as well as a drip irrigation system in each section controlled by a clock which is totally independent of the installed sensors.

The electricity consumption is measured by three network analyzers, a general one to record the total consumption, including the consumption of the Programmable Logic

Controller (PLC) and the data concentrator; and one for each of the greenhouse sections. Thus, each section contains the equipment shown in Figure 2:

- Temperature, relative humidity and CO_2 sensor.
- PAR radiation sensor.
- Fan to ventilate/extract.
- Heater for warming up with a power of 1800 W.
- 2-nozzle telescopic nebulizer.
- Drip irrigation system.
- Network analyzer for energy consumption analysis.

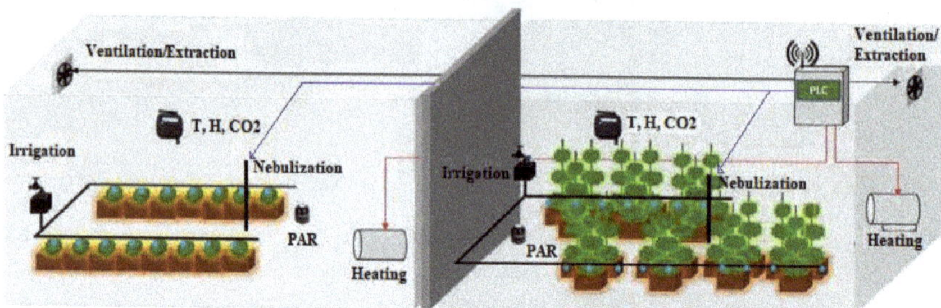

Figure 2. Sensors and actuators.

The modular monitoring system of the prototype consists of the following elements:

- Power: power module made up of power supplies and transformers, responsible for feeding all the components of the system.
- Sensors: room sensorization module.
- Data concentrator: information storage module.
- Wireless gateway: wireless communication module for receiving data from wireless sensors.
- PLC: control module.
- Actuators: module made up of the actuators or control elements of the system.

Figure 3 shows a summary block diagram of the monitoring system. The sensors communicate with the data concentrator through a wireless platform, while the actuators and the irrigation system do it directly by buried cable.

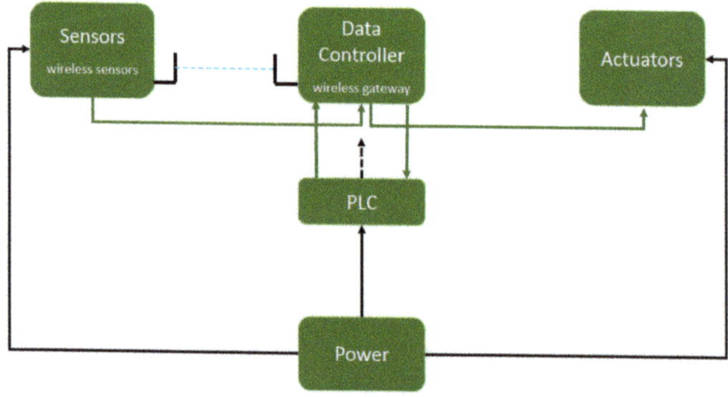

Figure 3. Block diagram of the general operation of the prototype.

The main element of the control system is the PLC. This device is in charge of establishing the rules and sending orders to the actuators so that the control parameters are kept within the optimal value ranges for horticultural production. The control is based on threshold values of the parameters of control which should not be exceeded inside of the greenhouse: maximum temperature (T_{max}), minimum temperature (T_{min}) maximum relative humidity (H_{max}), minimum relative humidity (H_{min}) and minimum concentration of CO_2 ($CO_{2\,min}$). In this way, the PLC receives information from the sensors, with a frequency of 1 min, and acts by sending commands to the actuators so that the control parameters do not exceed the previously programmed thresholds. These thresholds can be modified based on the real needs of each crop. Figures 4–8 show the control loops that the PLC uses to determine the actions that must be performed.

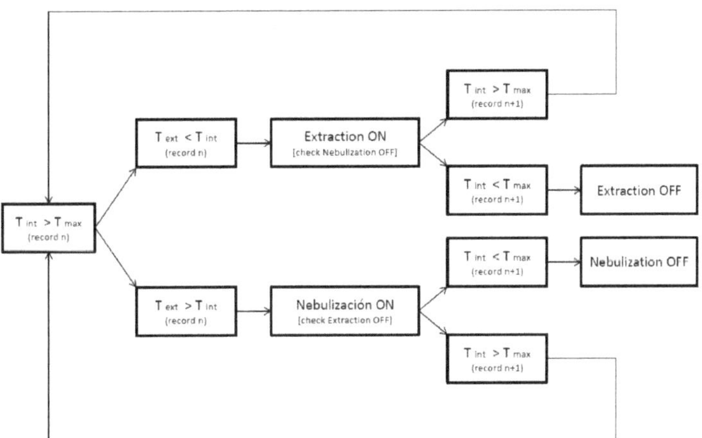

Figure 4. Control loops when $T_{int} > T_{max}$.

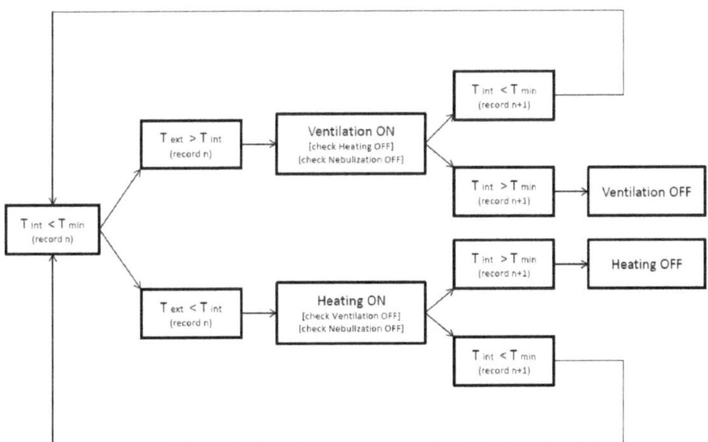

Figure 5. Control loops when $T_{int} < T_{min}$.

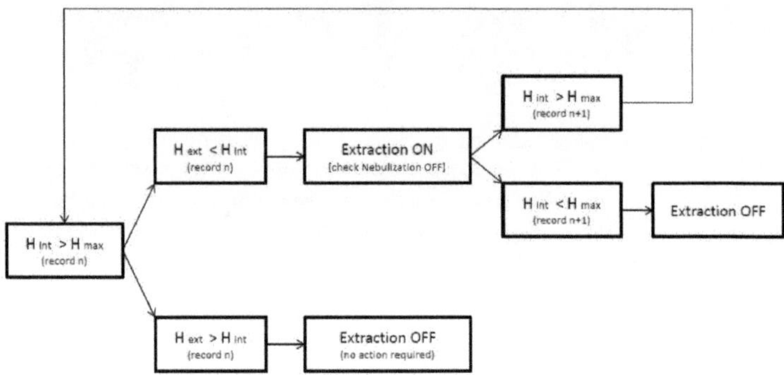

Figure 6. Control loops when $H_{int} > H_{max}$.

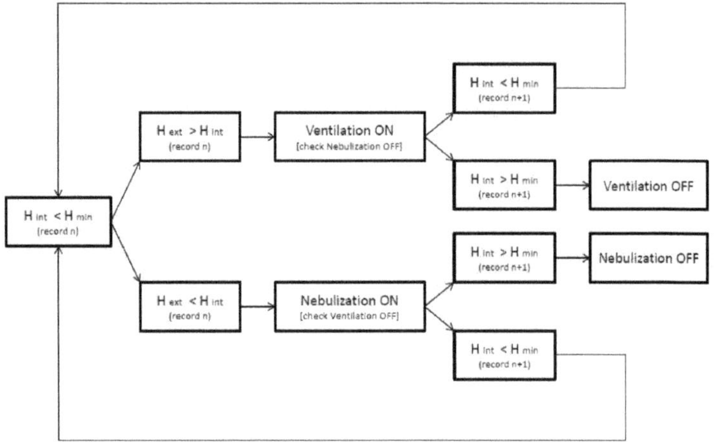

Figure 7. Control loops when $H_{int} < H_{min}$.

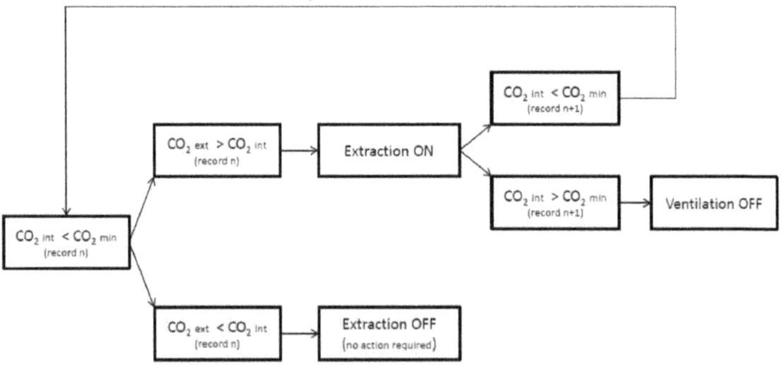

Figure 8. Control loops when $CO_{2\ int} < CO_{2\ min}$.

Both the data concentrator and the PLC are housed in the same electrical control panel that also includes:

- Protection against electrical overloads.
- The system's power supplies.
- The power supply transformers of the solenoid valves.
- Network analyzers.

- The communication gateway with the wireless sensors.
- Contactors to control the loads of the control system of both greenhouses.
- Connectors for sensors.
- Connectors for the actuators.

As previously detailed, three of the four sides of the PV greenhouse section are covered with PV glass: the southwest side of the greenhouse, the northeast side of the greenhouse and the roof. By having three PV production zones, the PV devices were grouped into three different arrays. The southwest façade (12° to the West from South) is made up of 15 vertical PV modules in an array of five columns and three rows with an installed power of 330 Wp. The northeast façade (12° to the East from North) is made up of 25 vertical modules in an array of five columns and five rows with an installed power of 550 Wp. The roof has a southwest orientation (12° to the West from South), it is composed of 20 modules inclined 35° (with respect to the terrain) in an array of five columns and four rows with an installed power of 440 Wp. In this way, the total power of the installation is 1320 Wp.

Figure 9 shows the configuration and interconnection of the three zones. For design reasons of the PV regulator, each glass is connected to its own fuse, to later make the appropriate groups according to the array. The diagram also shows the interconnection with the PV regulator, and from this device to the available electrical storage set (batteries). Two batteries are installed in series and two branches, of this association, in parallel. In this way, one has a 24 VDC bus, since each battery is 12 V, and C 100 is 250 Ah. Finally, an inverter is placed to feed the alternating loads, constituted by the consumption of the greenhouse itself.

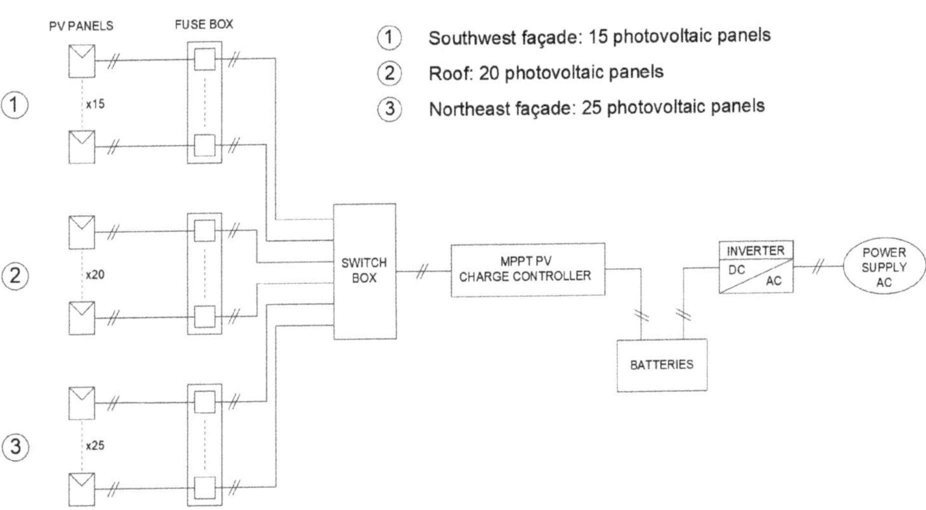

Figure 9. Connection diagram of the PV system and connected to batteries by regulator and inverter.

The devices selected for this project have different technologies and therefore they generate different outputs. The MPC-374 data concentrator is capable of simultaneously acquiring signals from sensors that produce analog outputs, as well as from sensors that work under communication protocols. In the following Table 2, the inputs and outputs of the system devices are shown.

Table 2. Inputs and outputs of system devices.

Sensor	Departure
Outside temperature, humidity and CO 2	Modbus RTU
Indoor temperature, humidity and CO 2	Modbus RTU
PAR radiation	0–10 VDC
Network analyzer	Modbus RTU
Actuators	**Entry**
Extractor fan	0–230 VAC
Heater	0–230 VAC
Nebulizer	0–230 VAC
Others	**Entry**
Drip irrigation	0–230 VAC

Digital sensors and network analyzers are connected to an RTU Modbus communication port, specifically to the UART2. The data is acquired by a physical RS485 interface. A client RTU Modbus is configured on this interface. The sensors are interrogated according to the standard set by this protocol. Actuators are connected to the analog outputs port of the MPC-374. It is a port formed by potential-free control outputs. These outputs reach the coil of each of the contactors that turn on/off the actuators as ordered by the PLC. The coils are attacked with 24 VDC. This turning on/off process is carried out according to the control rules programmed in the PLC. The nebulizers are managed directly by the digital output port included in the ILC-131 PLC.

As previously mentioned, drip irrigation is not part of the actuators. It is controlled directly by the discrete output port of the MPC-374, with an hourly programming time configured in the MPC-374.

Once the measurements of the sensors are stored in the MPC-374, they are already available to be interrogated by the PLC, SCADA or any software that makes the function of master of the Modbus network. The MP C-374, in addition to being able to be interrogated, has been configured to periodically send the information stored by FTP.

The relationship between the energy demanded from all devices of the greenhouse and the PV glass generation was studied at the end of the period of horticultural production by analyzing net balance, comparing energy consumption and actual production.

2.2. Type of Crop

The planting framework depends on the type of crop, and this, in turn, on whether the cycle is spring-summer or autumn-winter. In this project, the study period began on 18 October 2019, looking for an autumn-winter cycle in which lettuce (*Lactuca sativa* L.) and broad beans (*Vicia faba* L.) were planted. The study was concluded on 29 November 2019, coinciding with the collection of lettuces. The complete vegetative cycle of the broad beans had not ended by the time the study concluded, but the recorded growth measurements were sufficient to obtain the first results. Broad beans are common in the Mediterranean diet and are usually grown in greenhouses during the autumn-winter cycle. In addition, broad beans are very compatible with lettuces, so they are often grown together. However, as already stated the main aim of this research was to verify the feasibility of amorphous silicon semitransparent glass technology, rather than a study of the optimum type of crop.

Beans are considered crops improvers because their cultivation has the additional advantage of fixing atmospheric nitrogen in the soil when their roots are symbiotic with *Rhizobium* bacteria. Beans are best cultivated in mild climates. Temperatures above 30 °C, between flowering and fruit setting, can cause flowers and immature pods to fall off, increasing their hardness with the consequent loss of quality. They tolerate moderate frosts, and even strong ones of short duration, as long as they do not occur while in flower. Beans have no tendrils, or terminals or leaf, so they are not a branch plant, i.e., their angular and strong stems keep the plants upright without support. The cultivation is simple, without tutors or support, as long as one takes care that the long pods are not in contact with the ground.

Lettuces prefer uniform warm-temperate temperatures. The seeds do not germinate above 20 °C. Temperatures above 30 °C, during the period between flowering and fruit setting of the pods, can cause abortions of both flowers and immature pods, increasing their fibrosity. They are very sensitive to a lack of water, especially from flowering to pod filling. It is common to associate lettuce with other horticultural species, given its tolerance to shading and its rapid growth, which allows it to be cultivated among larger plants without significant competition for nutrients or causing other harmful effects.

As for the planting frame, a central corridor was set to allow passage and the manipulation of the plants. Beans seeds were spaced about 40 cm apart, as well as lettuce seedlings. To exploit to the maximum the available space, the quincunx schema shown in Figure 10 was created.

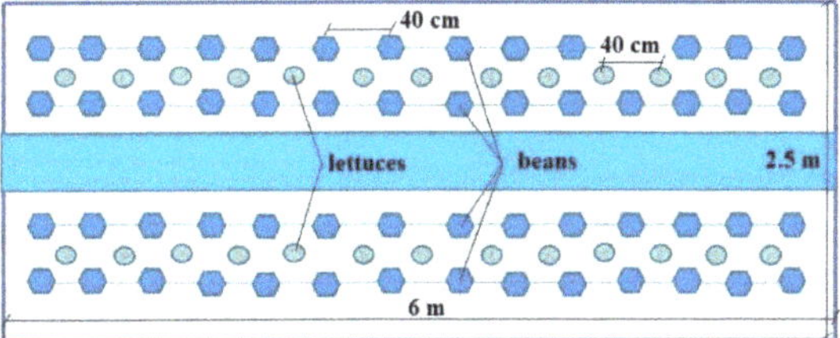

Figure 10. Planting frame in each section of the greenhouse.

Considering the characteristics of both types of crops and to ensure optimal horticultural production, the following thresholds were programmed: T_{min} = 10 °C, T_{max} = 30 °C, H_{min} = 50%, H_{max} = 90%, $CO_{2\,min}$ = 300 ppm.

After harvesting lettuces, about 1.5 months after planting, they were weighed to characterize the growth of both sections and to discern if there were any statistically significant differences between them.

The entire vegetative cycle of beans is much longer than that of lettuces, so it made more sense to monitor their growth rather than their plant production during the study period. Thus, the heights of the seedlings were periodically measured, making the first measurement on 25 October, and the subsequent ones on 6, 22 and 29 November 2019.

3. Results

In order to subsequently be able to study the behavior of the prototype during the growth of the crops, it was necessary to know the behavior and the relationship between the variables that were measured (T, H, CO_2 and PAR) without the corrective action of the actuators (ventilation/extraction, heating and fogging), and without the presence of horticultural crops, that is, before planting. On the other hand, with a minute recording frequency, the amount of data that would be needed to be handle since the greenhouse implementation was immense. Therefore, two representative intervals were selected for data treatment, one before and one after planting. The interval before planting started on 27 August and finished 2 September 2019. The interval after planting started on 25 November and ended 29 November 2019.

3.1. Monitoring Before Planting

The chosen interval corresponds to the course of warm days characterized, at this location, by reaching high temperatures and high radiation. Figure 11 shows the evolution of the control parameters and the PAR radiation, in both sections of the greenhouse. In the figure, 1 refers to the section with PV glass, and 2 to the conventional glass section.

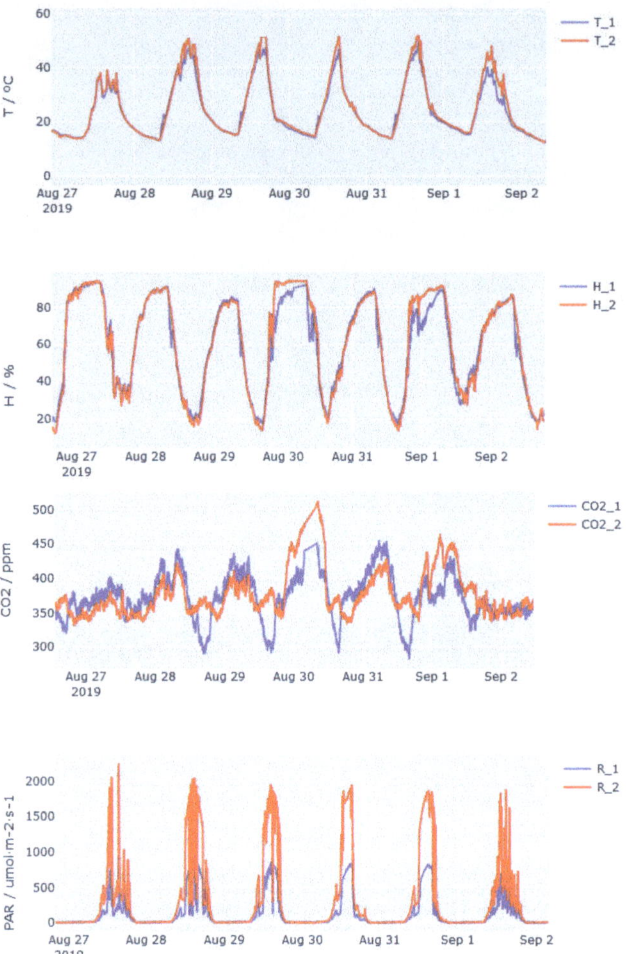

Figure 11. Evolution before planting of temperature, relative humidity, concentration of CO_2 and PAR radiation (1: PV glasses, 2: conventional glasses).

It is observed how the section with PV glass slightly delays the increase in temperature and reaches slightly lower maximum values than the section with conventional glass. Other authors report similar results [29]. This fact is reasonable and desired since PV glass filters a large part of the solar radiation received, allowing less energy to pass into the greenhouse.

Considering that at higher temperature, absolute humidity in the air per unit volume is higher too, the relative humidity follows a reverse curve to that of the temperature, i.e., when the temperature is high, the relative humidity is minimum, and vice versa. The minima of relative humidity occur in the conventional glass section, a logical result because they coincide with the moments of maximum temperatures. Sudden changes in relative humidity are also observed, decreasing a lot during the day and increasing a lot overnight. This is due to the fact that inside the greenhouse there is almost no contribution or extraction of humidity, and the temperature also varies a lot, by up to 30 °C, between day and night.

Regarding the concentration of CO_2, the evolution has displays a reasonable behavior. During the day, it is reduced due to the photosynthesis of some herbs that had grown spontaneously inside the greenhouse, and at night, it increases with their respiration process. It is curious that in the section with PV glass there were, in general, lower levels of CO_2 during the day. This is due to the fact that a greater amount of spontaneous plant mass was developed due to the damping effect of PAR radiation and temperature, which are clearly excessive on those dates and in the chosen location.

PAR radiation is a more complex parameter to evaluate since the sensors are fixed and the Sun moves on the horizon throughout the day, varying the angle of incidence on the sensor. It can be seen that the PAR radiation that the PV glass let through is never greater than 50% of the PAR radiation present in the other section of the greenhouse. This data is consistent with the data provided by the manufacturer (30% transparency).

Figure 12 shows the evolution of the control parameters and PAR radiation on 30 August 2019, in both sections of the greenhouse. Of course, all the variables are related to each other and to the amount of radiation that penetrates the interior.

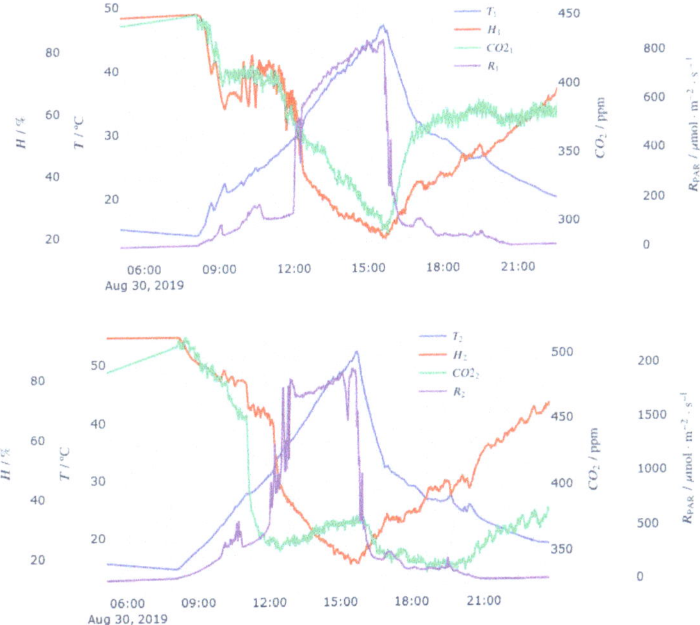

Figure 12. Relationship between temperature, relative humidity, CO_2 concentration and PAR radiation before planting (1: PV glasses, 2: conventional glasses).

In both sections, logical behaviors of all variables are obtained. At night, with the abrupt cessation of radiation, the temperature decreases and the relative humidity and concentration of CO_2 increase. At the beginning of the day, with the onset of radiation, the temperature rises while the relative humidity and CO_2 decrease. However, it is observed how the CO_2 concentration in the conventional glass section rises more slowly at night. This is because, as discussed above, in this section less herbs grew spontaneously due to the high values of radiation and temperature.

3.2. Monitoring after Planting

The chosen interval corresponds to the course of cold days characterized, in this location, by reaching low temperatures and moderate radiation. Figure 13 shows the evolution of the control parameters and PAR radiation in both sections of the greenhouse. The section with PV glass has been labelled as 1, and the section with conventional glass as 2.

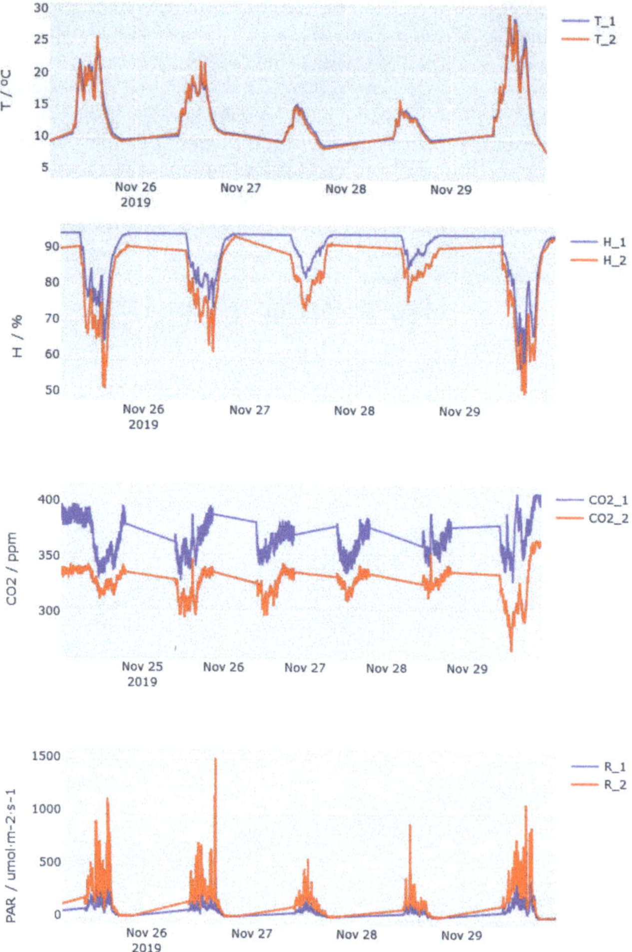

Figure 13. Evolution after planting of temperature, relative humidity, CO_2 concentration and PAR radiation (1: PV glass, 2: conventional glass).

It is seen that practically all the time the action of actuators maintains the temperature at values between 10 °C and 30 °C, the relative humidity between 50% and 90%, and the CO_2 concentration above 300 ppm. This means that the programming of the PLC and its communication with the sensors and actuators is adequate.

The evolution of the temperature curves is practically the same in both sections of the greenhouse due to the temperature correction of the actuators, however, the relative humidity is constantly lower in the part with conventional glass, as in the CO_2 concentration. The explanation, in both cases, must be found in the PAR radiation. It should be noted that the maximum radiation in this period of cold days is around half of that in the warm days period analyzed. This reduction in radiation, in intensity and duration, implies a significant reduction in the photosynthetic capacity of plants, which is why it is a limiting factor that should influence the behavior of the rest of the variables.

Obviously, PAR radiation is higher in the conventional glass section, especially during the central hours of the day. After receiving radiation, plants start the photosynthesis process using water absorbed by the roots and capturing the CO_2, which they incorporate into their plant structures. Consequently, the higher the PAR radiation, the lower the CO_2 concentration in the air.

However, the fact that the humidity is always lower in the conventional glass section does not have such an obvious immediate explanation. It is due to the effect of plant perspiration or loss of water in the form of water vapor. A large amount of water absorbed by the roots reaches the leaves, but only a small part is used in photosynthesis, the rest is lost through transpiration. The plants are being drip watered for their growth, in the same proportion in both sections of the greenhouse, so that the section with the least photosynthetic activity will also be the one that loses the most water through transpiration and, therefore, has the highest relative humidity in the air. As the section with PV glasses receives less PAR radiation, it will have lower photosynthetic activity and higher relative humidity.

Similarly, Figure 14 shows the evolution of the control parameters and PAR radiation on 26 November 2019, in both sections of the greenhouse. The same behavior pattern can be seen in both sections. CO_2 and relative humidity increase at night, when radiation ceases. With the beginning of the day, they decrease again, while the temperature rises with the increase of radiation.

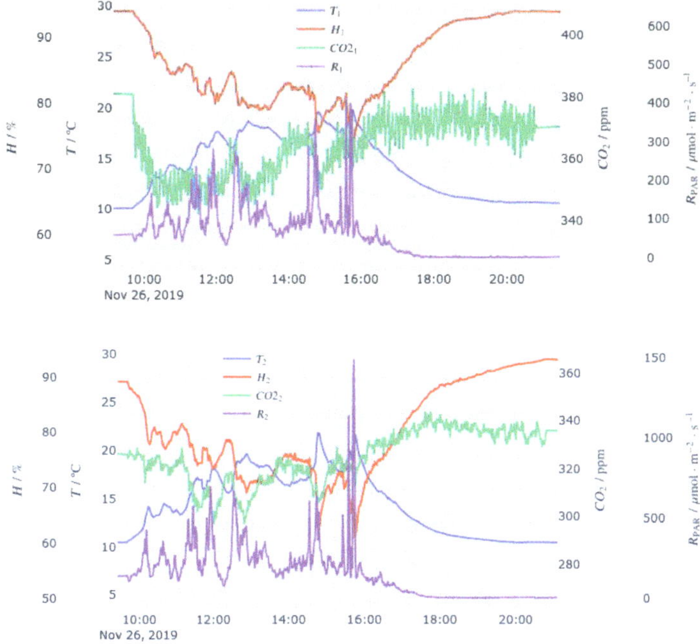

Figure 14. Relationship between temperature, relative humidity, CO_2 concentration and PAR radiation after planting (1: PV glass, 2: conventional glass).

3.3. Horticultural Production

The average mass of the lettuce in the section with conventional glass (mean weight: 510.6 g, CoV: 0.24) was slightly higher than the average mass in the section with PV glass (weight mean: 423.6 g, CoV: 0.37). The mass of lettuces in both sections after harvesting are shown in Figure 15. This is a reasonable result because in the autumn-winter cycle dates there is not much Sun available and the reduced radiation can result in a lower photosynthetic capacity and, therefore, production of plant tissues. Although the difference in means seems significant, statistically it does not reach a value that allows us to affirm with a 95% probability that it is not due to mere chance.

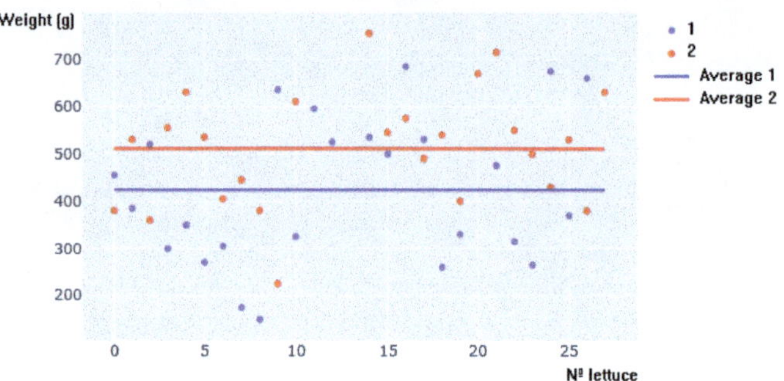

Figure 15. Mass of lettuces after harvesting (1: PV glasses, 2: conventional glasses).

Regarding the production of beans, in Table 3 the heights of the plants is measured vertically from the ground to the upper end thereof. In all measurements made it is observed that the plants are higher in the section with PV glass. It is also seen that the difference in height between both sections is increasing, due to an increase in the growth speed in the section with PV glass. This result is logical and it is a consequence of the lower amount of incident radiation. If not enough radiation is received, the natural survival mechanisms of the plant react as if there were a competition for light in the environment, giving priority to growth in height.

Table 3. Height of broad bean plants.

	25 October 2019		06 November 2019		22 November 2019		29 November 2019	
	Average Height (cm)	CoV	Average Height (cm)	CoV	Average Height (cm)	CoV	Average Height (cm)	CoV
PV glass section	30	0.23	49	0.2	52	0.2	57	0.19
Conventional glass section	26	0.23	44	0.25	46	0.22	50	0.2

In research carried out by other authors on the growth of lettuces with semitransparent monocrystalline silicon PV panels that occupied 20% of the roof surface and that reduced radiation by 35–40%, similar results were obtained with the same level of significance ($\alpha = 0.05$) [30]. In other studies carried out in southern China on the cultivation of tomatoes with integrated semi-transparent solar panels, it was determined that the reduction of radiation surface reduces the diameter and weight of the tomatoes produced, but not significantly. Furthermore, the tomato plants that were in the shaded area had a greater height and a greater number of leaves to compensate the loss of solar radiation, as well as a higher chlorophyll content [31].

Therefore, it is interesting to note that the crops, both lettuce and broad beans, have been able to develop correctly despite the fact part of the incident PAR radiation was subtracted, with part of the light being used for electricity generation.

3.4. Net Energy Balance

In addition to the environmental variables, the monitoring system also recorded the electrical variables related to the consumption of each of the sections separately, and of the sum of both plus the necessary consumption of the control panel electronics (PLC and data concentrator).

During the study period (18 October 2019–29 November 2019), the installed PV system produced 90.15 kWh. In the same period, the energy demanded by the PV section was

56.21 kWh, while the energy demanded by the complete installation was 133.07 kWh. Comparing the produced and the consumed energies, exclusively in the PV section, a positive net balance is obtained which means that PV panels are able to provide 100% of the energy demand. It shows, therefore, that the greenhouse is completely self-sufficient for the studied site. The energy balance also indicates that there is a surplus of excess energy (33.94 kWh), which means that it would not be necessary for the entire greenhouse surface to be covered by PV glasses.

This result is very interesting because in different systems analyzed by other researchers, where radiation was reduced by 35–40%, the PV installation was able to provide only 20% of the energy demanded by the greenhouse [30].

4. Discussion

The use of shading techniques is a very economical solution when trying to protect crops from excessive sunlight and reduce the temperature inside the greenhouse [30]. In research carried out in the south of Spain, with flexible integrated solar panels that covered 9.8% of the surface, the payback of the investment was calculated at 18 years [32]. In other research carried out in southern China, with integrated semi-transparent panels covering 20% of the surface, the payback was 9 years [31]. Based on the indicated references, the payback of investment for the semitransparent amorphous silicon (a-Si) panels, covering 100% of the surface of the roof and major sides of the greenhouse is attractive, which would make the implementation of this technology in those sites with sufficient resources feasible.

The semitransparent PV cell devices provide good light conditions for photosynthesis and plant growth [33]. However, depending on the type of crop, the vegetative period in which it is found, the location or the time of year, the shading caused by semi-transparent glasses can be beneficial or detrimental to horticultural development. The shade used in this article may be a more suitable solution to crops that do not have high solar radiation requirements (leaf crops: lettuce, chard, endive or spinach; root crops: beet, carrot, celery, leek or radish; crop fruits: broad bean, pea or strawberry; aromatic crops: mint, spearmint or parsley). However, it must be taken into account that when working with living beings the data has a lot of variability even within the same section of the greenhouse. For this reason, it is important to continue the investigation, collecting more data from different crops and in different cycles, in order to draw more relevant conclusions. The authors expect to analyze different crop types and cycles with this technology in subsequent studies.

On the other hand, the analysis of the transmittance of PV glass as a function of the wavelength constitutes a key field of research for the plants' growth, depending on whether or not the maximum absorption coincides with the wavelengths at which the photosynthetic molecules absorb the most. Thus, depending on the plant species, there are different photosynthetic molecules, which have different absorption spectra for the conversion of sunlight photons into the generation of organic matter.

5. Conclusions

BIPV applied to horticultural production constitutes a research and development completely in line with the latest trends in sustainable building, which advocate the incorporation of urban gardens in the building for the consumption of unprocessed ecological products.

The PV glass semitransparent amorphous silicon (a-Si) placement, covering the whole surface of the roof (PV_R of 100%) and of the main sides of the greenhouse, influences the growth of horticultural crops, slightly reducing the biomass production of the plants and accelerating the apical growth mechanism of heliophilic plants. However, from a statistical point of view, this influence is negligible so the studied technology is viable for horticultural production. Nevertheless, to obtain results with a greater scope and practical application, it is necessary to extend the study to other crops and growing seasons.

The energy balance carried out indicates that the energy produced by the PV section is greater than the energy demand, which shows that the greenhouse is completely viable and self-sufficient when installed at a site with the adequate solar resource.

Author Contributions: Conceptualization, J.-R.A. and L.H.-C.; methodology, J.-R.A., L.H.-C., S.G.-S., M.E.-G., V.A.-G. and M.-Á.M.-G.; software, V.A.-G.; validation, L.H.-C. and M.E.-G.; formal analysis, M.E.-G. and V.A.-G.; investigation, J.-R.A., L.H.-C., S.G.-S., M.E.-G., V.A.-G. and M.-Á.M.-G.; resources, J.-R.A., L.H.-C. and M.-Á.M.-G.; data curation, M.E.-G. and V.A.-G.; writing—original draft preparation, L.H.-C., S.G.-S. and V.A.-G.; writing—review and editing, J.-R.A., S.G.-S. and L.H.-C.; visualization, L.H.-C., M.E.-G. and M.-Á.M.-G.; supervision, J.-R.A. and L.H.-C.; project administration, L.H.-C.; funding acquisition, L.H.-C. and M.-Á.M.-G. All authors have read and agreed to the published version of the manuscript.

Funding: The project "Aplicación de la tecnología fotovoltaica integrada en edificios (BIPV) para invernaderos agrícolas" with file number 04/16/AV/0001 has been carried out within the framework of the "Proyectos de I + D en PYMES" program. This project has been funded by the Fondo Europeo de Desarrollo Regional (FEDER) of the European Union and the Junta de Castilla y León, through the Institute for Business Competitiveness of Castilla y León (ICE), with the aim of promoting research, technological development and innovation. The project has been led by Onyx Solar and the UVa has participated in this project.

Institutional Review Board Statement: Not applicable.

Informed Consent Statement: Not applicable.

Data Availability Statement: Confidential data.

Acknowledgments: Authors thank the company Onyx Solar Energy SL. The financing of this project.

Conflicts of Interest: The authors declare no conflict of interest. The funders had no role in the design of the study; in the collection, analyses, or interpretation of data; in the writing of the manuscript, or in the decision to publish the results.

Abbreviations

a-Si	amorphous silicon.
BIPV	building-integrated photovoltaic technology.
$CO_{2\,ext}$	outdoor CO_2 concentration.
$CO_{2\,int}$	indoor CO_2 concentration.
$CO_{2\,min.}$	minimum CO_2 concentration.
DC	direct current.
DG	distributed generation.
FTP	file transfer protocol.
H_{ext}	outdoor air humidity.
H_{int}	indoor air humidity.
H_{max}	maximum air humidity.
H_{min}	minimum air humidity.
MPPT	maximum power point tracker.
PAR	photosynthetically active radiation.
PLC	programmable logic controller.
PV	photovoltaic.
PVR	ratio of horizontal surface covered by solar panels placed on the roof.
SCADA	supervisory control and data acquisition.
T_{ext}	outdoor temperature.
T_{int}	indoor temperature.
T_{max}	maximum temperature.
T_{min}	minimum temperature.
UVa	University of Valladolid.

References

1. Aryanpur, V.; Atabaki, M.S.; Marzband, M.; Siano, P.; Ghayoumi, K. An overview of energy planning in Iran and transition pathways towards sustainable electricity supply sector. *Renew. Sustain. Energy Rev.* **2019**, *112*, 58–74. [CrossRef]
2. Locker, C.R.; Torkamani, S.; Laurenzi, I.J.; Jin, V.L.; Schmer, M.R.; Karlen, D.L. Field-to-farm gate greenhouse gas emissions from corn stover production in the Midwestern U.S. *J. Clean. Prod.* **2019**, *226*, 1116–1127. [CrossRef]
3. WBCSD. *Energy Efficiency in Buildings Transforming the Market*; WBCSD: Washington, DC, USA, 2009; ISBN 978-3-940388-44-5.
4. Lu, Y.; Zhang, X.P.; Huang, Z.; Lu, J.; Wang, D. Impact of introducing penalty-cost on optimal design of renewable energy systems for net zero energy buildings. *Appl. Energy* **2019**, *235*, 106–116. [CrossRef]
5. Hernández-Callejo, L.; Gallardo-Saavedra, S.; Alonso-Gómez, V. A review of photovoltaic systems: Design, operation and maintenance. *Sol. Energy* **2019**, *188*, 426–440. [CrossRef]
6. Foster, R.; Cota, A. Solar water pumping advances and comparative economics. *Energy Procedia* **2014**, *57*, 1431–1436. [CrossRef]
7. Chang, R.; Cao, Y.; Lu, Y.; Shabunko, V. Should BIPV technologies be empowered by innovation policy mix to facilitate energy transitions?-Revealing stakeholders' different perspectives using Q methodology. *Energy Policy* **2019**, *129*, 307–318. [CrossRef]
8. Bot, G.; Van De Braak, N.; Challa, H.; Hemming, S.; Rieswijk, T.; Straten, G.V.; Verlodt, I. The solar greenhouse: State of the art in energy saving and sustainable energy supply. In *Acta Horticulturae: International Society for Horticultural Science*; International Society for Horticultural Science: Leuven, Belgium, 2005; Volume 691, pp. 501–508.
9. Hassanien, R.H.E.; Li, M.; Lin, W.D. Advanced applications of solar energy in agricultural greenhouses. *Renew. Sustain. Energy Rev.* **2016**, *54*, 989–1001. [CrossRef]
10. Harjunowibowo, D.; Ding, Y.; Omer, S.; Riffat, S. Recent active technologies of greenhouse systems—A comprehensive review. *Bulg. J. Agric. Sci.* **2018**, *24*, 158–170.
11. Yano, A.; Furue, A.; Kadowaki, M.; Tanaka, T.; Hiraki, E.; Miyamoto, M.; Ishizu, F.; Noda, S. Electrical energy generated by photovoltaic modules mounted inside the roof of a north-south oriented greenhouse. *Biosyst. Eng.* **2009**, *103*, 228–238. [CrossRef]
12. Yano, A.; Kadowaki, M.; Furue, A.; Tamaki, N.; Tanaka, T.; Hiraki, E.; Kato, Y.; Ishizu, F.; Noda, S. Shading and electrical features of a photovoltaic array mounted inside the roof of an east-west oriented greenhouse. *Biosyst. Eng.* **2010**, *106*, 367–377. [CrossRef]
13. Kadowaki, M.; Yano, A.; Ishizu, F.; Tanaka, T.; Noda, S. Effects of greenhouse photovoltaic array shading on Welsh onion growth. *Biosyst. Eng.* **2012**, *111*, 290–297. [CrossRef]
14. Castellano, S. Photovoltaic greenhouses: Evaluation of shading effect and its influence on agricultural performances. *J. Agric. Eng.* **2014**, *45*, 168–175. [CrossRef]
15. Cossu, M.; Murgia, L.; Ledda, L.; Deligios, P.A.; Sirigu, A.; Chessa, F.; Pazzona, A. Solar radiation distribution inside a greenhouse with south-oriented photovoltaic roofs and effects on crop productivity. *Appl. Energy* **2014**, *133*, 89–100. [CrossRef]
16. Cossu, M.; Cossu, A.; Deligios, P.A.; Ledda, L.; Li, Z.; Fatnassi, H.; Poncet, C.; Yano, A. Assessment and comparison of the solar radiation distribution inside the main commercial photovoltaic greenhouse types in Europe. *Renew. Sustain. Energy Rev.* **2018**, *94*, 822–834. [CrossRef]
17. Cossu, M.; Yano, A.; Solinas, S.; Deligios, P.A.; Tiloca, M.T.; Cossu, A.; Ledda, L. Agricultural sustainability estimation of the European photovoltaic greenhouses. *Eur. J. Agron.* **2020**, *118*, 126074. [CrossRef]
18. Yano, A.; Onoe, M.; Nakata, J. Prototype semi-transparent photovoltaic modules for greenhouse roof applications. *Biosyst. Eng.* **2014**, *122*, 62–73. [CrossRef]
19. Cossu, M.; Yano, A.; Li, Z.; Onoe, M.; Nakamura, H.; Matsumoto, T.; Nakata, J. Advances on the semi-transparent modules based on micro solar cells: First integration in a greenhouse system. *Appl. Energy* **2016**, *162*, 1042–1051. [CrossRef]
20. Subhani, W.S.; Wang, K.; Du, M.; Wang, X.; Yuan, N.; Ding, J.; Liu, S.F. Anti-solvent engineering for efficient semitransparent CH3NH3PbBr3 perovskite solar cells for greenhouse applications. *J. Energy Chem.* **2019**, *34*, 12–19. [CrossRef]
21. Gupta, N.; Tiwari, A.; Tiwari, G.N. A thermal model of hybrid cooling systems for building integrated semitransparent photovoltaic thermal system. *Sol. Energy* **2017**, *153*, 486–498. [CrossRef]
22. Zisis, C.; Pechlivani, E.M.; Tsimikli, S.; Mekeridis, E.; Laskarakis, A.; Logothetidis, S. Organic photovoltaics on greenhouse rooftops: Effects on plant growth. In *Materials Today: Proceedings*; Elsevier Ltd.: Amsterdam, The Netherlands, 2020; Volume 21, pp. 65–72.
23. Song, W.; Fanady, B.; Peng, R.; Hong, L.; Wu, L.; Zhang, W.; Yan, T.; Wu, T.; Chen, S.; Ge, Z. Foldable Semitransparent Organic Solar Cells for Photovoltaic and Photosynthesis. *Adv. Energy Mater.* **2020**, *10*. [CrossRef]
24. Emmott, C.J.M.; Röhr, J.A.; Campoy-Quiles, M.; Kirchartz, T.; Urbina, A.; Ekins-Daukes, N.J.; Nelson, J. Organic photovoltaic greenhouses: A unique application for semi-transparent PV? *Energy Environ. Sci.* **2015**, *8*, 1317–1328. [CrossRef]
25. Marucci, A.; Cappuccini, A. Dynamic photovoltaic greenhouse: Energy efficiency in clear sky conditions. *Appl. Energy* **2016**, *170*, 362–376. [CrossRef]
26. Li, Z.; Yano, A.; Cossu, M.; Yoshioka, H.; Kita, I.; Ibaraki, Y. Electrical Energy Producing Greenhouse Shading System with a Semi-Transparent Photovoltaic Blind Based on Micro-Spherical Solar Cells. *Energies* **2018**, *11*, 1681. [CrossRef]
27. Moretti, S.; Marucci, A. A photovoltaic greenhouse with variable shading for the optimization of agricultural and energy production. *Energies* **2019**, *12*, 2589. [CrossRef]
28. Aira, J.R. Aprovechamiento y Valorización de la Madera. In *Parques Naturales y Espacios Naturales Protegidos, La Gestión del Parque Natural de la Serranía de Cuenca*; Universidad de Castilla-La Mancha: Toledo, Spain, 2016; pp. 17–56. ISBN 978-84-9044-224-1.

29. Hernández-Callejo, L.; Alonso-Gómez, V.; Eugenio-Gozalbo, M.; Rico-Rodríguez, E.; Huerta-Illera, I.; del Caño-González, T. Invernadero Fotovoltaicoes. In *Efficient, Sustainable, and Fully Comprehensive Smart Cities. II Ibero-American Congress of Smart Cities (ICSC-CITIES 2019)*; Leite, V., Callejo, L.H., Prieto, J., Cañón, C.L.Z., Ferreira, Â., Nesmachnow, S., Peña, F.C., Eds.; Editorial Universidad Santiago de Cali: Soria, Spain, 2019; pp. 434–455.

30. Hassanien, R.H.E.; Li, M. Influences of greenhouse-integrated semi-transparent photovoltaics on microclimate and lettuce growth. *Int. J. Agric. Biol. Eng.* **2017**, *10*, 11–22. [CrossRef]

31. Hassanien, R.H.E.; Li, M.; Yin, F. The integration of semi-transparent photovoltaics on greenhouse roof for energy and plant production. *Renew. Energy* **2018**, *121*, 377–388. [CrossRef]

32. Ureña-Sánchez, R.; Callejón-Ferre, Á.J.; Pérez-Alonso, J.; Carreño-Ortega, Á. Greenhouse tomato production with electricity generation by roof-mounted flexible solar panels. *Sci. Agric.* **2012**, *69*, 233–239. [CrossRef]

33. Shi, H.; Xia, R.; Zhang, G.; Yip, H.L.; Cao, Y. Spectral Engineering of Semitransparent Polymer Solar Cells for Greenhouse Applications. *Adv. Energy Mater.* **2019**, *9*. [CrossRef]

 *agronomy*

Article

Techno-Economic Viability of Agro-Photovoltaic Irrigated Arable Lands in the EU-Med Region: A Case-Study in Southwestern Spain

Guillermo P. Moreda [1], Miguel A. Muñoz-García [1,*], M. Carmen Alonso-García [2] and Luis Hernández-Callejo [3]

1   LPF-TAGRALIA, Departamento de Ingeniería Agroforestal, ETS de Ingeniería Agronómica, Alimentaria y de Biosistemas, Universidad Politécnica de Madrid, 28040 Madrid, Spain; guillermo.moreda@upm.es
2   Unidad de Energía Solar Fotovoltaica, CIEMAT, 28040 Madrid, Spain; carmen.alonso@ciemat.es
3   Departamento de Ingeniería de Sistemas y Automática, Universidad de Valladolid, 42004 Soria, Spain; luis.hernandez.callejo@uva.es
*   Correspondence: miguelangel.munoz@upm.es; Tel.: +34-9-1067-0968

Abstract: Solar photovoltaic (PV) energy is positioned to play a major role in the electricity generation mix of Mediterranean countries. Nonetheless, substantial increase in ground-mounted PV installed capacity could lead to competition with the agricultural use of land. A way to avert the peril is the electricity-food dual use of land or agro-photovoltaics (APV). Here, the profitability of a hypothetical APV system deployed on irrigated arable lands of southwestern Spain is analyzed. The basic generator design, comprised of fixed-tilt opaque monofacial PV modules on a 5 m ground-clearance substructure, featured 555.5 kWp/ha. Two APV shed orientations, due south and due southwest, were compared. Two 4-year annual-crop rotations, cultivated beneath the heightened PV modules and with each rotation spanning 24 ha, were studied. One crop rotation was headed by early potato, while the other was headed by processing tomato. All 9 crops involved fulfilled the two-fold condition of being usually cultivated in the area and compatible with APV shed intermitent shading. Crop revenues under the partial shading of PV modules were derived from official average yields in the area, through the use of two alternative sets of coefficients generated for low and high crop-yield shade-induced penalty. Likewise, two irrigation water sources, surface and underground, were compared. Crop total production costs, PV system investment and operating costs and revenues from the sale of electricity, were calculated. The internal rates of return (IRRs) obtained ranged from a minimum of 3.8% for the combination of southwest orientation, early-potato rotation, groundwater and high shade-induced crop-yield penalty, to a maximum of 5.6% for the combination of south orientation, processing-tomato rotation, surface water and low shade-induced crop-yield penalty.

Keywords: agrophotovoltaic; agrivoltaic; dual-land use; solar sharing; solar photovoltaic energy; water–food–energy nexus

check for **updates**

Citation: Moreda, G.P.; Muñoz-García, M.A.; Alonso-García, M.C.; Hernández-Callejo, L. Techno-Economic Viability of Agro-Photovoltaic Irrigated Arable Lands in the EU-Med Region: A Case-Study in Southwestern Spain. *Agronomy* **2021**, *11*, 593. https://doi.org/10.3390/agronomy11030593

Academic Editor: Ajit Govind

Received: 1 February 2021
Accepted: 17 March 2021
Published: 20 March 2021

Publisher's Note: MDPI stays neutral with regard to jurisdictional claims in published maps and institutional affiliations.

1. Introduction and Objectives

Nowadays, most countries worldwide are aware of the importance of preserving nature. Environment protection includes, amongst others, measures to limit the use of non-recyclable materials and to reduce the emission of greenhouse-effect gases (GHGs). Reduction in GHGs emission entails burning less fossil fuels and increasing the share of renewable energies in the electricity generation mix. The major renewable sources of electricity, wind and solar, intrinsically non-dispatchable due to the intermittency of their resource, could be backed-up by hydrogen fuel cells in the future. European Union (EU) member states are promoting increases in wind and solar installed capacity. Hereof, Spain and Italy planned national levels of 42% and 30%, respectively, of energy from renewable sources in their gross final energy consumption in 2030 [1,2].

Solar power plants produce electricity based on the photovoltaic (PV) effect or by concentrating solar energy onto a heat-transfer-fluid which produces the steam that drives a turbine-generator set. In many countries, ground-mounted solar PV power plants have become familiar in the rural landscape, being popularly known as solar farms. By the end of 2019, installed solar PV power capacity in Spain stood at 8913 MW, representing about 8% of the total installed power capacity in Spain [3].

Substantial further increase in ground-mounted PV power capacity could eventually lead to conflict of interest with the agricultural use of land. This could jeopardize the stability of agricultural produce prices, which should be carefully considered. In 1982, Goetzberger and Zastrow [4] analyzed the possibility of combining agricultural and electric-energy production in the same plot. This dual use of land was later coined in the literature as agrivoltaic or agrophotovoltaic. Here, we use the latter portmanteau, abbreviated as APV. The main differences between a conventional ground-mounted PV power plant and an APV system are:

i.   the spacing between PV module rows in APV systems is greater, to let more irradiation pass through and hit the crop; in conventional ground-mounted PV power plants, row spacing is kept to the minimum compatible with tolerable row self-shading.

1.   the PV modules in APV systems are substantially heightened above the ground, to decrease shade intensity and also to allow agricultural machinery operate beneath; thus, while in conventional ground-mounted PV power plants the vertical distance of the modules bottom edge to the ground is 0.5–1 m, in APV it is 5–6 m.

In 2010, a first APV prototype was erected by Dupraz et al. in Montpellier, France [5], using fixed-tilt PV modules. In their experiments, the main crop cultivated was lettuce [6]. Valle et al. [7] reported on the extension of the Montpellier 2010 prototype with sun-tracking PV modules. In 2016, a fixed-tilt 194.4 kWp APV array with bifacial PV modules and a ground clearance of 5 m was erected in Herdwangen, Germany [8]. The reason for using bifacial PV modules was two-fold: First, to harness snow reflectivity to produce more electricity; and second, to decrease crop shading, thanks to higher transparency of bifacial modules compared to monofacial counterpart. Schindele et al. [8] concluded that their system was profitable for potato but not for wheat. Dinesh and Pearce [9] concluded that PV installed capacity could be increased between 40 and 70 GW if lettuce cultivation alone were converted to APV systems in the United States of America. They recommended exploring the outputs for different crops and geographic areas, to determine the potential of APV farming worldwide. Recently, the consortium SolarPower Europe proposed to integrate a "European Agri-PV strategy" within the future Common Agricultural Policy (CAP) [10]. Hitherto, APV projects in Europe have been of limited acreage. To our knowledge, the largest APV complex, with 2.67 MW, spans 3.2 ha of raspberry near Arnhem, The Netherlands [11].

In APV systems, both the PV array and the understorey crop benefit mutually. For instance, in a watermelon field in the EU-Med region, the shade casted by the heightened PV modules could circumvent the need for anti-fruit-cracking solar protector spraying of the fruits. Apart from the economic saving for the farmer, this is beneficial for the environment. Another example of synergy is the soil moisture condition favored by APV sheds [12] that can save irrigation water. The latter is important for several reasons: First, environmental benefit; second, reduced cultivation cost; third, limited crop yield decrease in case of irrigation water allocation restricted due to drought; fourth, possibility of irrigating an acreage only slightly smaller than that of a non-drought year.

Albeit in Spain average yield per unit area of irrigated crops is 6.5 times greater than that of rainfed agriculture [13], drought episodes make granted water allocations not always deliverable. Irrigation blue water shortage is partly responsible for the difference between irrigable and irrigated area in many countries. Thus, in 2016 the share of irrigable and irrigated areas in the total Utilized Agricultural Area of Spain were of 15.7% and 13.2% respectively [14]. In the same year 2016, the corresponding shares were 32.6% and 20.2% for Italy and of 29.7% and 23.6% for Greece.

The objectives of this work were:

1. to design two irrigated annual crop rotations whereof crops are usually cultivated in the area of study and compatible with partial shading imposed by APV sheds.
2. to thoroughly determine the stream of expenditure and revenues for both agricultural and electricity production, with the final aim of analyzing the profitability of APV system for each combination of APV shed orientation (due south/ southwest), source of irrigation water (surface/underground), shade-induced crop yield penalty (low/ high) and crop rotation (early potato/processing tomato).

2. Materials and Methods

A hypothetical case-study was arranged with annual irrigated crops cultivated under APV sheds in the municipality of Brenes, close to the city of Seville, in southwestern Spain. The centroid of the site sits at 37°33′22″ N and 5°50′8″ W (Datum ETRS89), standing on average altitude of 40 m a.s.l. Some major woody and arable crops cultivated in the area are: Olive, citrus, almond, peach, alfalfa, early potato, maize, processing tomato, cotton and sunflower. Amongst the annual arable crops, we selected potato to be rotation-head, since under-shading yield data were found in the literature [15] for this crop. Based on agronomical considerations detailed in the next sub-section, a four-year rotation headed by early-potato was designed. Taking into account the 6 ha average size of the agricultural unit plot in Brenes, a total acreage of 24 ha was analyzed. Lettuce, a shade-tolerant crop with documented under-shade yield data, was disregarded in view of its limited cultivation in the area [16]. Conversely, cotton, a traditional local annual crop, was discarded because according to Weselek et al. [17] it does not thrive in shade. For the sake of universality, we considered as if land consolidation had not been implemented in the area of study. Thereby, the spatial distribution considered is fragmented, i.e., the four 6 ha plots are not adjacent to each other (Figure 1).

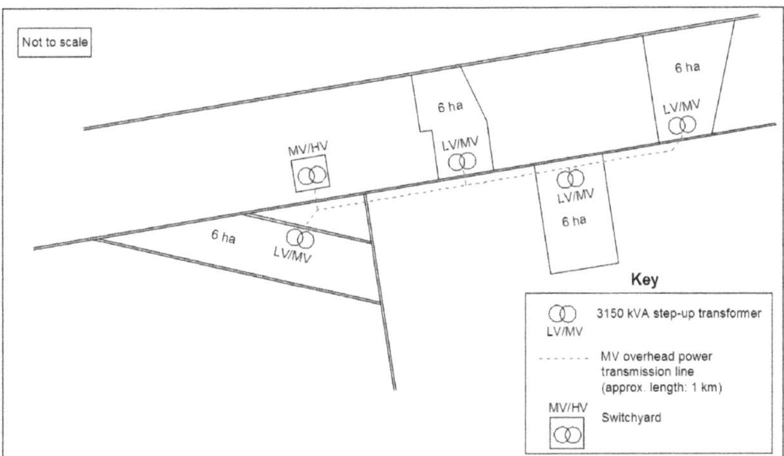

Figure 1. Spatial distribution of the four plots totalizing 24 ha, the medium voltage power transmission line budgeted and the pre-existing grid-connection switchyard.

2.1. PV System

The basic APV shed considered consisted of 22 non-tracking and heightened supporting structures aligned in two parallel swaths of 11 supporting structures each Figure 2. The PV modules are arranged in groups of 24 modules on each supporting structure. The fixed-tilt angle is of 27° (the local latitude minus 10°). The PV module considered was the opaque-monofacial-polycrystalline CSP290-60, of 290 Wp [18]. The dimensions of this module are 1640 mm × 992 mm and it weighs 18.2 kg. The supporting structures are equispaced 9.5 m. The mechanical configuration of the basic APV shed, including lateral (Northeast-Southwest direction for the due SW shed orientation depicted) lattice bracing that leave a ground-clearance of 5 m, is similar to the one reported by Schindele et al. [8]. This substructure is of known cost and would be valid for the location of Seville, where snow and wind loads are less or equal than in Herdwangen. The pillars of the substructure are fixed to the ground by means of a so-called spider-shaped anchor made of an anchoring bush plate with long threaded rods assembled in a circular fashion [19].

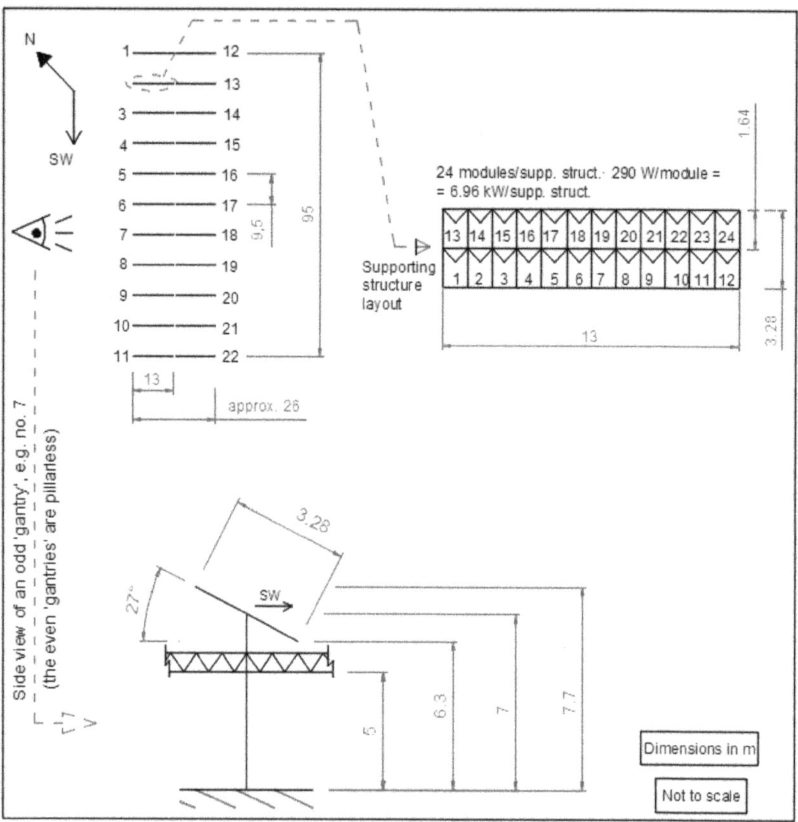

Figure 2. Basic Southwest-oriented agrophotovoltaic shed featuring an installed PV capacity of 153.1 kWp and connected to a 150 kW inverter.

With regard to the PV modules azimuthal angle, two orientations were compared: due southwest and due south. Although the latter is the one that maximizes electric production in the northern hemisphere, Beck et al. [20] concluded that either southeast or southwest is preferable for the crop cultivated beneath the APV shed, since ground radiation distribution is more uniform. The increased radiation uniformity favors crop plants isochronous ripening, which is particularly important for arable crops, usually harvested in mechanized or semi-mechanized one-single pass. Here, we used SAM 3D scene shade calculator [21] to compare due South (Figure A1, Appendix A) and due Southwest (Figure A2) orientations.

The number of PV modules in the basic APV shed is

$$24\frac{\text{PV modules}}{\text{Supporting structure}} \cdot 22\frac{\text{Supporting structures}}{\text{Basic APV shed}} = 528\,\text{PV modules/Basic APV shed}$$

The corresponding peak power is

$$528 \, \frac{\text{PV modules}}{\text{Basic APV shed}} \cdot 290 \frac{\text{Wp}}{\text{PV module}} = 153120 \, \text{Wp} \cong 153.1 \, \text{kWp/Basic APV shed}$$

To calculate the ground area covered by the basic APV shed, we have to multiply its overall width (26 m (Figure 2)) by its overall length. The latter, in turn, is the sum of 95 m (10 × 9.5 m, Figure 2) plus 1.5 m (the horizontal projection of the last module row cantilever 1.64 m to an angle of 27°) plus 9.5 m, i.e., 106 m. Hence,

$$26 \, \text{m} \times 106 \, \text{m} = 2756 \, \text{m}^2 = 0.2756 \, \text{ha}$$

To calculate the ground coverage ratio (GCR), the overall module surface area has to be first computed as:

$$528 \, \text{PV modules} \cdot \frac{1.68 \, \text{m} \cdot 0.992 \, \text{m}}{\text{PV module}} = 880 \, \text{m}^2$$

Then, the GCR is

$$\frac{880 \, \text{m}^2}{2756 \, \text{m}^2} = 32\%$$

Power density is

$$\frac{153.1 \, \text{kWp}}{0.2756 \, \text{ha}} \cong 555.5 \, \text{kWp/ha}$$

Considering an effective land area of 5.7 ha, obtained by reducing by 5% the size of the agricultural unit plot to account for mismatch between plot legal boundaries and APV shed orientation, the PV capacity installed in each of the four 6 ha plots is

$$555.5 \frac{\text{kWp}}{\text{ha}} \cdot 5.7 \, \frac{\text{ha}}{\text{plot}} = 3166 \, \text{kWp/plot}$$

The inverter considered is the SMA SHP Peak 3, which features a nominal AC power of 150 kW [22]. The number of inverters in the agricultural unit plot would be

$$\frac{3166 \, \text{kWp /agricultural unit plot}}{153.1 \, \text{kWp/inverter}} \cong 21 \, \text{inverters /agricultural unit plot}$$

The lifespan considered for the APV system is 25 years, the conventional lifespan of the PV modules. The service life assigned to the inverters is 13 years, which entails inverters replacement in the year 14. Module degradation was computed by means of a degradation coefficient, assigned a value of 1 until year 11 and annually decreased by 0.5% from the year 12 onwards. A simulation was run to estimate annual income from the sale of the energy generated by the PV system. Simulation was done using SISIFO [23], an online free simulation tool for the quality and bankability of PV systems. Table 1 presents a compilation of the main data fed to the SISIFO PV simulator.

2.2. Irrigated Crops

In the EU-Med countries, irrigation water use represents an average 70% of total water withdrawals [24]. On the other hand, more than 80% of the irrigated land-acreage in Spain, involving $7 \cdot 10^5$ irrigators and $2 \cdot 10^6$-hectare, is serviced by an irrigation district [25]. According to Masia et al. [26], surface water from reservoirs represents most of the water stewardshipped and distributed by irrigation districts in the EU-Med region. Concurrently, underground water abstractions in Spain irrigate over one-third of the country irrigable area [27]. Our study is on the arable lands of the Guadalquivir river valley near Seville, where the climate is Mediterranean-oceanic [28]. According to the site coordinates, water would be served by the Comunidad de Regantes del Valle Inferior del Guadalquivir irrigation district, hereinafter abbreviated as Valle Inferior Irrigation District (VIID). Their irrigation scheme provides surface water with a pressure head (p_h) of 441 kPa of water measured at the pumping station [29].

Table 1. Input data for PV simulation of a 5.7 ha agro-photovoltaic plant oriented due Southwest, in Brenes (Seville, Spain).

SISIFO Simulator Input Data	
Site Geographical Latitude	37.557351° N
Site geographical longitude	5.834142° W
Local altitude (m)	40
Meteorological data type	TMY [a]
PV system peak power (kWp)	3166
PV system peak power per inverter (kWp)	153.1
Inverter nominal power (kW)	150
Real power/peak power (dimensionless)	0.98
PV system peak power per transformer (kWp)	3166
Generator inclination or PV modules tilt angle (°)	27
Generator orient. or azimuth angle (°)	−45
Generator height at supporting structure center (m)	7
Separation among structures (dimensionless)	3 [b]
PV generator width (dimensionless)	8 [c]
Deviation of back structure (dimensionless)	0 [d]
LV/MV transformer power (kVA)	3150
LV/MV transformer iron losses (kW)	32
LV/MV transformer copper losses (kW)	32
DC wiring losses (% of peak power)	2.0
AC wiring losses between inverter and LV/MV transformer (% of peak power)	2.0
Soiling impact (%)	1.0

[a] Typical meteorological year. [b] 9.5 m/3.28 m ≈ 3. [c] 26 m/3.28 m ≈ 8. [d] 0 m/26 m = 0.

Instead of continuous mono-cropping, a crop rotation was designed to submit to the principles of sustainable farming. The crop selected as rotation head was early-potato. To do the study more comprehensive, we extended it to an alternative four-year crop rotation. The head-of-rotation in this case was processing-tomato, which in the last years competes with early-potato in local farmers preferences. It is worthy of note that processing-tomato is also mainstream in other EU-Med countries such Italy. The main difference between the two crop rotations is that the early-potato rotation is symmetrical, unlike the processing-tomato counterpart. The latter is asymmetrical because tomato withstands well—or indeed "'prefers'"—to be cultivated up to thrice on the same plot, whereas potato is required to be cultivated in a different plot each year.

The crops included in the early-potato rotation, apart from the potato (Solanum tuberosum L.) itself, were: Canola (*Brassica napus*), onion (*Allium cepa* L.), faba bean (*Vicia faba* L.) and forage maize (*Zea mays* L.). The latter two are cultivated on the same plot one after the other in the same year, practice known as sequential cropping or double cropping (Table 2). This scheduling was designed following the sustainable agriculture principles of: (i) avoid cultivating two demanding crops one after the other in the same plot; and (ii) for every plot, avoid repeating the botanic family of the previous year.

Table 2. The early-potato rotation designed, wherein potato returns to each plot every four years.

Year	Plot 1	Plot 2	Plot 3	Plot 4
1	FB-FM [a]	Canola	Potato	Onion
2	Canola	Potato	Onion	FB-FM
3	Potato	Onion	FB-FM	Canola
4	Onion	FB-FM	Canola	Potato

[a] Sequential cropping of faba bean (FB) in first harvest and forage maize (FM) in second harvest.

The crops comprising the processing-tomato rotation (Table 3), apart from tomato (*Lycopersicum esculentum* or *Solanum lycopersicum*) itself, were: Melon (*Cucumis melo* L.), carrot (*Daucus carota* L.), onion and dry peas (*Pisum sativum* L.). Due to the asymmetry of this rotation, its full 25-year scheduling derivation is tedious and is relegated to Table A1 (Appendix B).

Table 3. Tomato rotation designed, wherein processing-tomato is cultivated in the same plot for three consecutive years.

Year	Plot 1	Plot 2	Plot 3	Plot 4
1	Melon	Onion	Carrot	Tomato
2	Onion	Carrot	Melon	Tomato
3	Carrot	Melon	Onion	Tomato
4	Tomato	Onion	Pea	Melon

2.3. Profitability Analysis

From an entrepreneurship point of view, an APV farm is a business comprised of two activities, namely crop production and electric power generation. To determine the profitability of an APV system, the stream of annual income and expenditure during the project lifetime has to be computed. Expenditure comprise initial investment cost (ascribed to the PV activity, since all the crops considered are annual), plus annual operation and maintenance costs (due to both agricultural and electric energy generation activities). Revenues originate from both activities and go from the second year onwards, since the first year is unproductive and carries only investment costs.

With regard to the APV system investment cost or capital expenditure (CapEx), the main items are the acquisition and installation of the substructure and mounting structures, PV modules, inverters, transformer and the so-called balance-of-system (cables, switch-boards, etc.). Table A2 (Appendix C) includes a breakdown of the cost items involved. With regard to the operating expenditures (OpEx), they can be split in two: First, the annual PV OpEx. Second, the annual costs incurred for crop cultivation beneath APV sheds. Table 4 includes estimated crop production costs under the partial shading of an APV shed.

Table 4. Crop production cost under full sunlight and under APV partial shading (in both cases assuming irrigation with surface water).

	Crop Production Cost [a] under Full Sunlight (€/ha)	Savings Due to Synergetic APV Partial Shading				Crop Production Cost under APV Partial Shading (€/ha)
		Irrigation Water Saving (%)	Fertilization Saving (%)	Hail in Surance Saving (%)	Fruit Solar Protector Saving (%)	
Canola	934	11.5 [b]	–	–	–	931
Carrot	8978	11.5 [b]	–	–	–	8964
Forage maize	1826	11.5 [b]	–	–	–	1813
Dry faba bean	544	11.5 [b]	–	–	–	541
Melon	7725	14.0 [c]	–	2.5	1.5	7697
Onion	7899	11.5 [b]	–	–	–	7885
Dry pea	631	11.5 [b]	–	–	–	628
Early potato	4701	9.0 [d]	–	–	–	4694
Processing tomato	4430	9.0 [d]	2.0 [e]	2.5	–	4403

[a] See Table A4 (Appendix D) for details. [b] Due to scarcity of bibliographic data on water saving for some crops under shade, they are assigned the mean between 14% (c) and 9% (d). [c] Assimilated to cucumber as both melon and cucumber belong to the botanical family of *Cucurbitaceae*; data for cucumber available at [30]. [d] Data for cherry tomato available at [31]. In addition, processing-tomato, the same value is assigned to early-potato, since the latter belongs to the same botanical family as tomato, namely, *Solanaceae*. [e] [32].

The irrigation district is liable for pressurizing the pipeline network. Therefore, the electricity generated by the APV system would not be partially self-consumed by a pump station, but entirely sold in the electricity wholesale market. Hence, the management of electricity binomial tariff cost items (contracted power, energy consumption, capacity charge, meter-gauge leasing, irrigators partially exempted electric tax, etc.) is on the irrigation district behalf.

Following the recommendations/prescriptions of the EU Water Framework Directive, many irrigation districts have abandoned the traditional per-hectare flat-rate pricing. The installation of volumetric metering valves has enabled irrigation districts to change to binomial tariffs. These consist of a fixed per-hectare component that is proportional to the area with irrigation rights and a variable or volumetric component that is proportional to the volume of water used [33,34].

The irrigation district determines the volumetric component, based on energy cost. This cost depends on the p_h of the irrigation network, the acreage irrigated and the type of crop, since the required hydraulic power is equal to the p_h multiplied by the flow rate (in turn, p_h depends on whether the water source is underground or surface, the irrigated area topography (plot elevations, size and shape) and the irrigation system service pressure, e.g., sprinklers demand more pressure than drippers) For the VIID, current energy cost is of 0.012 EUR/m^3 [35]. Rodríguez-Díaz et al. [36] measured energy consumption and power required per unit of irrigated area for several surface-water irrigation districts in southern Spain. One of them, the Bembézar Margen Derecha (BMDID), featured a p_h of 461 kPa, almost identical to the VIID p_h of 441 kPa. The crops irrigated are similar in both irrigation districts and similar to those of our study. In a sequel work, Fernández-García et al. [37] reported an energy cost of 0.02 EUR/m^3 and a total irrigation cost of 283 EUR/ha for the BMDID. Here, we took the energy cost of 0.02 EUR/m^3 [37] instead of the lower 0.012 EUR/m^3 [35]. Total irrigation cost in sites where underground water is used is usually two-fold (600 EUR/ha) and sometimes it can reach 900 EUR/ha [38].

Apart from the energy cost, irrigation cost includes water as a fixed-cost levied upon the land. The fixed per-hectare component in turn splits in two: The royalty of the River Basin Organism (Confederación Hidrográfica del Guadalquivir), a public incumbrance for upstream public civil works that allow water disposal to irrigation schemes (42.47 EUR/ha·annum); and the irrigation district fee, for management and maintenance of downstream irrigation district proprietary pipeline network and facilities (63.23 EUR/ha·annum in the case of the VIID [35]). The sum of both amounts equals 105.8 EUR/ha·annum. Further, VIID subscribers are currently, subjected to surcharge disbursement of 84.30 EUR/ha·annum in concept of amortization of pipeline, reservoirs and pumping stations upgrading works commissioned in 2008. Herein, this cost item is not included, since the expected remaining surcharge payment period is shorter than our study lifespan. In addition, regarding water use, Table 5 compiles the volumetric component and the total irrigation cost under full sunlight for the 9 crops considered.

With regard to annual revenues, the main entries are: First, the income from the sale of electricity generated by the PV modules; and second, the income originated from the sale of agricultural produce. With regard to the first, the two variables that intervene are: The annual specific energy yield (kWh/kWp) and the price perceived for the energy generated (EUR/kWh). The annual yield will decrease from year 12 onwards, due to the module degradation coefficient abovementioned. With regard to the wholesale electricity market price perceived for the energy sold, we proceeded as follows: From the future solar contracts due 2026 (FTS YR-26) published in the OMIP 2019 sessions market bulletins [42], one day per month of the year 2019 was selected for averaging. The mean of the 12 prices was 45.02 EUR/MWh (specifically, we took the 11th day of every month, except for May and August, where the 13th and 12th day, respectively, were picked, to skip the eventual Sunday effect). We deliberately dropped the 2020 sessions of the future solar market, because concerns arose about the prices thereof being convoluted with the COVID-19 effect. Afterwards, we divided the 25-year lifespan into three periods: The first one encompassing the first 9 years, while the second and third period spanning the following eight-year each. Finally, we assigned the OMIP FTS mean price previously calculated, 45.02 EUR/MWh, to the first period of 9 years; a price diminished by 5% to the second period (0.95·45.02 = 42.77 EUR/MWh); and a price diminished by 10% to the last 8-year period (0.90·45.02 = 40.52 EUR/MWh). The reason to assign the foregoing reduced prices to the second and third period is that foreseeably—and unfortunately for generators—increasing solar PV capacity installed in the forthcoming years will lead to decline of wholesale electricity prices [43].

Table 5. Irrigation costs under full sunlight.

		Surface Water		Groundwater	
	Water Use (m³/ha)	Energy Cost [b] (EUR/ha)	Total Irrigation Cost [c] (EUR/ha)	Energy Cost [e] (EUR/ha)	Total Irrigation Cost [f] (EUR/ha)
Canola	1200	24	130	48	291
Carrot	6000	120	226	240	483
Maize	5600	112	165 [d]	224	346
Faba bean	1300	26	79 [d]	52	174
Melon	4300	86	192	172	415
Onion	5900	118	224	236	479
Pea	1300 [a]	26	132	52	295
Potato	4000	80	186	160	403
Tomato	4951	99	205	198	441

[a] Following Karkanis et al. [39] and ITACYL [40]. [b] Energy cost translates into a volumetric component of 0.02 EUR/m³ [37]. [c] Fixed component: The sum of the River Basin Organism royalty plus the irrigation district fee equals 105.8 EUR/ha [35]. [d] The fixed cost is halved because faba bean and forage maize share the same field in one year. [e] Assuming a well depth of 100 m, energy cost translates into a volumetric component of 0.04 EUR/m³ [41]. [f] The fixed component is related to the capacity factor charged to the irrigation district (ID) by the electric utility; in turn, the ID apportions this charge to irrigators. It is estimated as 2.3 × 105.8 ≅ 243 EUR/ha; with 105.8 EUR/ha taken from this same table footnote (c) as representative of a surface water ID. The rationale behind the 2.3 coefficient is that, as a rule of thumb, more powerful pumps are required in groundwater IDs compared to surface water IDs; this has a direct effect on the capacity factor charge.

Table 6 is a compilation of estimated decreases in crop yield due to APV shed shading compared to full-sunlight cultivation. In Table 7, values are five-year (2014–2018) averages calculated from official data of Spain Department of Agriculture [16] compiles crop yields (kg/ha) and prices (EUR/t) under full sunlight.

Table 6. Crop yield variation under APV partial shading with respect to full sunlight.

	Crop Yield Variation under Shading [a], High Crop-Yield Penalty (%)	Source [b]	Crop Yield Variation under Shading [c], Low Crop-Yield Penalty (%)
Canola	−20	[44]	−5
Carrot	−10	[45]	+5
Maize	−7	[46]	+8
Faba bean	0	[47]	+15
Melon	−17	[48]	−2
Onion	−6	[49]	+9
Pea	−15	[50]	0
Potato	−23	[15]	−8
Tomato	−5	[32]	+10

[a] Rounded to the closest integer. [b] See Appendix E for details on the *uncertainty factors* applied. [c] Assumption: Δ+15% over the values in the second column.

Table 7. Crop revenues under full sunlight conditions.

	Crop Yield under Full Sunlight (t/ha) [a]	Produce Price Paid to the Farmer (EUR/t)	Farmer Income from Produce Sale under Full Sunlight (EUR/ha)	EU-CAP Direct Payment to the Farmer (EUR/ha) [b]	Total Income under Full Sunlight (EUR/ha)
Canola	3.10	326.9	1013	35	1048
Carrot	49.22	303.4	14,933	NA	14,933
Forage maize	59.37	41.3	2452	NA	2452
Dry faba bean	1.79	223.6	400	45	445
Melon	34.60	337.7	11,684	NA	11,684
Onion	44.74	211.4	9458	NA	9458
Dry pea	1.79	220.6	395	45	440
Potato	30.98	246.2	7627	NA	7627
Processing tomato	85.00	72.5	6162	200	6362

[a] Values calculated as 5-year (2014–2018) averages from [16]. [b] NA: Not applicable.

3. Results

A well-established metric to assess the performance of dual-land use systems like agroforestry [51] and also APV, is the land equivalent ratio (LER). However, the LER exclusively accounts system revenues and not the expenditure. On the other hand, the benchmark yardstick for energy generation systems, the levelized cost of electricity (LCoE), computes cost relative to electricity yield, but does not incorporate the crop production activity. Here, the indicator wherethrough profitability was evaluated was the internal rate of return (IRR). Once income and expenditure from both agricultural and energy generation activity were accounted, their aggregation to obtain the IRR was straightforward.

Figure 3 shows the annual specific yield of 1628 kWh/kWp predicted by the PV simulator for the southwest-oriented APV shed. This value was introduced in Table 8 to compute annual PV income throughout the foreseen lifespan. Likewise, the annual specific yield was introduced in Table A3 to compute the annual PV OpEx. After the entire flow of APV income and expenditure was computed, the IRR was calculated for both early-potato and processing-tomato rotation. The same procedure was followed with the annual specific yield of 1786 kWh/kWp predicted by the SISIFO PV simulator for the south-oriented APV shed.

Unlike Beck et al. [20], we did not find substantial differences between the two orientations. A mean shade factor of 30.6% was calculated for due south orientation (Figure A3), while the counterpart for due SW was of 29.2% (Figure A4). The small difference between both values prevented from matching orientation and shade-induced crop yield penalty. Therefore, we determined that in our case-study APV shed orientation only affects electricity production. With the dichotomist sources of variation considered, namely, crop rotation (potato/tomato), source of irrigation water (surface/underground), level of shade-induced crop yield penalty (low/high) and APV shed orientation (SW/S), the number of combinations analyzed was of 2^4.

The formula for the IRR is given by (Equation (1)):

$$0 = \sum_{t=1}^{26} \frac{C_t}{(1 + IRR)^t} - CapEx \tag{1}$$

where C_t = net cash flow during the year t (calculated as the sum of annual income from electricity sale—Table 8 plus annual income from the sale of agricultural produce harvested, with subtraction of annual PV-OpEx—Table A3 and annual crop production cost; for the early-potato rotation, annual agricultural flow is constant throughout the 25-year lifespan, due to rotation symmetry, whereas in the case of the processing-tomato rotation, the annual agricultural flows vary according to the pattern shown in Table A1); CapEx = total initial investment cost (calculated as 562770 EUR/ha in Table A2 multiplied by 22.8 ha, giving 12,831,156 EUR).

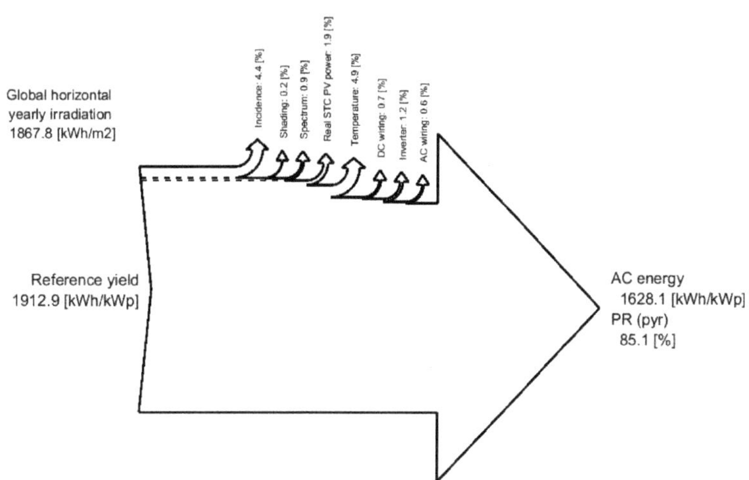

Figure 3. Annual AC energy output as predicted by PV simulator SISIFO [23] for the southwest-oriented APV system in Brenes, Seville, Spain.

Table 8. Annual PV income for 24 ha and due SW orientation, assuming a 5% loss as to PV productive land, due to plot dead corners (PV productive land of 22.8 ha).

Year	PV Module Degrad. Coeff.	Specific Yield (kWh/kWp)	(kWh/ha) [a]	Total Yield (kWh) [b]	(cEUR/kWh)	Energy Sale Income (EUR)	Total PV Income [c] (EUR)
1		0	0	0		0	0
2	1	1628	904,354	20,619,271	4.502	928,280	928,280
3	1	1628	904,354	20,619,271	4.502	928,280	928,280
4	1	1628	904,354	20,619,271	4.502	928,280	928,280
5	1	1628	904,354	20,619,271	4.502	928,280	928,280
6	1	1628	904,354	20,619,271	4.502	928,280	928,280
7	1	1628	904,354	20,619,271	4.502	928,280	928,280
8	1	1628	904,354	20,619,271	4.502	928,280	928,280
9	1	1628	904,354	20,619,271	4.502	928,280	928,280
10	1	1628	904,354	20,619,271	4.502	928,280	928,280
11	1	1628	904,354	20,619,271	4.277	881,886	881,886
12	0.995	1620	899,832	20,516,175	4.277	877,477	877,477
13	0.990	1612	895,310	20,413,078	4.277	873,067	873,067
14	0.985	1604	890,789	20,309,982	4.277	868,658	910,861
15	0.980	1595	886,267	20,206,886	4.277	864,249	864,249
16	0.975	1587	881,745	20,103,789	4.277	859,839	859,839
17	0.970	1579	877,223	20,000,693	4.277	855,430	855,430
18	0.965	1571	872,702	19,897,597	4.277	851,020	851,020
19	0.960	1563	868,180	19,794,500	4.052	802,073	802,073
20	0.955	1555	863,658	19,691,404	4.052	797,896	797,896
21	0.950	1547	859,136	19,588,308	4.052	793,718	793,718
22	0.945	1538	854,615	19,485,211	4.052	789,541	789,541
23	0.940	1530	850,093	19,382,115	4.052	785,363	785,363
24	0.935	1522	845,571	19,279,019	4.052	781,186	781,186
25	0.930	1514	841,049	19,175,922	4.052	777,008	777,008
26	0.925	1506	836,527	19,072,826	4.052	772,831	772,831

[a] (x kWh/kWp)·(555.5 kWp/ha) = y kWh/ha. [b] 0.95·24 ha = 22.8 ha; (y kWh/ha)·(22.8 ha) = z kWh. [c] Values in this column are equal to values in the adjacent-left column except for the due year of inverters replacement (year 14), where an income of 10% of inverter purchase price (18,505 EUR/ha, Table A2) is added in concept of old inverters residual value. Hence, the income added in the year 14 is of: 0.1·18,505·22.8 = 42,191 EUR.

Since the IRR is not explicit in Equation (1), it has to be solved by an iterative method, like the ad hoc function of Microsoft Excel®. Table 9 is a compilation of the IRR for each of the 2^4 combinations, wherein the minimum and maximum IRR are highlighted.

Table 9. Internal rate of return for the 16 combinations generated.

		IRR (%)			
		Due-Southwest Orientation		Due-South Orientation	
		Low Shade-InduceD Crop Yield Penalty	High Shade-InduceD Crop Yield Penalty	Low Shade-InduceD Crop Yield Penalty	High Shade-InduceD Crop Yield Penalty
Potato rotation	Surface Water	4.1	3.9	5.1	4.8
	Ground Water	4.0	3.8	5.0	4.8
Tomato rotation	Surface Water	4.7	4.3	5.6	5.2
	Ground Water	4.6	4.2	5.6	5.2

To elucidate the profitability associated to the foregoing IRRs, they were confronted with the private investor expected remuneration or annual cost of equity (r_e). According to Guaita-Pradas and Blasco-Ruiz [52], the cost of equity can be estimated through the capital asset pricing model, (Equation (2)):

$$r_e = r_f + (r_m - r_f) \cdot \qquad (2)$$

where r_e = annual cost of equity, i.e.,demanded rate of return on equity; r_f = annual risk-free rate of return; r_m = annual stock-exchange market rate of return; β = coefficient that reflects the sensitivity of the sector to market fluctuations.

A good representative for r_f in Spain is the interest rate of the 30-year maturity Public Treasury bonds, 1.31% [53]. Current market profitability, r_m, is of 4.5% [54]. In strict sense, in our case β would be somehow compounded, since the project economic activity sector is not only electric generation but also agricultural production. For the sake of simplicity, we took a PV β of 1.10 [55]. Substituting in (Equation (2)):

$$r_e = 0.0131 + (0.045 - 0.0131) \cdot 1.10 = 0.04819$$

Therefore, the threshold of profitability is 4.8%. The IRRs compiled in Table 9 indicate that some combinations would be profitable from the perspective of a private investor, whereas others would be not.

4. Discussion

Following the mainstream APV philosophy of prioritizing agricultural over power production and based on Beck et al. [20] conclusion, we initially performed calculations for a SW-oriented APV shed. Finally, a comparative shaded-fraction analysis between southwest and south orientation was undertaken. The small difference found (abovementioned values of 29.2% and 30.6%), together with the shape of histograms Figures A3 and A4 suggest little difference between both orientations. Perhaps the subtle difference in ground radiation uniformity in our case was due to the TMY data used. Edge effects could also play a role. This issue deserves more attention and should be further analyzed in a future work.

The result obtained for the early-potato rotation when the APV shed is oriented due Southwest is in line with Trommsdorff [56], who, for organic potatoes cultivated beneath the APV shed described by Schindele et al. [8], obtained an IRR 1.6% lower than WACC. In a broader sense, López Prol et al. [43] wondered if renewable energy generators like PV would ever be competitive considering the faster decline of the wholesale market price compared to the LCoE. From the inception of our study, it was envisaged that a negative factor for APV system profitability would be the high CapEx compared to conventional ground-mounted PV power plants. To restrain APV system CapEx, fixed-tilt PV modules were selected instead of single-axis trackers, which are 7% more expensive in average [57]. Here, the reason to select fixed-tilt PV generator was three-fold: First, to restrain system cost; second, to utilize the substructure described by Schindele et al. [8], which is of known cost; and third, to cast less shading on the understorey crop canopy. With regard to the latter, in an early stage a set of simulations was performed with the dual-use shading analysis tool, an on-line simulator promoted by the Massachusetts government to analyze the technical viability of APV layouts. Results indicated that single-axis tracking casted more shading, in w% per square meter than fixed-tilt. On the other hand, Amaducci et al. [58] concluded that reduction of global radiation beneath their APV shed was more affected by PV module array GCR than by tilt angle management (fixed tilt/sun-tracking).

In the case that in one of the 6 ha APV plots there existed an authorized underground water well equipped with a submersible pump, a reservoir to store the water abstracted by the well pump and a horizontal-axis pump to pressurize the whole 24 ha farmer irrigation network, the following management strategy could be analyzed: During PV productive hours, a small fraction of the energy produced would be self-consumed by the well pump to replenish the reservoir. To irrigate, i.e., to pressurize the irrigation network preferably during nocturnal hours, energy would be consumed from the grid, through the same HV power transmission line wherethrough PV production is injected. The energy consumed would be registered, for billing purposes, by a bi-directional metering gauge. This management strategy would remove the cost of water delivery charged by the irrigation district. Concurrently, the income from the sale of electricity would be diminished in the amount of the energy self-consumed—and therefore, not sold—by the well pump. Likewise, the farmer would incur in the cost of the nocturnal energy consumed from the grid by the irrigation pump. According to IDAE [59], average installed power pump for pressurized irrigation in Spain is of 2 kW/ha. For a flat topography like the area of study, assuming low pressure drip irrigation and for moderate water depth in the well, the share of installed power could be e.g., 70% for the submersible well pump and 30% for the irrigation pump (the well pump share would increase with increasing depth). Therefore, this results in 0.3 2 kW/ha = 0.6 kW/ha and 0.6 kW/ha 24 ha = 14.4 kW. This is significantly smaller than 450 kW, the minimum contracted power to benefit from the cheapest nocturnal electricity period of the Spanish tariff 6.1.

With regard to the possibility of reducing the 5 m clearance height of APV shed substructure and accordingly save in system CapEx, the following has to be considered: In our study, one head of rotation was first-early potato, planted in late December–January and harvested in late May–early June. In Spain, this type of potato is not harvested with the bulky and tall potato harvester, but with much smaller and shorter machines, namely, potato lifters and windrowers. These machines just dig-up and expose the tubers so that they can be afterwards hand-picked by manual workers. The reason to discard the potato harvester is to preserve tuber quality, since hand-picking is less aggressive. In other parts of Spain, where half-season potatoes are grown, the tubers spend more time within the ground, resulting in a thicker skin that withstands better the abrasions and impacts that occur inside potato harvester. Attending to the potato harvesting machinery used in the area, one could think of saving in substructure height, at least in the case of the early-potato rotation. However, a two-fold reason dissuade from this: First, canola and faba bean, two of the potato "partners" in the eponymous rotation, are harvested with the bulky combine harvester. Second, the main interest of a high substructure is not only to allow agricultural machinery work beneath, but to provide homogenous light distribution for the crop, casting shade of lower intensity. Analogous considerations apply for the processing-tomato rotation, whereinto processing tomato is harvested with a bulky-tall machine, similar in dimensions to both the potato and combine harvesters.

Among the circumstances that would yield lower IRR are: (i) APV plots remoteness from the grid-connection switchyard (distance higher than the 1 km assumed in Figure 1; (ii) re-activation of Spanish Law 15/2012, under which electricity generators must satisfy a tax of 7% on the value of the energy injected to the grid -this law was challenged before the Constitutional Court of Spain and the final judgment is pending [60].

Among the circumstances that could render higher IRR are: (i) Higher electricity yield due to favorable microclimate condition. Thus, the SISIFO PV simulator used here is not APV-specific but computes yield from site TMY climate data. Cooler temperatures on the back side of the modules, induced by the irrigated understory crop, would improve PV performance, especially in the summer months. (ii) Modification of the PV system electrical design: considering that while the peak power (153.1 kWp) to nominal power (150 kWp) ratio is of approximately 1.02 Wp/Wn; stronger oversizing of the PV array with respect to inverter is recommended [61].

Here, only the quantitative effect of shading on produce yield (t/ha) was considered. More research is needed to investigate the effect of APV shading on produce quality that ultimately affects revenues or even could be a limiting factor for the spread of APV. Nishizawa et al. [62] concluded that severe levels of shading negatively affected melon fruit firmness. Hernández et al. [32] measured not only higher concentration of lycopene, but also lower concentration of vitamin C and phenolic compounds, in tomatoes grown under partial shade compared to the full-sunlight counterpart.

In the authors' opinion, the lack of profitability in some of the combinations of the case-study analyzed herein does not tarnish the potential profitability of APV systems. Higher IRR is envisaged for specialty crops, thanks to extended synergies between the food generator—agricultural crop—and energy generator—PV modules. Savings in fruit orchard hail and bird netting allowed by APV sheds paddle in this direction [63]. Likewise, the utilization of semi-transparent PV modules could increase crop intercepted light without the need for the expensive 5 m ground-clearance substructure that supports conventional opaque PV modules. The fragility of specialties such as raspberry, blackberry and blueberry advises against their mechanized harvesting, contributing to the technical viability of cost-effective limited-height sheds and the subsequent increased profitability of APV systems.

5. Conclusions

In correspondence with objectives (1) and (2) indicated in Section 1, the following conclusions can be drawn:

1. two crop rotations, one of them headed by early-potato partnered with canola, faba bean, forage-maize and onion, and the other one headed by processing-tomato partnered with onion, dry-pea, carrot and melon were designed;
2. the stream of expenditure and revenues for both agricultural and electric energy production was determined for a lifespan of 25 years. The internal rates of return obtained ranged from a minimum of 3.8% for the combination of southwest orientation, early-potato rotation, groundwater and high shade-induced crop-yield penalty to a maximum of 5.6% for the combination of South orientation, processing-tomato rotation, surface water and low shade-induced crop-yield penalty.

Author Contributions: Conceptualization, M.A.M.-G., G.P.M. and M.C.A.-G.; methodology, G.P.M., M.A.M.-G. and M.C.A.-G.; validation, G.P.M., M.A.M.-G., M.C.A.-G. and L.H.-C.; formal analysis, G.P.M., L.H.-C. and M.C.A.-G.; investigation, G.P.M., M.A.M.-G., M.C.A.-G. and L.H.-C.; data curation, G.P.M.; writing—original draft preparation, G.P.M., M.A.M.-G. and M.C.A.-G.; writing—review and editing, G.P.M., M.A.M.-G., M.C.A.-G. and L.H.-C. All authors have read and agreed to the published version of the manuscript.

Funding: This research received no external funding.

Institutional Review Board Statement: Not applicable.

Informed Consent Statement: Not applicable.

Data Availability Statement: The data presented in this study are available on request from the corresponding author.

Acknowledgments: The authors wish to acknowledge the reviewers for their valuable comments.

Conflicts of Interest: The authors declare no conflict of interest.

Abbreviations

a.s.l.	Above sea level
CapEx	Capital expenditures (investment cost)
Med	Mediterranean
OpEx	Operating expenditures
AC	Alternating current
APV	Agrophotovoltaic
BMDID	*Bembézar Margen Derecha* Irrigation District
CAP	European Union Common Agricultural Policy
DC	Direct current
EU	European Union
FB	Faba bean
FM	Forage maize
FTS	Future solar contract
GCR	Ground coverage ratio
GHG	Greenhouse-effect gas
HV	High voltage
ID	Irrigation district
IRR	Internal rate of return
LCoE	Levelized cost of –electric- energy
LER	Land equivalent ratio
LV	Low voltage
MV	Medium voltage
PE	Polyethylene
PV	Photovoltaic
SW	Southwest
TMY	Typical meteorological year
VIID	*Valle Inferior* Irrigation District

Symbols

p_h	Pressure head
r_e	Cost of equity (demanded rate of return on equity)
r_f	Risk-free rate of return
r_m	Stock market rate of return
C_t	Cash flow in the year t

Appendix A

Comparison between due south and due southwest orientation of APV shed.

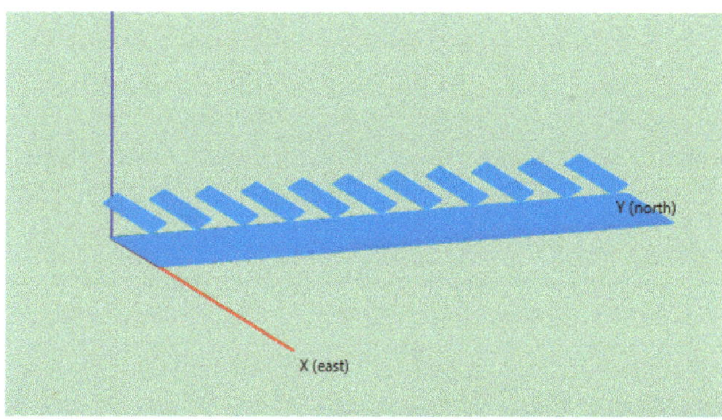

Figure A1. Due south APV shed. Ground area: 26 m × 106 m. Dimensions and number of PV module supporting structures: 26 m × 3.28 m × 11. Equidistance between supporting structures axes: 9.5 m, as in Figure 2.

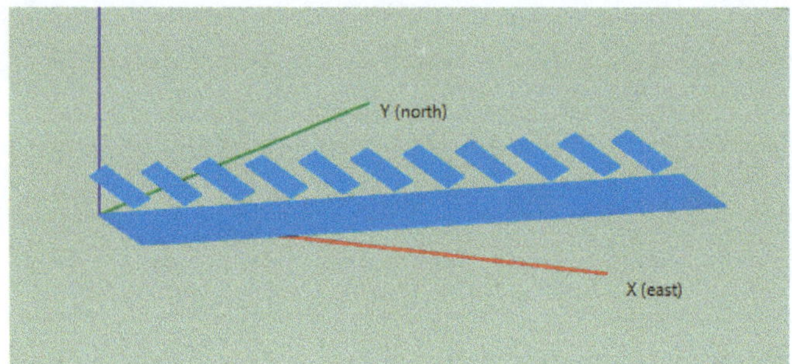

Figure A2. Due southwest APV shed.

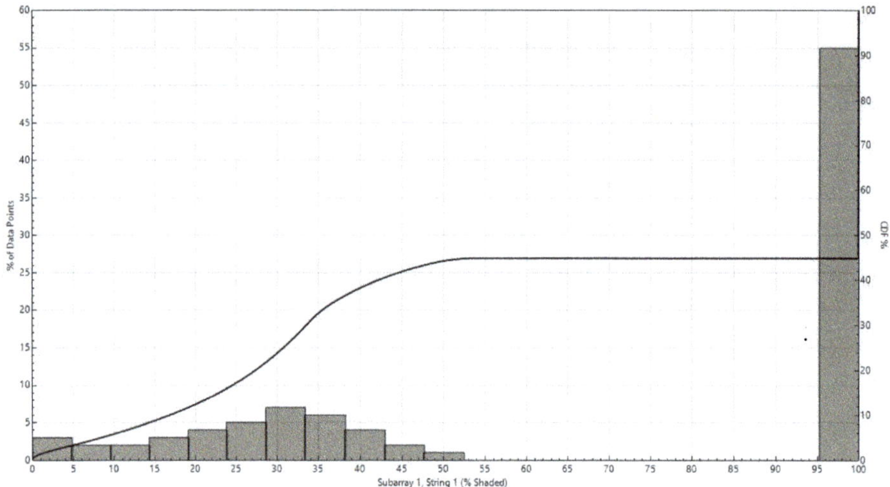

Figure A3. Time series (1 min step) shaded fraction histogram (zero values excluded) for due south orientation.

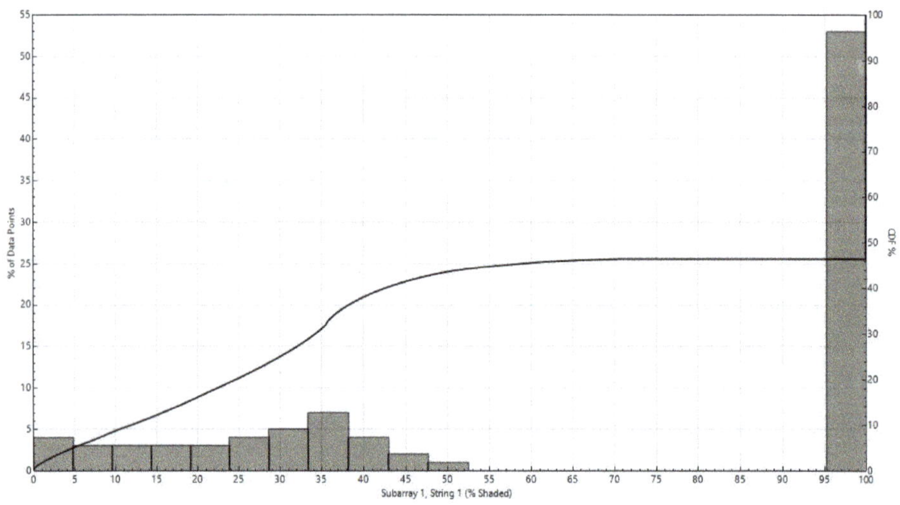

Figure A4. Time series (1 min step) shaded fraction histogram (zero-values excluded) for due southwest orientation.

Appendix B

Table A1. Processing-tomato rotation scheduling for 25-year lifespan.

Year	Plot 1	Plot 2	Plot 3	Plot 4
2	Melon	Onion	Carrot	Tomato
3	Onion	Carrot	Melon	Tomato
4	Carrot	Melon	Onion	Tomato
5	Tomato	Onion	Pea	Melon
6	Tomato	Onion	Onion	Onion
7	Tomato	Carrot	Carrot	Carrot
8	Melon	Tomato	Melon	Melon
9	Carrot	Tomato	Onion	Onion
10	Melon	Tomato	Carrot	Carrot
11	Onion	Melon	Tomato	Melon
12	Pea	Onion	Tomato	Onion
13	Onion	Carrot	Tomato	Pea
14	Carrot	Melon	Melon	Tomato
15	Melon	Onion	Onion	Tomato
16	Onion	Carrot	Carrot	Tomato
17	Tomato	Melon	Melon	Melon
18	Tomato	Onion	Onion	Onion
19	Tomato	Pea	Carrot	Carrot
20	Melon	Tomato	Melon	Melon
21	Onion	Tomato	Onion	Onion
22	Carrot	Tomato	Pea	Carrot
23	Melon	Melon	Tomato	Melon
24	Onion	Onion	Tomato	Onion
25	Carrot	Carrot	Tomato	Pea
26	Melon	Melon	Melon	Tomato

Appendix C

Table A2. APV system initial investment cost or capital expenditure, CapEx (the delayed CapEx item of inverters replacement incurred in year 14 is included in Table A3).

	EUR/unit	EUR/kWp	EUR/ha [a]	No. Units Per ag. Plot of 6(5.7) ha
(1) PV modules	60.9	210.0 [b]	116,655	10,919 [c]
(2) Galvanized steel mounting structure		378.6 [d]	210,312	
(3) Earthing		0.6 [e]	333	
(4) Lightning protection system			9000 [f]	
(5) DC switchboards (combiner boxes)	584.0 [g]	3.8	2111	21
(6) DC cables		35.0 [h]	19,443	
(7) Inverters	5100.0 [i]	33.3 [j]	18,505	21
(8) AC low voltage cables		18.4 [e]	10,221	
(9) LV/MV Transformer	80,500.0 [e]	25.4 [k]	14,124	1
(10) MV overhead power transmission line		1.9 [l]	1055	
(11) Monitoring and communications		0.9 [e]	500	
(12) Security		2.1 [e]	1167	
(13) Installation works			132,490 [m]	
(14) Subtotal 1 {=Σ(1) … (13)}			535,916	
(15) Administration costs (1%)			5359	
(16) Designer and construction manager fees (4%)			21,437	
(17) Subtotal 2 {= (14) + (15) + (16)}			562,712	
(18) Subsoiling			58 [n]	
(19) **TOTAL** {=(17) + (18)}			**562,770**	

[a] (x EUR/kWp)·(555.5 kWp/ha) = y EUR/ha. [b] [64]. [c] (528 modules/153.1 kWp)·3166 kWp = 10,919 modules. [d] Calclated from [8]. [e] Calculated from [65]. [f] Adapted from [66]. [g] From [67], (531 £)·(1.10 EUR/£) ≅ 584 EUR. [h] From [68] with consideration of 1000 VDC inverter wire section saving. [i] From [69], (0.034 EUR/W_{AC})·(150 kW_{AC}) = 5100 EUR. [j] 5100 EUR/153.1 kWp ≅ 33.3 EUR/kWp. [k] 80500 EUR/3166 kWp ≅ 25.4 EUR/kWp. [l] Calculated from [70]. [m] Calculated by introducing 155,241 EUR/ha [8] in the following breakdown model: Construction work, 65% of the installation works cost; electrical installation work, 35% of the installation works cost; labor share within construction work, 40%; labor share within electrical installation work, 70%; ancillary equipment (cranes, welding machines, tools, etc.) share within construction work, 60%; ancillary equipment share within electrical installation work, 30%. Price of construction labor in Spain relative to Germany: 54%; price of electrician labor in Spain relative to Germany: 89% [71]. [n] [72].

Table A3. Twenty-four hectare (22.8 ha effective) due SW-oriented PV system annual operating expenditure, OpEx, plus the delayed CapEx of inverters replacement in the year 14.

Year	Total Annual Yield [a] (MWh)	Grid Access Toll [b] (EUR)	Brokerage P.W.M. Agent [c] (EUR)	Maintenance and Repair [d] (EUR)	Insurance and Video-Surv. [e] (EUR)	Internet Fee [f] (EUR)	TOTAL (EUR)
1	0	0	0	0	0	0	0
2	20,619	10,310	4124	26,597	22,798	10,132	73,961
3	20,619	10,310	4124	26,597	22,798	10,132	73,961
4	20,619	10,310	4124	26,597	22,798	10,132	73,961
5	20,619	10,310	4124	26,597	22,798	10,132	73,961
6	20,619	10,310	4124	26,597	22,798	10,132	73,961
7	20,619	10,310	4124	26,597	22,798	10,132	73,961
8	20,619	10,310	4124	26,597	22,798	10,132	73,961
9	20,619	10,310	4124	26,597	22,798	10,132	73,961
10	20,619	10,310	4124	26,597	22,798	10,132	73,961
11	20,619	10,310	4124	26,597	22,798	10,132	73,961
12	20,516	10,258	4103	26,597	22,798	10,132	73,889
13	20,413	10,207	4083	26,597	22,798	10,132	73,817
14	20,310	10,155	4062	26,597	22,798	10,132	474,562 [g]
15	20,207	10,103	4041	26,597	22,798	10,132	73,672
16	20,104	10,052	4021	26,597	22,798	10,132	73,600
17	20,001	10,000	4000	26,597	22,798	10,132	73,528
18	19,898	9949	3980	26,597	22,798	10,132	73,456
19	19,795	9897	3959	26,597	22,798	10,132	73,384
20	19,691	9846	2954	19,948	17,099	7599	57,445
21	19,588	9794	2938	19,948	17,099	7599	57,378
22	19,485	9743	2923	19,948	17,099	7599	57,311
23	19,382	9691	2907	19,948	17,099	7599	57,244
24	19,279	9640	2892	19,948	17,099	7599	57,177
25	19,176	9588	2876	19,948	17,099	7599	57,110
26	19,073	9536	2861	19,948	17,099	7599	57,043

[a] From Table 8. [b] 0.5 EUR/MWh [73]. [c] p.w.m., power whosale market. Brokerage fee applied: 0.2 EUR/MWh [74], years 2 through 19; years 20 through 26: assumption of 25% price decrease applied [69], resulting in 0.15 EUR/MWh. [d] Years 2 through 19, (2.1 EUR/kWp)·(555.5 kWp/ha)·(22.8 ha) = 26,597 EUR; years 20 through 26: assumption of 25% price decrease applied [32], 0.75 × 26,597 = 19,948 EUR. [e] Years 2 through 19, (1.8 EUR/kWp)·(555.5 kWp/ha)·(22.8 ha) = 22,798 EUR; years 20 through 26: assumption of 25% price decrease applied [32], 0.75 × 22,798 = 17,099 EUR. [f] Years 2 through 19, (0.8 EUR/kWp)·(555.5 kWp/ha)·(22.8 ha) = 10,132 EUR; years 20 through 26: assumption of 25% price decrease applied [32], 0.75 × 10,132 = 7599 EUR. [g] Inverter's replacement cost included (0.95 × 18,505 EUR/ha × 22.8 ha = 400,818 EUR; a 5% price decrease is assumed with respect to the year zero; 18505 EUR/ha taken from Table A2).

Appendix D

Table A4. Annual production costs for the 9 crops under full sunlight. Values adapted from [75,76].

Costs in EUR/ha	Canola	Carrot	Forage Maize	Dry Faba Bean	Melon	Onion	Dry Pea	Early Potato	ProCessing Tomato
Seed	60	3900	170	60	3000	3925	55	1400	820
Fertilizer	205	860	610	10	800	750	10	600	590
Plant Protection products	115	650	50	60	505	230	60	250	510
Externalized works (mechanized harvest, etc.)	67	1390	85	55	80	75	55	110	850
Tractor fuel	60	420	120	60	330	180	60	105	110
Tractor & mach. Repair & Maint.	45	230	106	40	110	105	40	80	80
Tractor & mach. Shed costs	30	60	45	30	60	60	30	55	60
Amortization of tractor & mach.	20	170	105	17	150	145	17	120	140
Hired labor (manual harvest)	0	0	0	0	1125 [a]	1035 [b]	0	900 [c]	0
Soc. Sec. contrib.for hired labor (25%)	0	0	0	0	281	259	0	225	0
Own labor	65	480	200	50	500	350	50	290	580
Soc. Sec. contrib.for own labor (25%)	16	120	50	13	125	88	13	73	145
Insurances (crop, tractor)	15	57	15	15	100	100	15	57	100
Land property tax	70	70	35 [d]	35 [d]	70	70	70	70	70
Irrigation total cost	130	226	165 [d]	79 [d]	192	224	132	186	205
Subtotal	898	8633	1756	524					4260
Working capital interest (4%)	36	345	70	21					170
Total	**934**	**8978**	**1826**	**544**	**7725**	**7899**	**631**	**4701**	**4430**

[a] Considering a required harvest labor of 25 *labor units*/ha and a *labor unit* regulated price of 45 EUR/labor unit [77]; *labor unit* represents the work done by one worker in one day. [b] Considering a required harvest labor of 23 *labor units*/ha and a *labor unit* regulated price of 45 EUR/ labor unit. [c] Considering a required harvest labor of 20 *labor units*/ha and a *labor unit* regulated price of 45 EUR/ labor unit. [d] Since faba bean and forage maize are cultivated in sequential cropping, they are each ascribed with half the cost of land-property-tax and half the fixed component of irrigation cost.

Appendix E

Uncertainty factors applied on literature references to obtain percentage yield variation under shading for each crop (high crop yield penalty).

Appendix E.1. Canola

We took -20% yield value straightforwardly from Figure 1 [44], as the average of their experiments shading at flowering (2011) and shading at pod filling (2011). We disregarded yield value of shading at flowering (2010) because that year was extremely dry and we are analyzing irrigation farming.

Appendix E.2. Carrot

We pay attention to the marketable yield column of Table 2 [45]. From the different shading nets listed there, we select white polyethylene (PE) as the closer to our APV-shed configuration (at first glance, one could think that due to monofaciality of our PV modules, black PE would be more similar, but the shading intensity decrease due to modules height plays a role). With respect to no-shade, the variation is of 9.6%, which we rounded to 10%.

Appendix E.3. Maize

We paid attention to Table 2 [46], biomass of corn stover and Table 3, grain yield. To be conservative, for both tables we focused on the higher PV GCR. We took data from both tables because forage maize crop harvest is a mix of chopped stover, ears and grains. From Table 2 [46], we obtained -3% under shading, whereas from Table 3 [46], we obtained -3.6%. The average of both values is 3.3%. To be conservative, we applied an uncertainty factor of 2, which multiplied by 3.3 equals 6.6%, and finally, rounded to 7%. The reason underlying the uncertainty factor of 2 is that Sekiyama and Nagashima [46] experiments were conducted at latitude 35 °N, while our latitude is higher (37 °N).

Appendix E.4. Faba Bean

Table 2 [47] shows higher yield under shading than under full sunlight. To be conservative, we assume zero variation with respect to full sunlight.

Appendix E.5. Melon

In Figure A3 [48], we took marketable yields corresponding to control (full sunlight) and aluminet shading net, which to our understanding is more similar to our APV shading than the other two types of shading net categorized in Figure 3 [48]. The difference between them is approximately 8.4 t/ha, which divided by the control equals 16.5%, which we rounded to 17%.

Appendix E.6. Onion

From Table 7 [49], we calculated an average yield variation of 2.3% between full sunlight and shading conditions. Then, we applied an uncertainty factor of 2.5, for a three-fold reason: First, the latitude of Khan et al. [49] experiment was tropical, unlike ours; second, their shade was not generated by an inert artificial screen, but by a plant canopy which entails a competition not only for sunlight, but also for soil nutrients. In third place, the onion yield (t/ha) reported [49] are much lower than the common in our area, most probably because spacing between plants was rather large. Finally, we calculated $2.3\% \cdot 2.5 = 5.8\%$, which we rounded to 6%.

Appendix E.7. Pea

In Table 2 [50], we took the yield values of *lighter shading*—one layer of screen—which, to our understanding, reflects better the light conditions under our APV shed and compare them to the no-shade conditions. For the year 1973, we obtained 19.4%, whereas for 1974 we obtained 10.5%. The average of both is approximately 14.9%, which we rounded to 15%.

Appendix E.8. Potato

We took years 2015 and 2017 from Figure 3 [15]. To be conservative, we assumed an average shading level of 38%, the mean of 26% and 50%, two of the shading intensities shown in Figure 3 [15]. Reading in the graph the pertinent values and calculating, a shading level of 38% delivered an average tuber yield variation of 23.4%, which we rounded to 23%. We decided to apply no further uncertainty factor due to the following: although, with respect to the availability of the solar resource, our latitude of Seville is more advantageous than the latitude of Germany [15], this is cancelled-out by the fact that our early potato crop season is shifted towards winter.

Appendix E.9. Tomato

The data compiled in Table 1 [32] indicate no tomato yield variation between full-sunlight and shade (60% light). To be conservative, we considered a -5% in yield, to account for the fact that Hernández et al.'s experiment [32], although at the same latitude than ours, was conducted inside a greenhouse.

References

1. Assessment of the Draft National Energy and Climate Plan of Spain. *Commission Staff Working Document 2019–262 Final*; European Commission: Brussels, Belgium, 2019.
2. Assessment of the Draft National Energy and Climate Plan of Italy. *Commission Staff Working Document 2020–911*; European Commission: Brussels, Belgium, 2020.
3. Red Eléctrica de España. Renewable Energy in the Spanish Electricity System-2019. 2020. Available online: www.ree.es/en (accessed on 19 March 2021).
4. Goetzberger, A.; Zastrow, A. On the coexistence of solar-energy conversion and plant cultivation. *Int. J. Sol. Energy* **1982**, *1*, 55–69. [CrossRef]
5. Dupraz, C.; Marrou, H.; Talbot, G.; Dufour, L.; Nogier, A.; Ferard, Y. Combining solar photovoltaic panels and food crops for optimising land use: Towards new agrivoltaic schemes. *Renew. Energy* **2011**, *36*, 2725–2732. [CrossRef]
6. Marrou, H.; Wery, J.; Dufour, L.; Dupraz, C. Productivity and radiation use efficiency of lettuces grown in the partial shade of photovoltaic panels. *Eur. J. Agronomy* **2013**, *44*, 54–66. [CrossRef]
7. Valle, B.; Simonneau, T.; Sourd, F.; Pechier, P.; Hamard, P.; Frisson, T.; Ryckewaert, M.; Christophe, A. Increasing the total productivity of a land by combining mobile photovoltaic panels and food crops. *Appl. Energy* **2017**, *206*, 1495–1507. [CrossRef]
8. Schindele, S.; Trommsdorff, M.; Schlaak, A.; Obergfell, T.; Bopp, G.; Reise, C.; Braun, C.; Weselek, A.; Bauerle, A.; Högy, P.; et al. Implementation of agrophotovoltaics: Techno-economic analysis of the price-performance ratio and its policy implications. *Appl. Energy* **2020**, *265*, 114737. [CrossRef]
9. Dinesh, H.; Pearce, J.M. The potential of agrivoltaic systems. *Renew. Sustain. Energy Rev.* **2016**, *54*, 299–308. [CrossRef]
10. SolarPower Europe. Agri-PV: How Solar Enables the Clean Energy Transition in Rural Areas. Briefing Paper. Available online: www.solarpowereurope.org (accessed on 29 December 2020).
11. Bellini, E. Special Solar Panels for Agrivoltaics. 2020. Available online: https://www.pv-magazine.com/2020/07/23/special-solar-panels-for-agrivoltaics/ (accessed on 19 March 2021).
12. Hassanpour Adeh, E.; Selker, J.S.; Higgins, C.W. Remarkable agrivoltaic influence on soil moisture, micrometeorology and water-use efficiency. *PLoS ONE* **2018**, *13*, e0203256. [CrossRef] [PubMed]
13. Expósito, A.; Berbel, J. Agricultural irrigation water use in a closed basin and the impacts on water productivity: The case of the Guadalquivir river basin (southern Spain). *Water* **2017**, *9*, 136. [CrossRef]
14. Eurostat. Agri-environmental indicator –irrigation. Statistics Explained. 16/4/2019 (data from February 2019). p. 10. Available online: https://ec.europa.eu/eurostat/statistics-explained/ (accessed on 19 March 2021).
15. Schulz, V.S.; Munz, S.; Stolzenburg, K.; Hartung, J.; Weisenburger, S.; Graeff-Hönninger, S. Impact of Different Shading Levels on Growth, Yield and Quality of Potato (*Solanum tuberosum* L.). *Agronomy* **2019**, *9*, 330. [CrossRef]
16. MAPA Avance del Anuario de Estadística 2019. 2020 Ministerio de Agricultura, Pesca y Alimentación. Available online: https://www.mapa.gob.es (accessed on 19 March 2021).
17. Weselek, A.; Ehmann, A.; Zikeli, S.; Lewandowski, I.; Schindele, S.; Högy, P. Agrophotovoltaic systems: Applications, challenges, and opportunities. A review. *Agron. Sustain. Dev.* **2019**, *39*, 35. [CrossRef]
18. Cell Solar. Polycristalline 60 Cells CSP270-290W. Available online: www.cellsolar-energy.com (accessed on 29 December 2020).
19. Oberhofer, A. Spinnanker (Spider-Shaped Anchor). European Patent EP 1 750 020 B2, 7 February 2007.
20. Beck, M.; Bopp, G.; Goetzberger, A.; Obergfell, T.; Reise, C.; Schindele, S. Combining PV and food crops to agrophotovoltaic-optimization of orientation and harvest. *EUPVSEC Proc.* **2012**, *1*, 4096–4100.
21. NREL. *System Advisor Model (SAM)*; 2020.11.29-R1, SSC 252.; National Renewable Energy Laboratory: Golden, CO, USA, 2020.

22. SMA. Sunny HighPower Peak 3. Available online: https://www.sma.de/en/products/solarinverters/sunny-highpower-peak3.html (accessed on 19 March 2021).

23. SISIFO. On-line Simulator of PV Systems. Solar Energy Institute of the Universidad Politécnica de Madrid. Web Service Supported by the European Commission with the H2020 Project MASLOWATEN. Available online: https://www.sisifo.info/en/datainput (accessed on 2 March 2021).

24. Berbel, J.; Borrego-Marín, M.M.; Expósito, A.; Giannoccaro, G.; Montilla-López, N.M.; Roseta-Palma, C. Analysis of irrigation water tariffs and taxes in Europe. *Water Policy* **2019**, *21*, 806–825. [CrossRef]

25. FENACORE. Dossier de Prensa. Federación Nacional de Comunidades de Regantes. 2017. Available online: http://www.fenacore.org/empresas/fenacoreweb/documentos/DOSSIER%20PRENSA%20FENACORE%202017.pdf (accessed on 19 March 2021).

26. Masia, S.; Susnik, J.; Marras, S.; Mereu, S.; Spano, D.; Trabucco, A. Assessment of irrigated agriculture vulnerability under climate change in Southern Italy. *Water* **2018**, *10*, 209. [CrossRef]

27. De Stefano, L.; Fornés, J.M.; López-Geta, J.A.; Villarroya, F. Groundwater use in Spain: An overview in light of the EU Water Framework Directive. *Intl. J. Water Resour. Dev.* **2015**, *31*, 640–656. [CrossRef]

28. Consejería de Medio Ambiente y Ordenación del Territorio. El Clima de Andalucía en el Siglo XXI. 2014 Junta de Andalucía. Available online: http://www.juntadeandalucia.es/medioambiente/site/portalweb/ (accessed on 29 December 2020).

29. TEXLA Renovables. Proyecto Planta Fotovoltaica Valle Inferior Solar 6 MWp. Comunidad de Regantes del Valle Inferior del Guadalquivir. 2018. Available online: www.valleinferior.es (accessed on 29 December 2020).

30. Marrou, H.; Dufour, L.; Wery, J. How does a shelter of solar panels influence water flows in a soil-crop system? *Eur. J. Agronomy* **2013**, *50*, 38–51. [CrossRef]

31. Barron-Gafford, G.A.; Pavao-Zuckerman, M.A.; Minor, R.L.; Sutter, L.F.; Barnett-Moreno, I.; Blackett, D.T.; Thompson, M.; Dimond, K.; Gerlak, A.K.; Nabhan, G.P.; et al. Agrivoltaics provide mutual benefits across the food–energy–water nexus in drylands. *Nat. Sustain.* **2019**, *2*, 848–855. [CrossRef]

32. Hernández, V.; Hellín, P.; Fenoll, J.; Flores, P. Interaction of nitrogen and shading on tomato yield and quality. *Sci. Hortic.* **2019**, *255*, 255–259. [CrossRef]

33. Berbel, J.; Expósito, A.; Gutiérrez-Martín, C.; Mateos, L. Effects of the Irrigation Modernization in Spain 2002–2015. *Water Resour. Manag.* **2019**, *33*, 1835–1849. [CrossRef]

34. Hernández-Mora, N.; Martínez Cortina, L.; Llamas Madurga, M.R.; Custodio Gimena, E. *Groundwater Issues in Southern EU Member States. Spain Country Report*; European Academies Science Advisory Council: Brussels, Belgium, 2007.

35. Comunidad de Regantes del Valle Inferior del Guadalquivir. Circular n° 1/2016. 2016. Available online: http://valleinferior.es/circular-n-1-2016_aa14.html (accessed on 19 March 2021).

36. Rodríguez Díaz, J.A.; Poyato, E.C.; Pérez, M.B. Evaluation of Water and Energy Use in Pressurized Irrigation Networks in Southern Spain. *J. Irrig. Drain. Eng.* **2011**, *137*, 644–650. [CrossRef]

37. Fernández García, I.; Rodríguez Díaz, J.A.; Poyato, E.C.; Montesinos, P.; Berbel, J. Effects of modernization and medium term perspectives on water and energy use in irrigation districts. *Agric. Syst.* **2014**, *131*, 56–63. [CrossRef]

38. Sanchis-Ibor, C.; García-Mollá, M.; Avellà-Reus, L. Effects of drip irrigation promotion policies on water use and irrigation costs in Valencia, Spain. *Hydrol. Res.* **2017**, *19*, 165–180. [CrossRef]

39. Karkanis, A.; Ntatsi, G.; Kontopoulou, C.-K.; Pristeri, A.; Bilalis, D.; Savvas, D. Field Pea in European Cropping Systems: Adaptability, Biological Nitrogen Fixation and Cultivation Practices. *Not. Bot. Horti Agrobot. Cluj-Napoca* **2016**, *44*, 325–336. [CrossRef]

40. ITACYL. Plan de Monitorización de los Cultivos de Regadío en Castilla y León: Resultados de la Encuesta de Cultivos de la Campaña Agrícola 2013–2014. Junta de Castilla y León. 2015. Available online: www.inforiego.org (accessed on 19 March 2021).

41. Murugarren, N. Precios de la energía en la campaña 2012. Importancia de su negociación en la gestión de las comunidades de regantes. *Navarra Agraria* **2013**, 7–11. Available online: www.navarraagraria.com (accessed on 19 March 2021).

42. OMIP. Market Bulletin MIBEL SPEL Solar Load (FTS). Available online: https://www.omip.pt/es/dados-mercado?date=2019-11-19&product=EL&zone=ES&instrument=FTS&maturity=YR (accessed on 19 March 2021).

43. López-Prol, J.; Steininger, K.W.; Zilberman, D. The cannibalization effect of wind and solar in the California wholesale electricity market. *Energy Econ.* **2020**, *85*, 104552. [CrossRef]

44. Zhang, H.; Flottmann, S. Source-sink manipulation in canola indicates that yield is source limited. In Proceedings of the 17th ASA Conference, Hobart, Australia, 20–24 September 2015.

45. Barmon, N.C.; Bala, P.; Roy, U.K.; Azad, A.K. Growth and yield of carrot influenced by shading characters. *Eco-friendly Agril. J.* **2012**, *5*, 13–16.

46. Sekiyama, T.; Nagashima, A. Solar Sharing for Both Food and Clean Energy Production: Performance of Agrivoltaic Systems for Corn, A Typical Shade-Intolerant Crop. *Environments* **2019**, *6*, 65. [CrossRef]

47. Nasrullahzadeh, S.; Ghassemi-Golezani, K.; Javanshir, A.; Valizade, M.; Shakiba, M.R. Effects of shade stress on ground cover and grain yield of faba bean (*Vicia faba* L.). *J. Food Agric. Environ.* **2007**, *5*, 337–340.

48. Pereira, F.; Puiatti, M.; Finger, F.; Cecon, P. Growth, assimilate partition and yield of melon charenthais under different shading screens. *Hortic. Bras.* **2011**, *29*, 91–97. [CrossRef]

49. Khan, M.A.A.; Larkin, A.; Singh, V.; Agarwal, Y.K. Impact of fertilizer levels on the growth and yield of onion (*Allium cepa* L.) under jatropha (*Jetropha curcas* L.) based agroforestry system. *Int. Arch. App. Sci. Technol.* **2019**, *10*, 107–113.
50. Gubbels, G.H. Quality, yield and seed weight of green field peas under conditions of applied shade. *Can. J. Plant Sci.* **1980**, *61*, 213–217. [CrossRef]
51. Seserman, D.M.; Veste, M.; Freese, D.; Swieter, A.; Langhof, M. Benefits of agroforestry systems for land equivalent ratio-case studies in Brandenburg and Lower Saxony, Germany. In Proceedings of the 4th European Agroforestry Conference—Agroforestry as a Sustainable Land Use, Nijmegen, The Netherlands, 28–30 May 2018; pp. 26–29.
52. Guaita-Pradas, I.; Blasco-Ruiz, A. Analyzing Profitability and Discount Rates for Solar PV Plants. A Spanish Case. *Sustain. J. Rec.* **2020**, *12*, 3157. [CrossRef]
53. Banco de España. 30-year Maturity Bonds. Spain. Financial Indicators. Daily Series. Available online: https://www.bde.es/webbde/en/estadis/infoest/sindi.html (accessed on 6 November 2020).
54. Bolsas y Mercados Españoles. Annual Report. 2019. Available online: https://www.bolsasymercados.es/docs/inf_legal/ing/economico/2019/IA-BME-2019-Eng.pdf (accessed on 19 March 2021).
55. Infrontanalytics. Levered/Unlevered Beta of SMA Solar Technology AG. Available online: https://www.infrontanalytics.com/fe-EN/40278ED/SMA-Solar-Technology-AG/Beta (accessed on 29 December 2020).
56. Trommsdorff, M. An Economic Analysis of Agrophotovoltaics: Opportunities, Risks and Strategies towards a More Efficient land Use. Master's Thesis, University of Freiburg, Freiburg, Germany, 2016.
57. Stein, A. Utility-Scale Fixed-Tilt PV vs. Single-Axis Tracker PV: NEMS Projections to 2050. Capstone Paper. Master's Thesis, Johns Hopkins University, Baltimore, ML, USA, 2018. Available online: https://jscholarship.library.jhu.edu/handle/1774.2/59881 (accessed on 26 November 2020).
58. Amaducci, S.; Yin, X.; Colauzzi, M. Agrivoltaic systems to optimize land use for electric energy production. *Appl. Energy* **2018**, *220*, 545–561. [CrossRef]
59. IDAE. Ahorro y Eficiencia Energética en la Agricultura de Regadío. 2005. Available online: www.idae.es (accessed on 19 March 2021).
60. González Ruiz, J.I.; Descalzo Benito, M.J. Electricity Regulations in Spain: Overview. Thomson Reuters Practical Law. Available online: https://uk.practicallaw.thomsonreuters.com/4-529-8116?transitionType=Default&contextData=(sc.Default)&firstPage=true#co_anchor_a808110 (accessed on 29 December 2020).
61. Wang, H.X.; Muñoz-García, M.A.; Moreda, G.P.; Alonso-García, M.C. Optimum inverter sizing of grid-connected photovoltaic systems based on energetic and economic considerations. *Renew. Energy* **2018**, *118*, 709–717. [CrossRef]
62. Nishizawa, T.; Ito, A.; Motomura, Y.; Ito, M.; Togashi, M. Changes in fruit quality as influenced by shading of netted melon plants (*Cucumis melo* L. 'Andesu' and 'Luster'). *J. Jpn. Soc. Hort. Sci.* **2000**, *69*, 563–569. [CrossRef]
63. Willockx, B.; Herteleer, B.; Cappelle, J. Techno-economic study of agrivoltaic systems focusing on orchard crops. In Proceedings of the EUPVSEC 2020 online Conference, online, 2 September 2020.
64. Schachinger, M. Module Price Index. 2020. Available online: www.pv-magazine.com/module-price-index (accessed on 4 December 2020).
65. Renovables Arlumi, S.L. Anteproyecto Planta Fotovoltaica 44.95 MWn La Torre 40 S.L. 2017. Available online: https://www.juntadeandalucia.es/export/drupaljda/tramite_informacion_publica/20/07/ANTEPROYECTO%20PFV%20LA%20TORRE%2040.pdf (accessed on 19 March 2021).
66. Sueta, H.E.; Mocelin, A.; Zilles, R.; Obase, P.F.; Boemeisel, E. Protection of photovoltaic systems against lightning. Experimental verifications and techno-economic analysis of protection. In Proceedings of the XII SIPDA-International Symposium on Lightning Protection, Belo Horizonte, Brazil, 7–11 October 2013; pp. 442–447.
67. Cclcomponents. SMA DC String Combiner Box for STP/SHP Inverters. Available online: https://www.cclcomponents.com/sma-solar-dc-string-combiner-box-16-f2-s (accessed on 4 December 2020).
68. Vartiainen, E.; Masson, G.; Breyer, C. PV LCOE in Europe 2014–30—Final Report. 2015 European PV Technology Platform Steering Committee. PV LCOE Working Group. Available online: www.eupvplattform.org (accessed on 19 March 2021).
69. Vartiainen, E.; Masson, G.; Breyer, C.; Moser, D.; Román Medina, E. Impact of weighed average cost of capital, capital expenditure, and other parameters on future utility-scale PV levelised cost of electricity. *Prog. Photovolt. Res. Appl.* **2020**, *28*, 439–453. [CrossRef]
70. Esgueva, N.A. Línea de Conexión a 20 kV, Centro de Trasformación y Planta de Generación Fotovoltaica en Valdeolivas (Cuenca). Bachelor's Thesis, Universidad Carlos III de Madrid, Madrid, Spain, 2016. Available online: https://e-archivo.uc3m.es/handle/10016/24446 (accessed on 4 December 2020).
71. Eurostat. Labour Cost Levels by NACE. Available online: https://ec.europa.eu/eurostat/web/labour-market/labour-costs (accessed on 29 December 2020).
72. Más que Máquinas Agrícolas. El Blog de Maquinaria de la Revista Agricultura. Costes de Externalización de Labores. Precios de Trabajos y Faenas Agrícolas. 2014. Available online: http://www.masquemaquina.com/2014/01/costes-de-externalizacion-de-labores-y.html (accessed on 29 December 2020).
73. Ministerio de Industria. *Turismo y Comercio RD 1544/2011, de 31 de octubre, por el que se Establecen los Peajes de Acceso a las Redes de Transporte y Distribución que Deben Satisfacer los Productores de Energía Eléctrica*; BOE 276; Ministerio de Industria, Turismo y Comercio: Madrid, Spain, 2011.

74. TOTAL Gas y Electricidad España SAU. Huertos Solares: Venta a Mercado o PPAs. 2019. Available online: https://www.totalenergia.es/es/pymes/blog/huerto-Solar-parque-fotovoltaico-venta-energia-mercados-ppa (accessed on 29 December 2020).
75. MAPA. *Resultados Técnico-Económicos de Cultivos Hortícolas 2016*; Ministerio de Agricultura, Pesca y Alimentación: Madrid, Spain, 2020.
76. MAPAMA. *Resultados Técnico-Económicos de Cultivos Herbáceos 2015*; Ministerio de Agricultura y Pesca, Alimentación y Medio Ambiente: Madrid, Spain, 2017.
77. Consejería de Economía-Consejería de Empleo. Texto Articulado del Convenio Colectivo Provincial de Sevilla Para las Faenas Agrícolas, Forestales y Ganaderas 2017–2021. Boletín Oficial de la Provincia de Sevilla, 267. 2018 Junta de Andalucía. Available online: http://www.juntadeandalucia.es/empleo/mapaNegociacionColectiva/descargarDocumento?uuid=d093aafe-ec99-11 e8-94d4-b70735927673 (accessed on 4 December 2020).

 *agronomy*

Article

Estimation of the Hourly Global Solar Irradiation on the Tilted and Oriented Plane of Photovoltaic Solar Panels Applied to Greenhouse Production

Francisco J. Diez [1,*], Andrés Martínez-Rodríguez [1], Luis M. Navas-Gracia [1], Leticia Chico-Santamarta [2], Adriana Correa-Guimaraes [1] and Renato Andara [3]

[1] Department of Agricultural and Forestry Engineering, University of Valladolid, Campus La Yutera, 34004 Palencia, Spain; andres.martinez.rodriguez@uva.es (A.M.-R.); luismanuel.navas@uva.es (L.M.N.-G.); adriana.correa@uva.es (A.C.-G.)

[2] International Department, Harper Adams University, Newport TF10 8NB, UK; lchico-santamarta@harper-adams.ac.uk

[3] Antonio José de Sucre National Experimental Polytechnic University, 3001 Barquisimeto, Venezuela; renatoandara@gmail.com

* Correspondence: x5pino@yahoo.es; Tel.: +34-979-108-342

Citation: Diez, F.J.; Martínez-Rodríguez, A.; Navas-Gracia, L.M.; Chico-Santamarta, L.; Correa-Guimaraes, A.; Andara, R. Estimation of the Hourly Global Solar Irradiation on the Tilted and Oriented Plane of Photovoltaic Solar Panels Applied to Greenhouse Production. *Agronomy* **2021**, *11*, 495. https://doi.org/10.3390/agronomy11030495

Academic Editors: Miguel-Ángel Muñoz-García, Luis Hernández-Callejo and Byoung Ryong Jeong

Received: 30 December 2020
Accepted: 3 March 2021
Published: 6 March 2021

Publisher's Note: MDPI stays neutral with regard to jurisdictional claims in published maps and institutional affiliations.

Abstract: Agrometeorological stations have horizontal solar irradiation data available, but the design and simulation of photovoltaic (PV) systems require data about the solar panel (inclined and/or oriented). Greenhouses for agricultural production, outside the large protected production areas, are usually off-grid; thus, the solar irradiation variable on the panel plane is critical for an optimal PV design. Modeling of solar radiation components (beam, diffuse, and ground-reflected) is carried out by calculating the extraterrestrial solar radiation, solar height, angle of incidence, and diffuse solar radiation. In this study, the modeling was done using Simulink-MATLAB blocks to facilitate its application, using the day of the year, the time of day, and the hourly horizontal global solar irradiation as input variables. The rest of the parameters (i.e., inclination, orientation, solar constant, albedo, latitude, and longitude) were fixed in each block. The results obtained using anisotropic models of diffuse solar irradiation of the sky in the region of Castile and León (Spain) showed improvements over the results obtained with isotropic models. This work enables the precise estimation of solar irradiation on a solar panel flexibly, for particular places, and with the best models for each of the components of solar radiation.

Keywords: extraterrestrial solar irradiation; global, beam and diffuse solar components; ground-reflected solar radiation; horizontal, tilted and oriented solar irradiation

1. Introduction

The solar irradiation incident on the surface of a solar panel is the fundamental parameter for the design of photovoltaic systems that are best integrated into greenhouses for agricultural production and for determining the amount of electrical energy that is produced by such a panel, as well as for the simulation of its operation with the required precision. The value of the solar radiation that affects the solar panels is the main variable needed to determine the performance of a photovoltaic (PV) system, together with the ambient temperature, humidity and the speed and direction of the wind (see Pérez-Alonso et al. [1]).

In modern agriculture, greenhouses are intended to increase the productivity, quality, and precocity of crops that are characterized by the intensive use of land and of other means of production and inputs (see Yano and Cossu [2]). Greenhouses, except those located in large, protected production areas, are usually located in rural off-grid areas, and connection to the grid can be very expensive for technical, economic, or environmental reasons; therefore, an autonomous power source is required (see Chaurey and Kandpal [3] and

Qoaider and Steinbrecht [4]). Thus, an efficient framework is needed to use solar/diesel systems in off-grid greenhouses (see Cai et al. [5]). On the other hand, the highest electrical consumptions in greenhouses correspond to ventilation, refrigeration, and pumping equipment (water and nutrients). These agricultural structures are usually located in open spaces where they receive large amounts of direct solar radiation. Hence, the largest demand for electricity occurs during periods in which solar irradiation is available in large quantities, thus matching the demand and supply, which makes the use of solar energy viable (see Al-Ibrahim et al. [6]).

To estimate the incident solar irradiation, a pyranometer can be used [7], installed in the same solar panel plane that is to be studied [8], and if such a sensor is not available, its value can be estimated with the measurements of pyranometers installed in nearby meteorological stations, from which such measurements are normally taken on the horizontal surfaces.

The solar irradiation received by a solar panel inclined at a certain angle with respect to the horizontal surface and oriented with a deviation towards the east or west with respect to the equator, with respect to the solar irradiation that reaches the horizontal surface, which is usually measured and recorded in meteorological stations, depends on various variables and parameters. Furthermore, this transformation is performed by treating the three components of solar radiation separately, namely the direct radiation received in the direction of the sun; the diffuse radiation, coming from all directions of the celestial vault when the plane of the panel is inclined horizontally; and the albedo, which is the solar radiation reflected from the surroundings of the earth's surface.

The effect of the inclination of the solar panel on its electrical production performance has been studied by different authors. Hafez et al. [9] detailed most of the design criteria for a solar collector, suggesting a low optimal angle of inclination for summer and spring and a high one for winter and autumn. In addition, photovoltaic systems show their best performance with an optimal annual angle of inclination for which the solar tracking system is not a necessary element. In addition, the solar irradiation incident on an inclined surface has been studied depending on the geographical location. In the Mediterranean Region, Darhmaoui and Lahjouji [10] found the optimal angle of inclination to achieve the maximum annual solar energy collection, starting from the latitude of the place and the values of the daily global solar irradiation on the horizontal surface, assuming a correct south orientation.

In India, Pandey and Katiyar [11] studied the variation of the hourly global solar irradiation for surfaces inclined at intervals of 15°, where the one received with an angle of inclination equal to the latitude of the place was the optimum throughout the year, by using, for its simplicity, the isotropic model of Liu and Jordan [12] to estimate the monthly mean hourly global solar irradiation on inclined surfaces. The same model developed by Liu and Jordan [12] was used by Klein [13] to calculate the monthly mean daily solar irradiation on inclined surfaces, regardless of the orientation of the collecting surface. In the same period, Temps and Coulson [14] estimated the values of solar irradiation on the inclined and oriented plane, using the solar flux model developed by Robinson [15].

From the beginning, the technological development of photovoltaic and thermal solar energy has included scientific and technical work to estimate the solar irradiation available on the horizontal surface from easily measurable parameters. More recently, Gómez and Casanovas [16] proposed a model that applied to Spanish conditions of solar irradiation on inclined surfaces arbitrarily oriented based on procedures of fuzzy logic. This model considers overlapping classes, thus allowing a better description of the sky situations in the transition zones between contiguous categories. Other studies have been published analyzing the performances of different models of global solar irradiation (e.g., Loutzenhiser et al. [17], Evseev and Kudish [18], El Mghouchi et al. [19], and Li et al. [20]).

The direct component of the solar irradiation incident on an inclined plane can be calculated trigonometrically, but it is also necessary to know the diffuse component of the available solar irradiation on the horizontal surface. In some places, the global and diffuse solar irradiation on the horizontal surface is measured but, generally, only global data are measured or inferred from satellite data. In South Korea, Yoon et al. [21] evaluated 20 cases (five solar radiation models for each of the four albedo models) and proposed the photographic method with two factors (sky view and ground view) acquired from the pyranometer; the precision was improved, mainly by increasing the angle of inclination (i.e., considering the influence of obstacles against solar radiation); this improved the prediction accuracy for diffuse irradiation. However, the prediction accuracy of direct radiation was not improved.

The most widely available solar energy data are the measurements of global solar irradiation on a horizontal surface and thus these are the main models used to estimate diffuse solar irradiation on the horizontal surface, utilizing the horizontal global solar irradiation. After the first studies, numerous models emerged utilizing a method that provided a relationship for solar irradiation (diffuse vs. global) on a horizontal surface. These models are generally expressed in terms of polynomials from the first to fourth degrees, relating the diffuse fraction to the clearness index. Validity is discussed in these studies in order to apply the findings at different locations from where the data have been used for their development and for different climatic conditions or other geographical latitudes. The original correlations were developed for daily values, but in this study the hourly diffuse fraction vs. hourly clearness index was used, as it is the hourly solar irradiation incident on the surface of the solar panel that is a fundamental input required in the simulation of a more comprehensive design of a photovoltaic system.

Due to the lack of data series for solar irradiation measured on an inclined surface, several models have been used to estimate the solar irradiation incident on the surface of the solar panel from the measurement of global irradiation on a horizontal surface. This estimation requires previous knowledge of the components (direct and diffuse) of the global horizontal irradiation. Normally, they are not recorded at measurement stations, so the search for these components is generally done through estimation models. In the case of diffuse irradiation, the most widely used models or correlations are those that refer to the diffuse fraction k_d and the clearness index k_t on an hourly, daily, or monthly average basis. For the case of hourly fractions k_d vs. k_t, state-of-the-art models can be classified as first-order models (e.g., Boland et al. [22]), second-order models (e.g., Hawlader [23]), third-order models (e.g., Karatasou et al. [24]), and fourth-order models (e.g., Soares et al. [25], in this case utilizing an artificial neural network technique). Muneer and Munawwar [26], with a wide network of stations in Europe and Asia, show that the conventional model (k_d vs. k_t) for solar irradiation diffusion produces a high dispersion and therefore it is not satisfactory. For Australia, Ridley et al. [27] developed multiple predictions, using the hourly and daily clearness indexes as predictors, along with the solar height, the apparent solar time, and a measure of the persistence of the global solar irradiation level, suggesting its use as a universal model. For Spain, Posadillo and López [28,29] studied the dependence of k_d and k_t on solar height for their generalization to different places. Other experimental studies concerning diffuse solar irradiation on the horizontal surface can be found, such as those of Elminir [30], Ruíz-Arias et al. [31], and Torres et al. [32].

The method required for modeling the components of solar irradiation (beam, diffuse, and ground-reflected) to estimate the incident on the solar panel is extensive and its application can be complicated, so this study intended to make its use easier by providing a methodology using Simulink-MATLAB blocks. This methodology was applied with hourly horizontal global solar irradiation data from an agrometeorological station near to a greenhouse, resulting in a better approximation thanks to the use of anisotropic models of the diffuse solar irradiation.

2. Materials and Methods

This section describes the databases used, the component models applied, and the methodology developed with Simulink-MATLAB.

2.1. Materials

The hourly horizontal global solar irradiation data used in this study were recorded in 2011 in an agrometeorological station that belongs to the Agroclimatic Information System for Irrigation (SIAR), located in Mansilla Mayor (León, Castile and León region, Spain) with the following geographical coordinates: 42°30′43″ N and 5°26′46″ W, altitude 791 mamsl and local time GMT-21.725555. SIAR is a project of the Ministry of Environment and Rural and Maritime Areas of Spain, managed by the Agricultural Technological Institute in Castile and León (ITACyL), which, through the InfoRiego service for irrigating information, provides farmers with management recommendations for the use of water for irrigation [33]. The sensor used was a silicon photocell that measures the solar irradiation incident in the spectrum band between 350–1100 nm in the Skye SP1110 photovoltaic pyranometer (Campbell Scientific, Inc., North Logan, UT, USA).

The hourly horizontal diffuse solar irradiation data used in this study were taken in 2011 from the State Meteorological Agency database (AEMet is its name in Spanish) of the Ministry for Ecological Transition of Spain [34], registered in the meteorological station located in La Virgen del Camino (León, Castile and León region, Spain) with the geographical coordinates: 42°35′18″ N and 5°39′04″ W, altitude 912 mamsl.

The solar irradiance data measured on the 45° inclined plane and oriented towards the equator, which were used for comparison with the results obtained by the estimates of the methodology proposed here, were recorded at the facilities of the University of León (León, Castile and León region, Spain) with the geographical coordinates: 42°36′50″ N and 5°33′39″ W, altitude 848 mamsl. The thermoelectric sensor used generated a voltage of 10 mV/(kW·m^2), with a measurement range between 0–2000 W/m^2 and a spectral field of 305–2800 nm, and was deployed as part of a 1st class LP PYRA 02 AC pyranometer (Delta OHM Srl, Padova, Italy), manufactured under the ISO 9060 standard following the recommendations of the World Meteorological Organization (WMO).

2.2. The Components of the Solar Irradiation Incident on an Inclined Plane

The evaluation of solar irradiation reaching an inclined plane is crucial because, usually, only solar irradiation data recorded on the horizontal surface is available. The methodology used for its estimation must determine the amount of received solar irradiation (direct and diffuse) and, for a good simulation of the photovoltaic system, it must be calculated with values for a minimum period of one hour. Methods mentioned in the literature to calculate each of the components of the solar irradiation that affect the solar panel are described below: directly from the sun; reflected from the ground; and diffused from the sky. These are generally deployed separately before their subsequent union into a global measurement.

2.2.1. The Beam Irradiation of the Sun Incident on an Inclined Plane

The direct solar irradiation incident on an inclined plane results from the relationship among the components of the solar beam irradiation (extraterrestrial, horizontal, and inclined), for which Iqbal [35] assumed that the direct irradiation on a surface (inclined vs. horizontal) is the same on the surface of the Earth as it is at the maximum height of the atmosphere, as shown in Equation (1) and also detailed in Equation (2), where r_b is the ratio of solar irradiation on a plane (inclined/horizontal) at the maximum height of the Earth's atmosphere $(I_{0\beta}/I_0) \approx (\cos \theta_0 / \cos \theta_z)$.

$$I_{b\beta\gamma} = I_b \frac{I_{0\beta\gamma}}{I_0} \tag{1}$$

$$I_{b\beta\gamma} = I_b \frac{\cos \theta}{\cos \theta_z} = I_b \, r_b \tag{2}$$

where:

I_b: direct hourly irradiation incident on a horizontal surface;

$I_{b\beta\gamma}$: direct hourly irradiation incident on an inclined and oriented plane;

r_b: ratio of irradiation on an inclined plane and the horizontal surface at the maximum of the earth's atmosphere $\left(\frac{I_{0\beta}}{I_0} \approx \frac{\cos\theta_0}{\cos\theta_z}\right)$.

2.2.2. The Radiation Reflected by the Earth Incident on an Inclined Plane

The solar radiation that reaches the ground has direct and diffuse components. The word "earth" here refers to the surface of the earth that the solar panel inclined plane sees. Depending on the type of land cover, the albedo of solar irradiation (direct and diffuse) is not the same, so the total irradiation reflected by the ground can be described, following Iqbal [35], by Equation (3). As a result, two cases of reflection (isotropic and anisotropic) can happen and are presented below.

$$I_r = (I_b\, \rho_b + I_d\, \rho_d)A_g \tag{3}$$

where:

I_r: diffuse hourly irradiation reflected by the earth incident on an inclined plane;

I_d: diffuse hourly irradiation incident on a horizontal surface;

ρ_b: albedo of the soil due to direct irradiation;

ρ_d: albedo of the soil due to diffuse irradiation;

A_g: total area of the terrain seen by the inclined plane.

- Albedo with Isotropic Reflection

The isotropic reflection albedo refers to the perfectly diffuse reflection that occurs when the global solar irradiation is mainly composed of diffuse irradiation (e.g., with a cloudy sky) and/or when the ground cover is a perfectly diffuse reflector (e.g., a floor of concrete). Then, by using the ratio of solar irradiation on an inclined plane to that on a horizontal surface, a configuration factor relating the ground to the inclined plane can be obtained, as developed by Iqbal [35] in Equation (4).

$$I_r = \frac{1}{2}I\,\rho(1 - \cos\beta) \tag{4}$$

where:

ρ: albedo of the ground (irradiation reflected from the ground/irradiation incident on the ground).

- Albedo with Anisotropic Reflection

The anisotropic reflection albedo refers to the imperfect diffuse reflection that occurs when global solar irradiation is mainly composed of direct irradiation (e.g., with a clear sky) and/or when the ground is wet or there are shiny surfaces. Then, the isotropic model can be corrected with the following factors, as described by Iqbal [35], in Equation (5).

$$I_r = \frac{1}{2}I\,\rho(1 - \cos\beta)\left[1 + \mathrm{sen}^2\left(\frac{\theta_z}{2}\right)\right](|\cos\Delta|) \tag{5}$$

where:

Δ: azimuth of the inclined surface to that of the Sun; this angle is reduced to ω for surfaces inclined towards the equator.

2.2.3. The Diffuse Irradiation of the Sky Incident on an Inclined Plane

The empirical formulations for the diffuse solar irradiation of the sky incident on an inclined surface are well-developed for each category of the sky (i.e., clear, cloudy, and partly cloudy).

- Circumsolar Model

The circumsolar model is applied with a clean and clear sky and assumes that all solar irradiation that reaches the horizontal surface comes from the direction of the Sun, and thus that it can be treated in the same way as direct irradiation, as in Equation (6) from Iqbal [35].

$$I_s = I_d \, r_b \tag{6}$$

where:

I_s: diffuse irradiation of the hourly sky incident on an inclined plane.

—   Isotropic Model

The isotropic model is applied with a cloudy sky and assumes that the diffuse solar irradiation of the sky is uniform throughout the celestial dome. The diffuse irradiation of the sky incident on an inclined plane can thus be obtained with the Liu and Jordan model [36], as in Equation (7).

$$I_s = \frac{1}{2}I_d(1 + \cos \beta) \tag{7}$$

—   Anisotropic Models

Anisotropic models are applied with a partially cloudy sky, which can vary between clear and cloudy sky. Below are three models for this case.

(a)   Temps and Coulson Model

The Temps and Coulson model of anisotropic distribution for clear skies was developed in [14] and can be calculated with Equation (8).

$$I_s = \frac{1}{2}I_d(1 + \cos \beta)\left[1 + \operatorname{sen}^3\left(\frac{\beta}{2}\right)\right]\left(1 + \cos^2 \theta \, \operatorname{sen}^3\theta_z\right) \tag{8}$$

This model includes correction factors for the isotropic diffuse irradiation model which take into account the zones of anisotropy in diffuse irradiation. Factor $[1 + \operatorname{sen}^3(\beta/2)]$ is included to explain the increase in skylight observed near the horizon on clear days, the factor $(1 + \cos^2\theta \, \operatorname{sen}^3\theta_z)$ approximates the brightness of the sky near the Sun, and the third factor represents the reflection of the earth in a better way.

(b)   Klucher Model

The Klucher model of anisotropic distribution for all of sky types [37] modifies the formulation of the Temps and Coulson model by including a factor $F = 1 - (I_d/I)^2$, as indicated in Equation (9).

$$I_s = \frac{1}{2}I_d(1 + \cos \beta)\left[1 + F \operatorname{sen}^3\left(\frac{\beta}{2}\right)\right]\left(1 + F \cos^2\theta \, \operatorname{sen}^3\theta_z\right) \tag{9}$$

When the sky is completely cloudy, $F = 0$ (i.e., the equation returns to the isotropic model), and when the sky is clear, $F = 1$ (i.e., the Temps and Coulson model is used), thus improving the estimates for all types of sky.

(c)   Hay Model

The Hay model utilizes a circumsolar component that comes directly from the direction of the Sun and another component of diffuse irradiation that is distributed isotropically from the rest of the celestial vault, as calculated by Hay [38,39] with Equation (10).

$$I_s = I_d\left\{\frac{I - I_d}{I_0}r_b + \frac{1}{2}(1 + \cos \beta)\left[1 - \frac{I - I_d}{I_0}\right]\right\} \tag{10}$$

These two components are weighted according to an isotropy index (i.e., a ratio of the direct horizontal solar irradiation on the earth and the extraterrestrial solar irradiation).

2.2.4. Global Solar Irradiation Incident on an Inclined Plane

The total amount of the solar radiation incident on an inclined plane is made up of three components (direct, reflected from the ground, and diffused from the sky) which are then combined once the individual values are known. In places where the hourly solar irradiation (global and diffuse) on horizontal surfaces is known or the latter can be estimated, the global irradiation on an inclined plane can be written, as in Iqbal [35], with Equations (11) or (12).

$$I_{\beta\gamma} = (I - I_d)r_b + I_r + I_s \tag{11}$$

$$I_{\beta\gamma} = I_{b\beta\gamma} + I_{d\beta} \tag{12}$$

where:

I: global hourly irradiation incident on a horizontal surface;
$I_{\beta\gamma}$: global hourly irradiation incident on an inclined and oriented plane;
$I_{b\beta\gamma}$: direct hourly irradiation incident on an inclined and oriented plane;
$I_{d\beta}$: diffuse hourly irradiation incident on an inclined and oriented plane ($I_r + I_s$).

2.3. Simulink-MATLAB Methodology for Estimating the Solar Irradiation Incident on the Inclined Plane

For the calculation of solar irradiation on an inclined and/or oriented surface, existing models described in the literature need the horizontal global solar irradiation, horizontal diffuse solar irradiation, solar height, and angle of incidence as input variables, with their values either measured or estimated. When other models are used for the calculation (e.g., horizontal diffuse solar irradiation), the value of extraterrestrial solar irradiation is provided through parameters such as the solar constant, a correction factor for Earth's eccentricity, the day of the year, the solar declination, and the solar time angle, as well as others pertaining to the location of the greenhouse, such as its geographical latitude and longitude, and the inclination and orientation of the solar panels.

In this study, the methodology was developed with models from different authors that have been accepted in the literature; however, to facilitate its use, the Simulink-MATLAB platform for programming with visual objects was deployed. The modular nature of its design provides the possibility of using different existing models for different case studies, depending on which is the most suitable, or creating new models to obtain the various variables required.

In this study, the value of the hourly global solar irradiation on an inclined plane, according to the scheme proposed in Figure 1, was calculated with six functional blocks:

- a block for calculating the hourly extraterrestrial solar irradiation on the horizontal surface, I_0;
- a block for calculating the solar height, α, for each hour of the day;
- a block for calculating the angle of incidence, Θ, of the solar rays on the inclined and/or oriented plane;
- a block for estimating the diffuse solar irradiation of the hourly sky on the horizontal surface, I_d;
- a block for the union of the three components on the inclined plane;
- a block for the conversion of irradiance to hourly mean solar irradiance.

This methodology received the following as input variables:

- the day of the year, J (i.e., 1 for January 1, . . . , 365 for December 31);
- the mean value of the hourly interval to study, t (i.e., the time of day + 0.5);
- the hourly global solar irradiation measured on the horizontal surface, I.

The rest of the parameters were set inside each block and identified the position of the solar panels of the photovoltaic system and the location of the greenhouse:

- the inclination of the solar panel, β;
- the orientation of the solar panel, γ;
- the solar constant, I_{sc};

- the albedo, ρ;
- the latitude, the geographical φ of the greenhouse;
- the geographical longitude of the greenhouse.

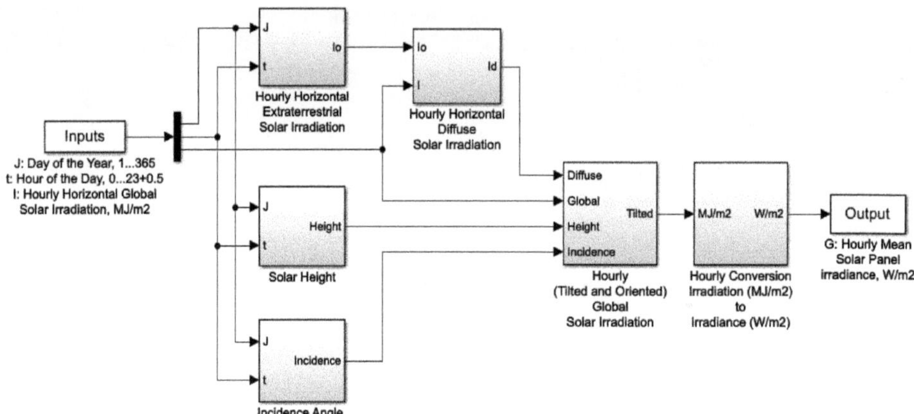

Figure 1. Methodology for estimating the measured values of global irradiance on a horizontal surface with the estimates received by the inclined plane of the solar panel.

The following sections detail the implementation of each block of the methodology developed in Simulink-MATLAB.

2.3.1. Hourly Extraterrestrial Solar Irradiation Block

The Sun emits a flow of energy which is considered to be almost constant except for small variations due to sun spots. This solar energy, when it reaches the top of the atmosphere, receives the name of extraterrestrial solar irradiation and it is this solar irradiation that would reach a certain point on Earth if the atmosphere that protects it did not exist.

Extraterrestrial solar irradiation for various latitudes can be estimated from the following parameters: the solar constant, the solar declination, and the time of year. For hourly or shorter periods, the solar angle at the beginning and the end of the period has to be considered (see the work of Allen [40], recommended by the FAO for the development of calculations for crop evapotranspiration, Duffie et al. [41], and Kalogirou [42]). This is done with Equation (13).

$$I_0 = \left(\frac{12\cdot60}{\pi}\right) I_{sc}\, E_0[(\omega_2 - \omega_1)\, \mathrm{sen}\, \phi\, \mathrm{sen}\, \delta + \cos\phi\, \cos\delta(\mathrm{sen}\, \omega_2 - \mathrm{sen}\, \omega_1)] \tag{13}$$

where:

I_0: hourly extraterrestrial solar irradiation incident on a horizontal surface, $MJ/(m^2 \cdot h)$;

I_{sc}: solar constant, $0.082\ MJ/(m^2 \cdot min)$;

E_0: correction factor for the eccentricity of the Earth $(r_0/r)^2$, with $E_0 = 1 + (0.033\cdot\cos(2\cdot\pi\cdot J/365))$;

r_0: average Sun–Earth distance, 1 ua;

r: current Sun–Earth distance, ua;

ua: astronomical unit, 1496×10^8 km;

J: day number of the year, 1 for January 1, . . . , 365 for December 31;

ϕ: geographic latitude (rad), north (+) and south (−): $-90° \leq \phi \leq 90°$ where $(rad) = \pi/180\cdot(°\text{decimal places})$;

δ: solar declination (rad), with $\delta = 0.409 \cdot \mathrm{sen}((2 \cdot \pi \cdot J/365) - 1.39)$. Defined as the angular position of the Sun at solar noon—that is, when the Sun is in the local meridian—in relation to the plane of the Earth's equator, north (+) and south (−): $-23.45° \leq \delta \leq 23.45°$;

ω_1: solar hour angle at the beginning of the period (rad), with $\omega_1 = \omega - ((\pi \cdot t_1)/24)$;

ω: solar hour angle at the moment when the midpoint of the considered period occurs (rad), with $\omega = (\pi/12) \cdot [(t + 0.06667 \cdot (L_z - L_m) + S_c) - 12]$;

t_1: duration of the period considered (hours), 1 for hourly periods, 0.5 for periods of 30 min;

t: standard time at the midpoint of the period considered (hours) (e.g., for a period between 2:00 p.m. and 3:00 p.m., $t = 14.5$);

L_z: geographic longitude of the center of the local time zone, degrees west of Greenwich: 75° East, 90° Central, 105° Rocky Mountain, 120° Pacific USA, 0° Greenwich, 330° Cairo, 255° Bangkok, 345° Spain (Iberian Peninsula);

L_m: geographic longitude of the measurement area, degrees west of Greenwich;

S_c: seasonal correction for solar time (hours), with $S_c = 0.1645 \cdot \mathrm{sen}(2b) - 0.1255 \cdot \cos(b) - 0.025 \cdot \mathrm{sen}(b)$ and $b = 2 \cdot \pi \cdot (J - 81)/364$;

ω_2: solar hour angle at the end of the period (rad), with $\omega_2 = \omega + ((\pi \cdot t_1)/24)$. If, in the morning, $\omega < -\omega_s$, or, in the afternoon, $\omega > \omega_s$, this indicates that the Sun is below the horizon such that $I_0 = 0$.

It is necessary to take into account the advance of the clock time in the official time (i.e., wintertime UTC+1 from the last Sunday in October and summertime UTC+2 from the last Sunday in March).

2.3.2. Horizontal Diffuse Solar Irradiation Block

The horizontal diffuse solar irradiation modeling from the hourly clearness index ($k_t = I/I_0$) and an hourly diffuse fraction ($k_d = I_d/I$) was carried out with a third-order model (that of Miguel et al. [43]) and was performed using data from several countries in the northern Mediterranean, resulting in Equation (14).

$$k_d = \begin{cases} 0.995 - 0.081\, k_t & k_t \leq 0.21 \\ 0.724 + 2.738\, k_t - 8.32\, k_t^2 + 4.967\, k_t^3 & \text{for} \quad 0.21 < k_t \leq 0.76 \\ 0.18 & 0.76 < k_t \end{cases} \tag{14}$$

2.3.3. Solar Height and Zenith Angle Block

Solar height, also called solar elevation, is the angular height of the Sun above the observer's celestial horizon, which is an angle between 0° and 90°. The zenith angle, also called the zenith distance, is the angle between the local zenith and the line that joins the observer and the Sun; the value of the angle is between 0° and 90°. The solar height is the complement of the zenith angle. For a given geographic position, in the absence of atmospheric refraction of the earth, the trigonometric relationship between the Sun and a horizontal surface provided by Iqbal [35] is defined by Equation (15).

$$\cos \theta_z = \mathrm{sen}\, \delta\, \mathrm{sen}\, \phi + \cos \delta\, \cos \phi\, \cos \omega = \mathrm{sen}\, \alpha \tag{15}$$

where:

α: solar height—the angle of elevation of the Sun above the true horizon;

θ_z: zenith angle—the angular position of the Sun in relation to the local vertical, $\theta_z = 90° - \alpha$.

2.3.4. Angle of the Incidence of Solar Irradiation on the Solar Panel Block

The angle of incidence is the angle formed between the normal to the inclined plane and the Sun–Earth vector. There are two cases: the inclined plane may be oriented towards the equator or with an arbitrary orientation towards east or west.

- Modeling the angle of incidence for an inclined solar panel oriented towards the equator

The angle of incidence for an inclined surface oriented towards the equator can be described, as detailed by Liu and Jordan [12], with Equation (16).

$$\cos\theta_0 = \operatorname{sen}\delta\,\operatorname{sen}(\phi-\beta) + \cos\delta\,\cos(\phi-\beta)\,\cos\omega \tag{16}$$

where:

θ_0: angle of incidence for an inclined surface oriented towards the equator;
β: inclination of the surface to the horizontal position.

 — Modeling the angle of incidence for an arbitrarily inclined and oriented solar panel

The relationship of the angle of incidence for a surface inclined and oriented in any direction with the local meridian is trigonometric (see Berod and Bock [44], Kondratyev [45], and Coffari [46]) and can be described with Equations (17) or (18).

$$\begin{aligned}
\cos\theta = {}&(\operatorname{sen}\phi\cos\beta - \cos\phi\operatorname{sen}\beta\cos\gamma)\operatorname{sen}\delta\\
&+(\cos\phi\cos\beta + \operatorname{sen}\phi\operatorname{sen}\beta\cos\gamma)\cos\delta\cos\omega\\
&+\cos\delta\operatorname{sen}\beta\operatorname{sen}\gamma\operatorname{sen}\omega
\end{aligned} \tag{17}$$

where:

θ: the angle of incidence for an arbitrarily inclined and oriented surface.

$$\cos\theta = \cos\beta\cos\theta_z + \operatorname{sen}\beta\cos\theta_z\cos(\psi-\gamma) \tag{18}$$

where:

γ: azimuth angle of the surface, orientation. Defined as the deviation of the normal to the surface of the solar collector from the local meridian in the directions west $(-)$, south (0), and east $(+)$;
ψ: solar azimuth with $\cos\psi = ((\operatorname{sen}\alpha\cdot\operatorname{sen}\phi - \operatorname{sen}\delta)/(\cos\alpha\cdot\cos\phi))$, with $0° \leq \psi \leq 90°$ for $\cos\psi \geq 0$, and with $90° \leq \psi \leq 180°$ for $\cos\psi \leq 0$. This is the angle at the local zenith between the plane of the observer's meridian and the plane of a great circle passing through the zenith and the Sun in the directions west $(-)$, south (0), and east $(+)$: $-180° \leq 0° \leq +180°$.

Measurement of the angle of incidence of the direct solar irradiation can be done by constructing a simple device. This consists of mounting a normal pointer to a flat surface, on which graduated concentric circles are marked. The length of the shadow cast by the pointer can be measured using concentric circumferences and can be used to determine the angle of incidence according to the international standard ISO 9806:2017 [47]. The device should be located on the plane and to one side of the solar panel.

2.3.5. Angle of the Incidence of the Solar Irradiation on the Solar Panel Block

This block was used to unify the three components of solar radiation that affect the solar panel in order to obtain the amount of the global solar radiation incident on the inclined surface.

2.3.6. Conversion from Hourly Global Irradiance to Hourly Mean Solar Irradiance Block

Finally, the conversion of the global hourly solar irradiance (i.e., energy value) obtained for the inclined and oriented plane to the corresponding hourly mean solar irradiance (i.e., power value) was undertaken. Through interpolation, these hourly values can be used to create a database with a time interval of one minute in order to carry out a more detailed simulation of the operation of the photovoltaic system under study (e.g., as input values for the simulation of a model of a greenhouse photovoltaic system).

3. Results

This section presents the results of the selected models for the estimation of the global solar irradiation value on the solar panel inclined plane, based on the global solar irradiation on the horizontal surface (which is usually recorded in meteorological stations) and using the following Simulink-MATLAB blocks:

- hourly extraterrestrial solar irradiation;
- hourly horizontal diffuse solar irradiation;
- the hourly solar height;
- the hourly angle of incidence on the solar panel;
- the hourly global solar irradiance and hourly average solar irradiance on the solar panel.

3.1. Result of the Hourly Extraterrestrial Solar Irradiation Block

The extraterrestrial solar irradiation modeling was applied to determine the temporal evolution of the extraterrestrial solar irradiation at the top of the atmosphere (which would be found to reach a certain point on Earth if the attenuation and scattering effects which are produced by the atmosphere when the sun's rays pass through it are not considered).

The results of the methodology for the calculation of hourly extraterrestrial solar irradiation, obtained for the 15th day of each month of the year at latitude 42° N and longitude 5.6° W, are shown in (Figure 2).

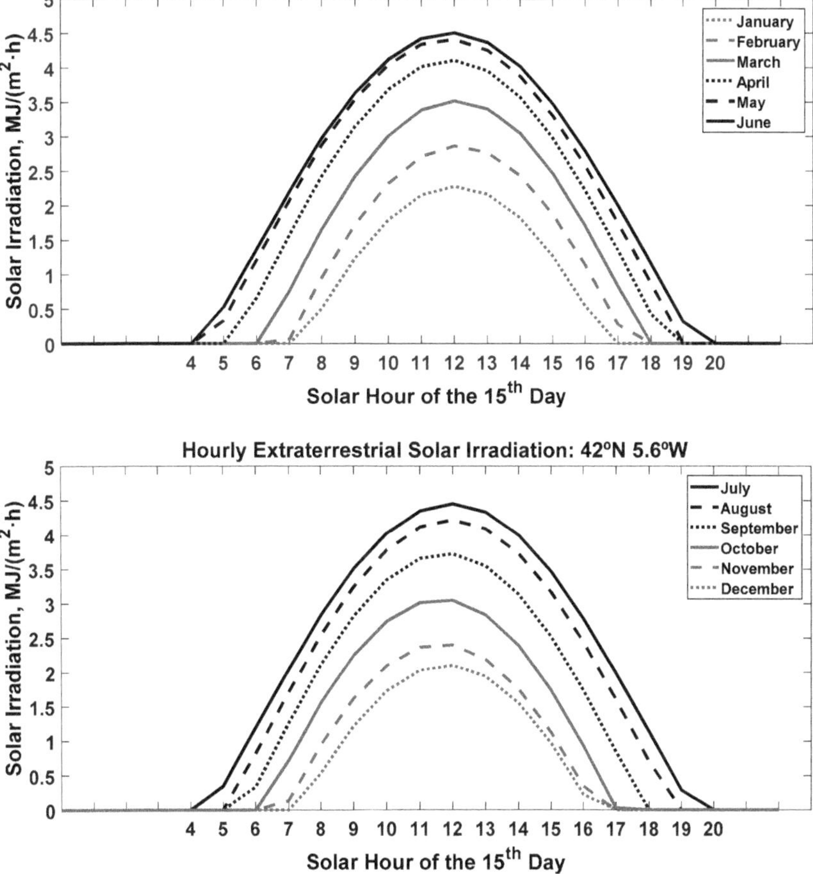

Figure 2. Hourly extraterrestrial solar irradiation calculated at latitude 42° N and longitude 5.6° W for the 15th day of each month. (**Top**) January to June; (**bottom**) July to December.

3.2. Result of the Hourly Horizontal Diffuse Solar Irradiation Block

The different correlations proposed for the hourly diffuse fraction in the introduction are shown in Figure 3. Generally, each model includes three relationships for each sky type, as detailed by Iqbal [35], that determine the daily clearness index in order to define sky conditions such that:

- for clear sky, $0.7 \leq k_t < 0.9$;
- for partly cloudy sky, $0.3 \leq k_t < 0.7$;
- for overcast sky, $0 \leq k_t < 0.3$.

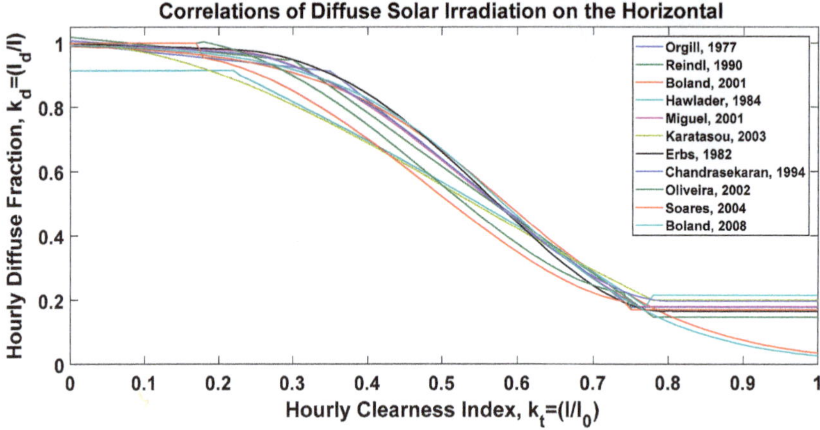

Figure 3. Different correlations of the diffuse solar irradiation on the horizontal surface proposed in the literature.

Figure 4 shows the results of the methodology used for the conversion of the hourly diffuse solar irradiation to the hourly global solar irradiation, using data recorded by the State Meteorological Agency (AEMet) at La Virgen del Camino station (León, Castile and León region, Spain), with regard to the relation between the hourly diffuse fraction and the hourly clearness index, with values recorded for the central eight hours of the day during 2011, along with the model selected (Miguel et al. [43]).

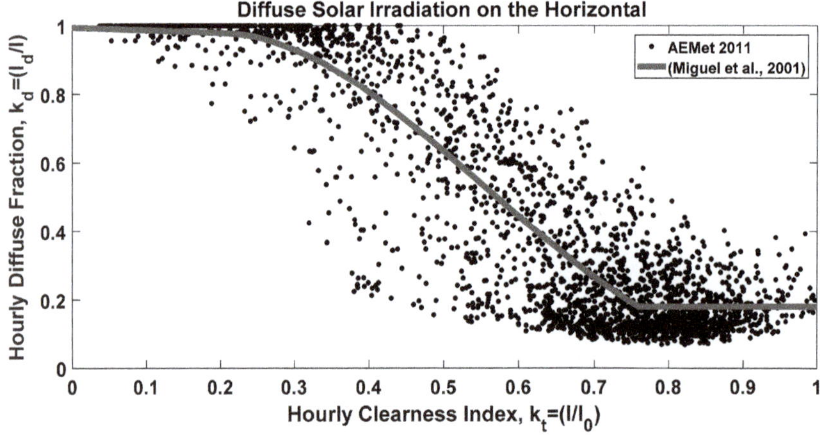

Figure 4. AEMet data from La Virgen del Camino (León, Castile and León region, Spain) for the hourly diffuse fraction vs. the hourly clearness index for the eight central hours of the day during the year 2011, along with the horizontal diffuse solar irradiation estimation model (Miguel et al. [43]).

3.3. Result of the Hourly Solar Height Block

The results of the methodology used for the modeling of the hourly solar height, obtained for the 15th day of each month of the year at latitude 42° N and longitude 5.6° W, are shown in Figure 5.

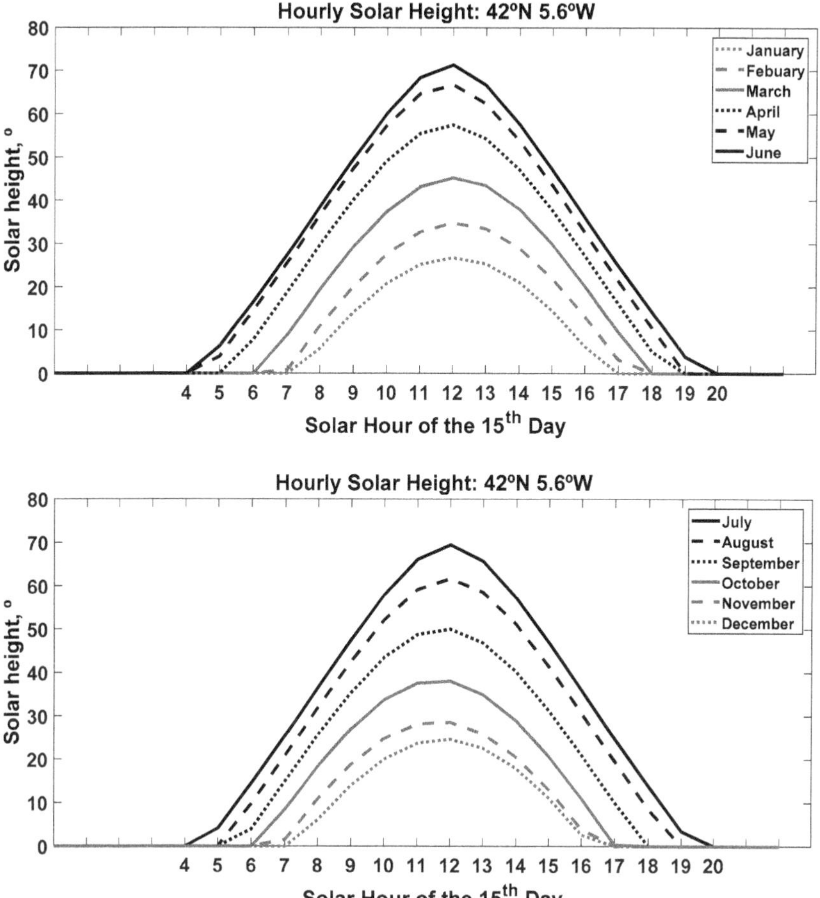

Figure 5. Hourly solar height calculated at latitude 42° N and longitude 5.6° W for the 15th day of each month. (**Top**) January to June; (**bottom**) July to December.

Figure 6 shows that the maximum values for the solar height were obtained at solar midday.

3.4. Result of the Hourly Incidence Angle Block

Figure 7 shows the results of the methodology used for the modeling of the hourly incidence angle obtained for the surface of a solar panel with an inclination of 45° and oriented towards the equator, for the 15th day of each month of the year at latitude 42° N and longitude 5.6° W.

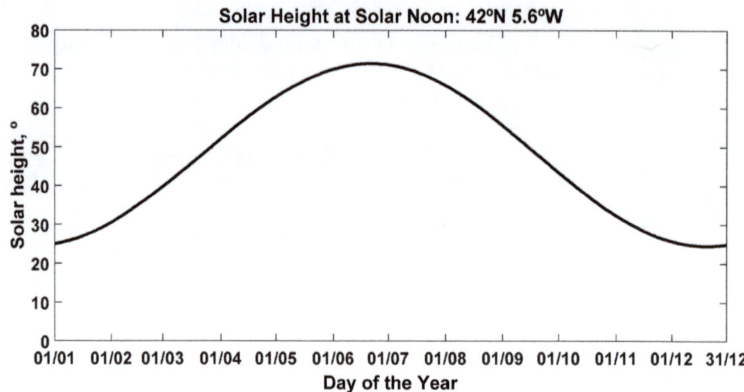

Figure 6. Solar height at solar noon calculated at latitude 42° N and longitude 5.6° W for each day of the year.

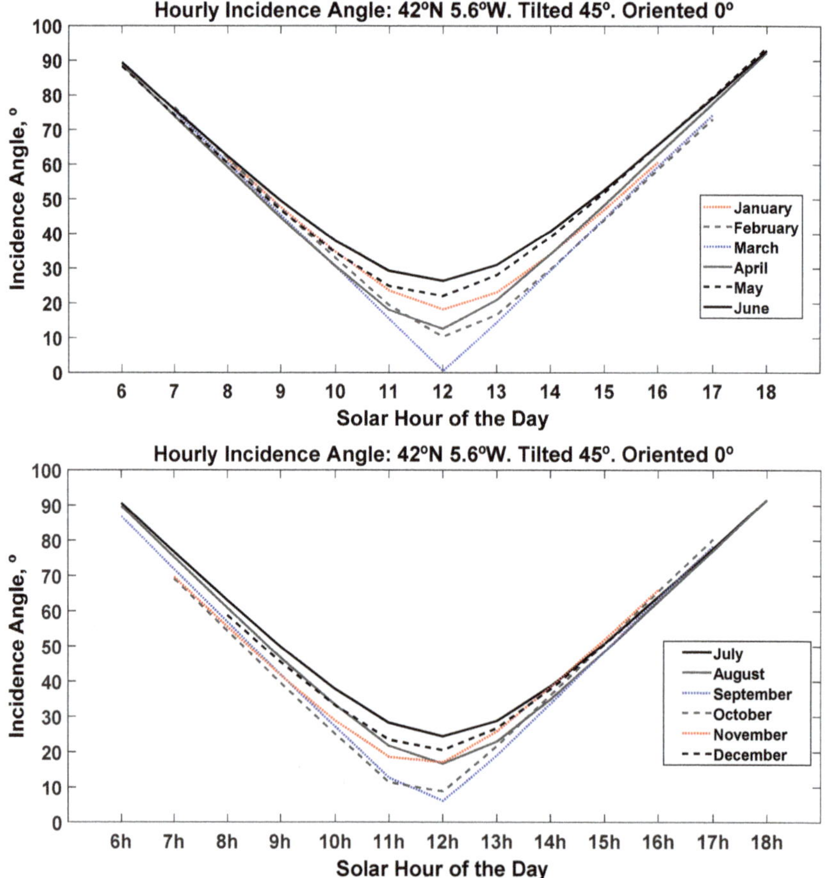

Figure 7. Hourly incidence angle calculated at latitude 42° N and longitude 5.6° W for the 15th day of each month. (**Top**) January to June; (**bottom**) July to December.

The minimum values of the angle of incidence, which were obtained at solar noon for each day of the year, are presented in Figure 8.

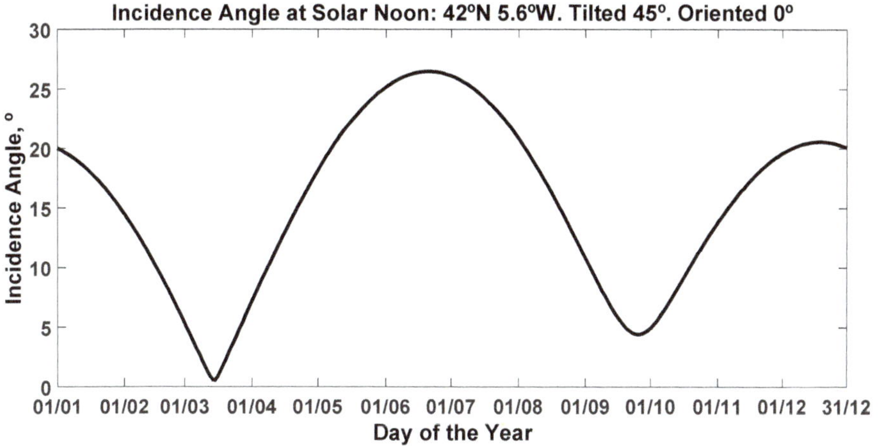

Figure 8. Incidence angle at solar noon calculated at latitude 42° N and longitude 5.6° W for all days of the year.

3.5. Results for Hourly Global Irradiance and Hourly Mean Solar Irradiance on the Solar Panel

The data recorded for the global solar irradiation on the horizontal surface during 2011 in the agrometeorological station located in the town of Mansilla Mayor (León, Castile and León region, Spain), as part of the SIAR network, were used to apply the methodology developed in Simulink-MATLAB for the estimation of the global solar irradiation value on a surface with an inclination of 45° and oriented towards the equator. The results of the different methods of obtaining diffuse solar irradiation are shown below.

Figure 9 shows the values of the daily global solar irradiation obtained from the horizontal SIAR network, together with the calculations carried out for the estimation of the daily global solar irradiation on a surface inclined at 45° and oriented to the equator using the following four models.

— Inclined Model 1

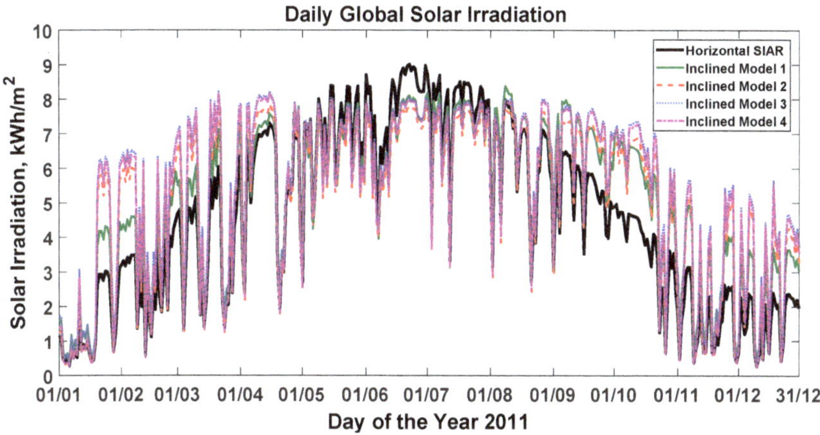

Figure 9. Global solar irradiation measured on the horizontal surface during the year 2011 and estimated for a the solar panel inclined at 45° and oriented to the equator with Inclined Models 1, 2, 3, and 4.

Inclined Model 1 used the CENSOLAR [48] method, which provides a coefficient to obtain the value of the global solar irradiation on the solar panel, depending on the latitude, the inclination, and the month of the year.

The following models (Inclined Models 2, 3, 4, and 5) used the methodology described in the previous sections for the values of the direct component and the albedo of the solar irradiation, while different models were used for the value of the diffuse component, therefore obtaining different results.

—　Inclined Model 2

Inclined Model 2 used the Liu and Jordan isotropic model, presented in Equation (7), for the estimation of the diffuse component.

—　Inclined Model 3

Inclined Model 3 used the anisotropic model of Temps and Coulson, presented in Equation (8), for the estimation of the diffuse component.

—　Inclined Model 4

Inclined Model 4 used the Klucher anisotropic model, presented in Equation (9), for the estimation of the diffuse component.

—　Inclined Model 5

Inclined Model 5 used Hay's anisotropic model, presented in Equation (10), to estimate the diffuse component.

Finally, the conversion module was applied to convert the global hourly solar irradiance values, given in energy units (MJ/m^2), on the inclined plane into the average hourly values of the solar irradiance, given in power units (W/m^2), on the inclined plane.

In Figure 10, the hourly variation of the solar irradiance is represented for one day in April (spring) and another day in October (autumn) in order to compare the solar irradiance on the inclined plane calculated with the five conversion methods with the evolution of the data from the global horizontal solar irradiance SIAR located in Mansilla Mayor (León, Castile and León region, Spain), specifically the solar irradiance recorded by the pyranometer of the solar panel. Table 1 lists the statistical results for four random sunny days, comparing the estimates made with the isotropic model and the three anisotropic models with the measured values for the inclined plane oriented towards the equator.

Table 1. Observed adjustment of the hourly solar irradiance on the plane inclined at 45° and oriented towards the equator, as estimated with Inclined Models 2, 3, 4, and 5 from data for the horizontal surface from the agrometeorological station SIAR in Mansilla Mayor (León, Castile and León, Spain), along with the values measured in León, for four days.

Day	Inclined Model 2		Inclined Model 3		Inclined Model 4		Inclined Model 5	
	RMSE	R^2	RMSE	R^2	RMSE	R^2	RMSE	R^2
11 April 2011	21.90	0.9751	25.09	0.9674	22.14	0.9746	<u>17.83</u>	<u>0.9835</u>
1 June 2011	38.51	0.8658	32.14	0.9065	<u>31.54</u>	<u>0.9100</u>	55.12	0.7250
30 June 2011	21.40	0.9415	40.11	0.7947	39.06	0.8052	<u>11.95</u>	<u>0.9817</u>
11 October 2011	62.36	0.7853	<u>31.44</u>	<u>0.9454</u>	32.51	0.9416	37.32	0.9230

Inclined Model 2 utilized the Liu and Jordan isotropic model; Inclined Model 3 utilized the Temps and Coulson anisotropic model; Inclined Model 4 utilized the Klucher anisotropic model; Inclined Model 5 utilized the Hay anisotropic model; RMSE, root mean square error (W/m^2); R^2, determination coefficient. The best results for each day are underlined.

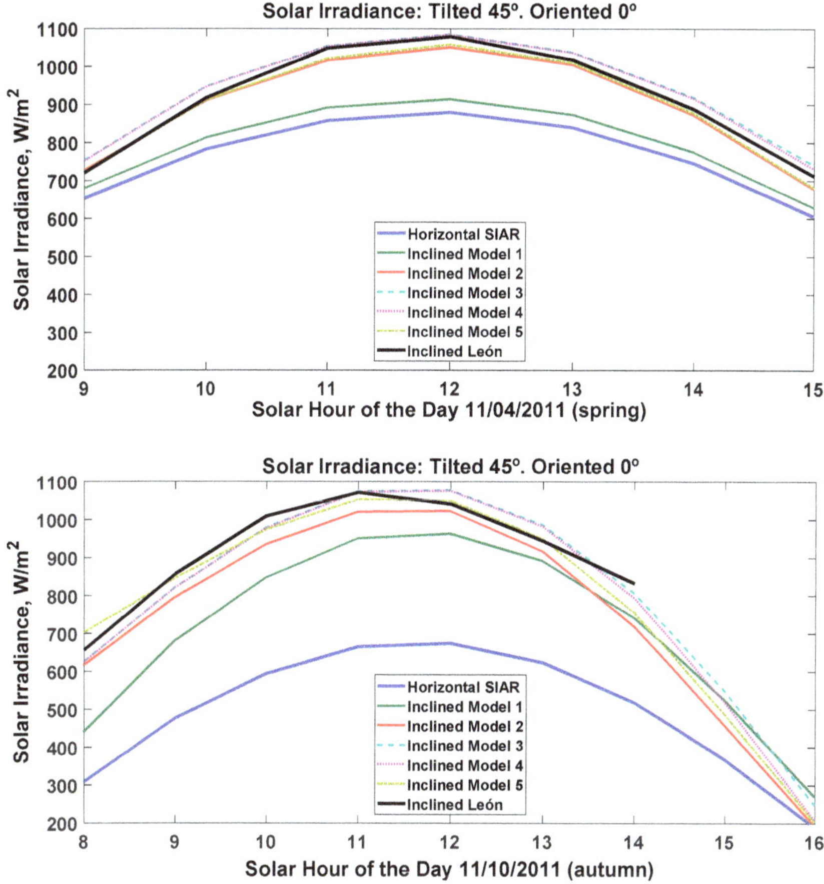

Figure 10. Hourly solar irradiance obtained for a single day from the horizontal SIAR in Mansilla Mayor (León, Castile and León region, Spain) compared with that estimated for a solar panel inclined at 45° and oriented to the equator with Inclined Models 1, 2, 3, 4, and 5. (**Top**) 11 April 2011 (spring); (**bottom**) 11 October 2011 (autumn).

4. Discussion

Once the estimation of the solar irradiation received by the inclined solar panel has been undertaken, the results obtained were analyzed using the Simulink-MATLAB blocks (Figure 1).

For the modeling of the extraterrestrial solar irradiation, symmetry can be observed throughout the day with regard to solar noon in Figure 2, along with a progressive increase in solar irradiation for the seasonal component from December to June and a decrease in solar irradiation from June to December (41.91-12.28 MJ/m^2 daily). The results were validated by comparing the sum of the hourly values of each day with those described by Allen [40] and resulted in a good approximation without the need for more statistics.

Three differentiated zones can be seen in the horizontal diffuse solar irradiation modeling (Figure 4). In one, a maximum hourly diffuse fraction (k_d) and a minimum hourly clearness index (k_t) were obtained (i.e., corresponding to the hours of the day with the sky covered); another zone was characterized by a minimum hourly diffuse fraction (k_d) and a maximum hourly clearness index (k_t) (i.e., corresponding to the hours of the day with clear skies); finally, there was a third intermediate zone with indices k_d and k_t inversely variable with each other, depending on the degree of cloudiness of the sky (i.e., corresponding to the hours of the partially cloudy day). The model provided in the study by Miguel et al. [43], which was developed in Mediterranean areas, shows a good approximation with the model obtained with the data recorded by the State Meteorological Agency (AEMet) at La Virgen del Camino station (León, Castile and León region, Spain) during the central eight hours of the day for the whole year of 2011. It can be observed that the data for clear sky days was different from the Miguel model (the mean diffuse fraction on clear days was lower than that of the Miguel model), which indicates a low dispersion of the solar irradiation of the sky on clear days (i.e., a very clear and clean sky in the place under study).

In the solar height modeling, in Figure 5 symmetry can be observed throughout the day with regard to solar noon, along with a progressive increase in solar height from December to June and a decrease in solar height from June to December. The results were validated by comparison with those described by Duffie et al. [41], showing a good approximation without the need for more statistics. The maximum values of solar height occurring at solar noon, which were observed throughout the year (Figure 6), reached their maximum during the last fortnight of June at the highest position of the Sun (72°), and the minimum solar height occurred during the last fortnight of December (24.5°).

For the results for the modeling of the angle of incidence, symmetry can be observed throughout the day with respect to solar noon, for which the minimum value was obtained, in Figure 7. In Figure 8, the progressive decrease from December to March (0.5°), the increase from March to June (26.5°), the decrease from June to September (4.5°), and the increase from September to December (20.5°) can be seen. The results were validated by comparison with those presented by Bérriz and Álvarez [49] and Duffie et al. [41], showing a good approximation without the need for more statistics. The most favorable months of the year, with the position of the solar panel inclined 45° and oriented towards the equator, for capturing solar energy were February, March, April, September, and October, as the inclined surface was in a more perpendicular position with regard to the path of the sun's rays on those dates.

The study of the variation of the angle of incidence is very important since it provides the time of year in which the use of solar energy is maximum, according to the inclination and orientation of the solar panel. The results of this model, if compared with the profile of annual energy needs of a photovoltaic system for a particular greenhouse, offer the possibility of finding the relationship (inclination–orientation) for each time of the year in which solar capture is optimal.

In M'Sila, Algeria, Ihaddadene et al. [50] theoretically searched for the best angle of inclination (monthly, seasonally, and annually), examining the Liu and Jordan model, the circumsolar model, the Hay model, and the Reindl model as the most appropriate models. They decided to change the angle of inclination of solar conversion systems monthly by 7272.27 MJ/m^2, or seasonally by 7184.94 MJ/m^2, instead of fixing them at 6836.83 MJ/m^2 in order to increase the amount of energy for a year; they recommended that this be studied using hourly data.

In Beijing, China, the annual optimal tilt angle shows a downward trend (Shen et al. [51]) compared to the optimum in the 1960s (i.e., 38°); the optimal tilt angle decreased by 2° from 2011-2015 (i.e., to 36°), caused mainly by the decrease in the direct irradiation ratio, which is highly related to atmospheric conditions (i.e., pollution that increases the proportion of diffuse irradiation and decreases the direct irradiation).

Kondratyev and Manolova [52] have stated that direct solar irradiation has been studied in detail, but diffuse and reflected solar irradiation on inclined planes are quite complicated, requiring study in relation to the distribution of the angular intensity. Therefore, special attention must be taken when selecting the most suitable diffuse irradiation model for a particular place; furthermore solar panels are more sensitive to electrical energy conversion in photovoltaic systems than solar collectors in thermal systems.

In light of these points, in this study, different models of diffuse irradiation were analyzed for a specific place where a greenhouse is located, taking into account that the diffuse solar irradiation of the sky that falls upon the inclined plane of the solar panel is estimated in a different way as a function of the model. The circumsolar model overestimated it in Equation (6); the isotropic model underestimated it on the opposite slope of the inclined surface in Equation (7); and in clear and partly cloudy sky conditions, which occur in many cases, the light of the sky is anisotropic, meaning that the Temps and Coulson model provided a good prediction for the clear sky in Equation (8) but overestimated the solar irradiance when used for the cloudy sky. In this case, the Klucher model made corrections by setting factor F to 0 when the sky was completely cloudy in Equation (9), thus returning to the isotropic model, and setting F to 1 when the skies were, resulting in the Temps and Coulson model; this improved the estimates for all types of the sky.

In Nakhon Pathom and Ubon Ratchathani, Thailand, Wattan and Janjai [53] investigated the performances of 14 models in estimating hourly diffuse solar irradiation on inclined (30°, 60°, and 90°) and oriented (north, south, east, and west) surfaces at two tropical sites, finding that the Muneer and Gueymard models performed better.

In Algiers and Ghardaia, Algeria, and Málaga, Spain, Takilalte et al. [54] estimated the inclined global irradiation in 5-min intervals, using global irradiation on the horizontal plane, geographical parameters, site albedo, and two cloudiness factors, based on a combination of two models (Perrin Brichambaut and Liu and Jordan) for which the parameters of the state of the sky transformed the isotropic models into an anisotropic model. The results of the proposed model for all sky conditions with regard to the normalized root mean square error (nRMSE), the relative percentage error (RPE), the normalized mean absolute error (nMAE), and the correlation coefficient (R^2) varied between 4.70–6.41%, 5.50–5.90%, 3.07–4.73%, and 0.97–0.99, respectively, which are very accurate results, especially for such short time scales in which there is no compensation or average effects, as occur when using monthly data.

Putting the three solar irradiation components together using Equation (11), with the solar panel at an inclination of 45° and oriented towards the equator, differences with regard to the incident on the horizontal surface shown in Figure 9 were observed according to the time of year:

— it was higher in the months of average solar irradiation (i.e., February, March, April, September, October, and November) because then the solar height produces a lower angle of incidence on the solar panel than in other months;
— it was moderate in the months of high solar irradiation (i.e., May and August);
— it was lower in the months of very high solar irradiation (i.e., June and July) due to the high solar height.

In Athens, Greece, Raptis et al. [55] studied the ideal inclination to maximize the capture of solar irradiation, which was determined by the latitude and the time of year, with the horizontal surface receiving more irradiance than the inclined surface during the summer months and on cloudy winter days, due in this case to an anisotropy of the diffuse light, with a greater diffuse contribution coming from angles closer to the zenith; however, the inclined surface reached higher values than the horizontal one in winter, with the optimum angle found to be around 30° during the year.

The result of the comparison of the recorded measurements of solar irradiance on the horizontal surface (i.e., the SIAR data) and the inclined plane of the solar panel (i.e., the pyranometer in León) with the results obtained with Inclined Models 1, 2, 3, 4, and 5 (Figure 10) show a better approximation with the anisotropic models for four random sunny days. Specifically: Inclined Model 5, which used the Hay corrections, obtained better results on 4 November 2011 and 30 June 2011 with regard to the RMSE (17.83 and 11.95 W/m^2) and R^2 (0.9835 and 0.9817), respectively. Inclined Model 4, which used the Klucher corrections, performed best on 6 January 2011, with RMSE = 31.54 W/m^2 and R^2 = 0.9100. Inclined Model 3, which used Temps and Coulson corrections, obtained better results on 10 November 2011, with RMSE = 31.44 W/m^2 and R^2 = 0.9454.

The solar panel global solar irradiation for 2011 was 1.84 MWh/m^2 with the CEN-SOLAR model, which used coefficients; 1.84 MWh/m^2 for the Liu and Jordan model; 1.99 MWh/m^2 for the Temps and Coulson model; 1.93 MWh/m^2 for the Klucher model; and 1.86 MWh/m^2 for the Hay model.

The global solar irradiation for the horizontal SIAR measured in Mansilla Mayor (León, Castile and León region, Spain) during 2011 was 1.67 MWh/m^2, thus resulting in a higher value on the solar panel using corrections of 10.17% for the model by Liu and Jordan, 19.16% for the Temps and Coulson model, 15.56% for the Klucher model, and 11.37% for the Hay model. However, this was distributed throughout the year, as shown in Figure 9, and as mentioned previously, was higher in the months with moderate solar irradiation (i.e., February, March, April, September, October, and November), moderate in high solar irradiation months (i.e., May and August), and lower in the months with very high solar irradiation (i.e., June and July).

In Adrar, Algeria, an increment in the performance of horizontally placed solar panels was achieved by Bailek et al. [56] with a fixed inclination of 20.61% monthly, 19.58% seasonally, 19.24% semi-annually, and 13.78% for annual adjustments, with the optimal tilt angles in each period.

5. Conclusions

Glasshouses are agricultural productive structures intended to increase the production and quality of early bloomer crops. They can be energetically characterized as follows:

1. they involve intensive use of soil and means of production, which requires a safe provision of all the supplies, including energy;
2. any of them are located in off-grid rural areas, so they need an autonomous energy supply;
3. they are located in open areas, with great availability of solar resources and time synchronization between the supply (i.e., the Sun) and the demand (i.e., ventilation, cooling, and ferti-irrigation).

Thus, distributed generation PV systems are ideal for connection to glasshouses, either on their own or together with power generators where the value of the solar irradiance which falls upon the solar panels is the main variable to determine the performance of the PV system.

The literature pertaining to the estimation of the incidental solar irradiance on the solar panel plane (at an angle and/or oriented) in relation to the irradiance received on the horizontal surface (data registered in meteorological stations) is highly diverse, especially with regard to the types of sky in particular places. However, the practical use of this diverse information is complex, which is what incentivized the present work, in which the following methodology was adopted.

1. Measured data of incidental solar irradiance on the horizontal surface in an agrometeorological station was used to obtain an estimation of the incidental solar irradiance on the plane of the glasshouse solar panels, where the verification pyranometer was located.
2. A flexible methodology was built with Simulink-MATLAB software blocks that could be adapted to the numerous existing models in the literature.
3. The application of components (beam, diffuse, and ground-reflected) was provided in order to ensure the use of the most adequate model for each type of sky in each location.
4. Irradiance on the solar panel was obtained with an hourly resolution for various days of the year and hours of the day, along with the hourly horizontal global solar irradiance, with the location coordinate fixed. This temporal resolution is more adequate for use in the simulation of PV systems.
5. The results obtained with models of diffuse anisotropic irradiance improved on those obtained with other models. As they are estimations on an hourly scale, when using data from stations close to the greenhouse, differences were observed for a few hours in the comparisons (e.g., at 12 h and 13 h on 10 November 2011 (autumn)),

which may have been due to some cloudiness or changes in the reflections of the surrounding light.

As a final conclusion to this paper, it should be noted that the solar estimation for the inclined plane can be used together with the daily prediction of solar irradiance, as detailed by Diez et al. [57], to obtain the value of available solar energy in the glasshouse PV system and thus to enable more efficient management of the glasshouse electric demand.

Author Contributions: Conceptualization, F.J.D., L.M.N.-G. and L.C.-S.; Methodology, F.J.D., L.M.N.-G. and A.M.-R.; Software, F.J.D., A.M.-R. and R.A.; Validation, F.J.D., L.C.-S. and A.C.-G.; Formal analysis, L.M.N.-G., A.M.-R. and R.A.; Investigation, F.J.D., L.M.N.-G. and L.C.-S.; Resources, L.M.N.-G. and A.C.-G.; Writing—original draft preparation, F.J.D., A.M.-R. and R.A.; Writing—review and editing, L.M.N.-G., L.C.-S., A.C.-G. and R.A.; Visualization, F.J.D. and A.M.-R.; Supervision, F.J.D., A.C.-G. and R.A.; Project administration, L.M.N.-G. and A.M.-R.; Funding acquisition, L.M.N.-G. All authors have read and agreed to the published version of the manuscript.

Funding: This research received no external funding.

Institutional Review Board Statement: Not applicable.

Informed Consent Statement: Not applicable.

Acknowledgments: The authors wish to acknowledge CYTED (the Ibero-American Program of Science and Technology for Development) for supporting this work through collaboration with the RITMUS network.

Conflicts of Interest: The authors declare no conflict of interest. The funders had no role in the design of the study; in the collection, analyses, or interpretation of data; in the writing of the manuscript, and in the decision to publish the results.

References

1. Pérez-Alonso, J.; Pérez-García, M.; Pasamontes-Romera, M.; Callejón-Ferre, A.J. Performance analysis and neural modelling of a greenhouse integrated photovoltaic system. *Renew. Sustain. Energ. Rev.* **2012**, *16*, 4675–4685. [CrossRef]
2. Yano, A.; Cossu, M. Energy sustainable greenhouse crop cultivation using photovoltaic technologies. *Renew. Sustain. Energ. Rev.* **2019**, *109*, 116–137. [CrossRef]
3. Chaurey, A.; Kandpal, T.C. Assessment and evaluation of PV based decentralized rural electrification: An overview. *Renew. Sustain. Energy Rev.* **2010**, *14*, 2266–2278. [CrossRef]
4. Qoaider, L.; Steinbrecht, D. Photovoltaic systems: A cost competitive option to supply energy to off-grid agricultural communities in arid regions. *Appl. Energy* **2010**, *87*, 427–435. [CrossRef]
5. Cai, W.; Li, X.; Maleki, A.; Pourfayaz, F.; Rosen, M.A.; Nazari, M.A.; Bui, D.T. Optimal sizing and location based on economic parameters for an off-grid application of a hybrid system with photovoltaic, battery and diesel technology. *Energy* **2020**, *201*, 117480. [CrossRef]
6. Al-Ibrahim, A.; Al-Abbadi, N.; Al-Helal, I. PV greenhouse system—System description, performance and lesson learned. In Proceedings of the International Symposium on Greenhouses, Environmental Controls and In-house Mechanization for Crop Production in the Tropics and Sub-Tropics, Cameron Highlands, Pahang, Malaysia, 30 June 2006; Rukunuddin, K., Hamid, A., Eds.; ISHS: Leuven, Belgium, 2006; Volume 710, pp. 251–264.
7. Brooks, D.R. *Bringing the Sun down to Earth: Designing Inexpensive Instruments for Monitoring the Atmosphere*; Springer: New York, NY, USA, 2008.
8. Vignola, F.; Michalsky, J.; Stoffel, T. *Solar and Infrared Radiation Measurements*, 2nd ed.; CRC Press/Taylor & Francis Group: Boca Raton, FL, USA, 2020.
9. Hafez, A.Z.; Soliman, A.; El-Metwally, K.A.; Ismail, I.M. Tilt and azimuth angles in solar energy applications—A review. *Renew. Sustain. Energ. Rev.* **2017**, *77*, 147–168. [CrossRef]
10. Darhmaoui, H.; Lahjouji, D. Latitude based model for tilt angle optimization for solar collectors in the Mediterranean Region. *Energy Procedia* **2013**, *42*, 426–435. [CrossRef]
11. Pandey, C.K.; Katiyar, A.K. Hourly solar radiation on inclined surfaces. *Sustain. Energy Technol. Assess.* **2014**, *6*, 86–92. [CrossRef]
12. Liu, B.Y.H.; Jordan, R.C. Daily insolation on surfaces tilted toward the equator. *Trans. ASHRAE* **1962**, 526–541.
13. Klein, S.A. Calculation of monthly average insolation on tilted surfaces. *Sol. Energy* **1977**, *19*, 325–329. [CrossRef]
14. Temps, R.C.; Coulson, K.L. Solar radiation incident upon slopes of different orientations. *Sol. Energy* **1977**, *19*, 179–184. [CrossRef]
15. Robinson, R. *Solar Radiation*; Elseiver: New York, NY, USA, 1966.
16. Gómez, V.; Casanovas, A. Fuzzy modeling of solar irradiance on inclined surfaces. *Sol. Energy* **2003**, *75*, 307–315. [CrossRef]

17. Loutzenhiser, P.G.; Manz, H.; Felsmann, C.; Strachan, P.A.; Frank, T.; Maxwell, G.M. Empirical validation of models to compute solar irradiance on inclined surfaces for building energy simulation. *Sol. Energy* **2007**, *81*, 254–267. [CrossRef]
18. Evseev, E.G.; Kudish, A.I. The assessment of different models to predict the global solar radiation on a surface tilted to the south. *Sol. Energy* **2009**, *83*, 377–388. [CrossRef]
19. El Mghouchi, Y.; Chham, E.; Krikiz, M.S.; Ajzoul, T.; El Bouardi, A. On the prediction of the daily global solar radiation intensity on south-facing plane surfaces inclined at varying angles. *Energ. Convers. Manag.* **2016**, *120*, 397–411. [CrossRef]
20. Li, D.H.W.; Lou, S.; Lam, J.C.; Wu, R.H.T. Determining solar irradiance on inclined planes from classified CIE (International Commission on Illumination) standard skies. *Energy* **2016**, *101*, 462–470. [CrossRef]
21. Yoon, K.; Yun, G.; Jeon, J.; Kim, K.S. Evaluation of hourly solar radiation on inclined surfaces at Seoul by Photographical Method. *Sol. Energy* **2014**, *100*, 203–216. [CrossRef]
22. Boland, J.; Scott, L.; Luther, M. Modelling the diffuse fraction of global solar radiation on a horizontal surface. *Environmetrics* **2001**, *12*, 103–116. [CrossRef]
23. Hawlader, M.N.A. Diffuse, global and extraterrestrial solar radiation for Singapore. *Int. J. Ambient. Energy* **1984**, *5*, 31–38. [CrossRef]
24. Karatasou, S.; Santamouris, M.; Geros, V. Analysis of experimental data on diffuse solar radiation in Athens, Greece, for building applications. *Int. J. Sustain. Energy* **2003**, *23*, 1–11. [CrossRef]
25. Soares, J.; Oliveira, A.P.; Božnar, M.Z.; Mlakar, P.; Escobedo, J.F.; Machado, A.J. Modeling hourly diffuse solar-radiation in the city of São Paulo using a neural-network technique. *Appl. Energ.* **2004**, *79*, 201–214. [CrossRef]
26. Muneer, T.; Munawwar, S. Potential for improvement in estimation of solar diffuse irradiance. *Energ. Convers. Manag.* **2006**, *47*, 68–86. [CrossRef]
27. Ridley, B.; Boland, J.; Lauret, P. Modelling of diffuse solar fraction with multiple predictors. *Renew. Energ.* **2010**, *35*, 478–483. [CrossRef]
28. Posadillo. R.; López, R. Hourly distributions of the diffuse fraction of global solar irradiation in Córdoba (Spain). *Energ. Convers. Manag.* **2009**, *50*, 223–231. [CrossRef]
29. Posadillo, R.; López, R. The generation of hourly diffuse irradiation: A model from the analysis of the fluctuation of global irradiance series. *Energ. Convers. Manag.* **2010**, *51*, 627–635. [CrossRef]
30. Elminir, H.K. Experimental and theoretical investigation of diffuse solar radiation: Data and models quality tested for Egyptian sites. *Energy* **2007**, *32*, 73–82. [CrossRef]
31. Ruíz-Arias, J.A.; Alsamamra, H.; Tovar-Pescador, J.; Pozo-Vázquez, D. Proposal of a regressive model for the hourly diffuse solar radiation under all sky conditions. *Energ. Convers. Manag.* **2010**, *51*, 881–893. [CrossRef]
32. Torres, J.L.; de Blas, M.; García, A.; de Francisco, A. Comparative study of various models in estimating hourly diffuse solar irradiance. *Renew. Energ.* **2010**, *35*, 1325–1332. [CrossRef]
33. InfoRiego. Información Meteorológica. Available online: http://www.inforiego.org (accessed on 16 March 2020).
34. AEMet. Agencia Estatal de Meteorología, Ministry for Ecological Transition of Spain. Available online: http://www.aemet.es (accessed on 16 March 2020).
35. Iqbal, M. *An Introduction to Solar Radiation*; Academic Press: Toronto, ON, Canada, 1983.
36. Liu, B.Y.H.; Jordan, R.C. The long-term average performance of flat-plate solar-energy collectors: With design data for the U.S., its outlying possessions and Canada. *Sol. Energy* **1963**, *7*, 53–74. [CrossRef]
37. Klucher, T.M. Evaluation of models to predict insolation on tilted surfaces. *Sol. Energy* **1979**, *23*, 111–114. [CrossRef]
38. Hay, J.E. Calculation of monthly mean solar radiation for horizontal and inclined surfaces. *Sol. Energy* **1979**, *23*, 301–307. [CrossRef]
39. Hay, J.E. A revised method for determining the direct and diffuse components of the total short-wave radiation. *Atmosphere* **1976**, *14*, 278–287. [CrossRef]
40. Allen, R.G.; Pereira, L.S.; Raes, D.; Smith, M. Crop evapotranspiration: Guidelines for computing crop water requirements. In *FAO Irrigation and Drainage Paper No. 56*; FAO: Rome, Italy, 1998; pp. 47–48.
41. Duffie, J.A.; Beckman, W.A.; Blair, N. *Solar Engineering of Thermal Processes, Photovoltaics and Wind*, 5th ed.; John Wiley & Sons: Hoboken, NJ, USA, 2020; pp. 37–41.
42. Kalogirou, S.A. *Solar Energy Engineering: Processes and Systems*, 2nd ed.; Elsevier/Academic Press: Amsterdam, The Netherlands, 2014; pp. 91–95.
43. Miguel, A.; Bilbao, J.; Aguiar, R.; Kambezidis, H.; Negro, E. Diffuse solar irradiation model evaluation in the North Mediterranean Belt area. *Sol. Energy* **2001**, *70*, 143–153. [CrossRef]
44. Benrod, F.; Bock, J.E. A time analysis of sunshine. *Trans. Am. Illum. Eng. Soc.* **1934**, *34*, 200–218.
45. Kondratyev, K.Y. *Radiation in the Atmosphere*; Elsevier/Academic Press: New York, NY, USA, 1969; pp. 342–346.
46. Coffari, E. The sun and the celestial vault. In *Solar Energy Engineering*; Sayigh, A.A.M., Ed.; Academic Press: Orlando, FL, USA, 1977; Volume 2, pp. 22–27.
47. ISO 9806:2017. *Solar energy–Solar thermal collectors–Test Methods*; International Organization for Standardization: Geneva, Switzerland, 2017.
48. Centro de Estudios de la Energía Solar (CENSOLAR). *Pliego de Condiciones Técnicas de Instalaciones de Baja Temperatura: Instalaciones de Energía Solar Térmica*; Instituto para la Diversificación y Ahorro de la Energía (IDAE): Madrid, Spain, 2009; pp. 102–108. Available online: https://www.idae.es/publicaciones/instalaciones-deenergia-solar-termica-pliego-de-condiciones-tecnicas-de-instalaciones-de-baja (accessed on 16 March 2020).

49. Bérriz, L.; Álvarez, M. *Manual Para el Cálculo y Diseño de Calentadores Solares*; Cubasolar: La Habana, Cuba, 2008.
50. Ihaddadene, N.; Ihaddadene, R.; Charik, A. Best Tilt Angle of Fixed Solar Conversion Systems at M'Sila Region (Algeria). *Energy Procedia* **2017**, *118*, 63–71. [CrossRef]
51. Shen, Y.; Zhang, J.; Guo, P.; Wang, X. Impact of solar radiation variation on the optimal tilted angle for fixed grid-connected PV array—case study in Beijing. *Global Energy Interconnect.* **2018**, *1*, 460–466.
52. Kondratyev, K.J.; Manolova, M.P. The radiation balance of slopes. *Sol. Energy* **1960**, *4*, 14–19. [CrossRef]
53. Wattan, R.; Janjai, S. An investigation of the performance of 14 models for estimating hourly diffuse irradiation on inclined surfaces at tropical sites. *Renew. Energ.* **2016**, *93*, 667–674. [CrossRef]
54. Takilalte, A.; Harrouni, S.; Yaiche, M.R.; Mora-López, L. New approach to estimate 5-min global solar irradiation data on tilted planes from horizontal measurement. *Renew. Energ.* **2020**, *145*, 2477–2488. [CrossRef]
55. Raptis, P.I.; Kazadzis, S.; Psiloglou, B.; Kouremeti, N.; Kosmopoulos, P.; Kazantzidis, A. Measurements and model simulations of solar radiation at tilted planes, towards the maximization of energy capture. *Energy* **2017**, *130*, 570–580. [CrossRef]
56. Bailek, N.; Bouchouicha, K.; Aoun, N.; EL-Shimy, M.; Jamil, B.; Mostafaeipour, A. Optimized fixed tilt for incident solar energy maximization on flat surfaces located in the Algerian Big South. *Sustain. Energy Technol. Assess.* **2018**, *28*, 96–102. [CrossRef]
57. Diez, F.J.; Navas-Gracia, L.M.; Chico-Santamarta, L.; Correa-Guimaraes, A.; Martínez-Rodríguez, A. Prediction of horizontal daily global solar irradiation using artificial neural networks (ANNs) in the Castile and León region, Spain. *Agronomy* **2020**, *10*, 96. [CrossRef]

 *agronomy*

Article

Improving the Power Supply Performance in Rural Smart Grids with Photovoltaic DG by Optimizing Fuse Selection

Santiago Pindado [1,*], Daniel Alcala-Gonzalez [2], Daniel Alfonso-Corcuera [1], Eva M. García del Toro [2] and María Isabel Más-López [3]

[1] Instituto Universitario de Microgravedad "Ignacio Da Riva" (IDR/UPM), ETSI Aeronáutica y del Espacio, Universidad Politécnica de Madrid, Pza. del Cardenal Cisneros 3, 28040 Madrid, Spain; daniel.alfonso.corcuera@alumnos.upm.es

[2] Departamento de Ingeniería Civil, Hidráulica y Ordenación del Territorio, ETSI Civil, Universidad Politécnica de Madrid, Alfonso XII, 3, 28014 Madrid, Spain; d.alcalag@upm.es (D.A.-G.); evamaria.garcia@upm.es (E.M.G.d.T.)

[3] Departamento de Ingeniería Civil, Construcción Infraestructura y Transporte, ETSI Civil, Universidad Politécnica de Madrid, Alfonso XII, 3, 28014 Madrid, Spain; mariaisabel.mas@upm.es

* Correspondence: santiago.pindado@upm.es

Abstract: The recent increase in the use of renewable sources in electrical systems has transformed the electrical distribution network with the subsequent implementation of the distributed generation (DG) concept. The high penetration level of photovoltaic units increases their injected fault current that may result in a lack of coordination of fuse reclosers in distribution networks. One of the main protection devices that is generally used in rural distribution networks is the fuse. A correct size selection is key for ensuring good operation and coordination with other protection devices. The DG implementation makes the selection above more difficult, as the current flow both in steady state and in case of short-circuit is subject to alterations. A new protection fuse selection method for distribution networks with implemented DG is proposed in this paper with the aim of ensuring an effective coordination between them, avoiding untimely behaviors. Different case studies have been analyzed (for diverse locations of DG in the network with various penetration levels which represent 25%, 50%, 75%, and 100% of the total installed load), using an IEEE 13-node test feeder. Besides, a new model to analyze fuse performance is proposed in this work. This model has proven to fit the manufacturer's data well, with a maximum error of 2% within the normal trip current values.

Keywords: fuses coordination; protection device; distributed generation; distribution network; photovoltaic; DIgSILENT

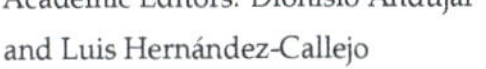

Citation: Pindado, S.; Alcala-Gonzalez, D.; Alfonso-Corcuera, D.; García del Toro, E.M.; Más-López, M.I. Improving the Power Supply Performance in Rural Smart Grids with Photovoltaic DG by Optimizing Fuse Selection. *Agronomy* **2021**, *11*, 622. https://doi.org/10.3390/agronomy11040622

Academic Editors: Dionisio Andújar and Luis Hernández-Callejo

Received: 9 February 2021
Accepted: 19 March 2021
Published: 25 March 2021

Publisher's Note: MDPI stays neutral with regard to jurisdictional claims in published maps and institutional affiliations.

1. Introduction

The presence of distributed generation (DG) of renewable energy sources in rural electrical distribution networks has caused an important change from the classic energy supply model. The DG concept aims to bring nearer power generation and users, in contrast with the traditional view of centralized power generation plants [1]. This involves a need for adapting the electrical infrastructure to allow a correct energy distribution that ensures an optimal end-point quality [2].

The traditional procedures for planning, managing, and operating a rural electrical distribution network with a radial typology is based on assuming a unidirectional power flow, with a transmission from higher voltage levels down to distribution levels. This assumption allows implementing relatively simple and cheap protection schemes, perfectly coordinated, which allows an effective operation of the protection system [3,4].

The implementation of distributed generation (DG) causes a structural change in the distribution network, which no longer acts as a radial network, compromising the adequate coordination between protection devices [3,5–7]. The impact level that DG can cause in the distribution network will depend on, among others, the generator size, type, and location in the network [8,9].

One of the most frequently used protection devices is the fuse. To this device, the implementation of DG can cause a lack of coordination and untimely tripping [10], as the steady-state currents are subject to alterations which can even involve currents going through the fuse in reverse direction [7]. This means currents going through the fuse can be generated from locations both downstream and upstream of the fuse. Studies by Hadjsaid et al. show how current values can be altered as a consequence of deploying DG [11]. Girgis and Brahma [12,13] described that the lack of coordination between fuses was an issue when implementing DG. Chaitusaney and Yokoyama have studied in detail the impact of the DG on the system stability, considering the lack of coordination between protection devices [14]. Finally, Boonyapakdee et al. analyzed the dynamic coordination of recloser fuses affected by synchronous distributed generators [15].

Besides, Razavi et al. addressed the voltage regulation methods in the presence of DG units and their impact on protection systems [16]. Bayati et al. proposed a local protection method without communication links. This methodology can be used in both digital and conventional protection devices installed in DC microgrids [17]. The proposed scheme formulates the fuse recloser switch coordination challenge as a curve-fitting problem and solves this problem to obtain the settings of the digital recloser switch and fuse. Finally, Alam et al. suggest the use of a new scheme of recloser fuse coordination for reconfigurable radial distribution networks (RDNs) with DG, to obtain the optimum recloser fuse settings [18].

As mentioned above, it is possible to state that DG in any feeder of an RDN might change the flow of currents through fuses during faults. Additionally, the magnitude of the fault current passing through the recloser placed at the substation is also modified. In some cases, the fault currents through fuses become larger than the ones through the recloser, while in some other cases, the direction of the flow of current through some fuses is reversed due to the presence of DG in the fault path [10,13,14,19,20]. Under these circumstances, it is difficult to provide appropriate protection using the conventional fuse–fuse coordination scheme [13,21]. Additionally, the presence of multiple DGs in the network makes the coordination of fuses very complex. Also, synchronous machine-based DGs contribute more to the fault and are more prone to causing lack of coordination.

To the best of our knowledge and based on the available literature, it is necessary to propose a new reconfiguration of protection devices based on fuses coordination, as the main interest in the mentioned available literature seems to be focused on fuse–relay coordination [7,22,23]. This procedure should reduce the negative effects of DG in power networks (e.g., lack of coordination between protection relays and fuses), which imply that electrical protection schemes become not valid or, at least, less effective.

With the aim to both ensure a good electrical supply quality, with adequate protection device functionality, and reach a more sustainable energy network, this paper proposes a new method for fuse selection that avoids untimely tripping and ensures effective coordination between protection devices. It should be emphasized the novelty of this approach, that is based on a completely new model of fuse performance, based on polynomial expansion of the inverse of the tripping time in relation to the current root. This fuse performance modeling leaves aside thermal behavior, which may be relevant [24].

The present work is organized as follows. In Section 2, the methodology and the network used to analyze the coordination are described, whereas the results are included in Section 3. Finally, conclusions are summarized in Section 4.

2. Materials and Methods

2.1. Coordination between Protection Fuses

The trip curve of protection fuses is an inverse time–current characteristic curve. The straight line I^2t log–log plot, in which I is the tripping current and t is the tripping time, is usually expressed for the minimum melting and total clearing times for fuses. From the fuse characteristic on the log–log curve, it is better to approximate it by the second-order polynomial function. However, the interested range of the curve approaches a straight line. Moreover, a linear equation can be simply applied to reduce the calculation task. The general equation describing the fuse characteristic curve can be expressed as the following equation [22,25,26]:

$$\log(t) = m \log(I) + n, \tag{1}$$

where m and n are the parameters of the selected tripping curve [27]. Nevertheless, it should be also said that other more complex equations have been suggested to describe the behavior of fuses. Santoso and Short [19] suggested the following one:

$$t = \exp\left(\sum_{n=0}^{4} a_n (\ln(I))^n \right), \tag{2}$$

which was reduced in one term by Tang and Ayyanar [28]. Additionally, Abdel-ghany et al. [29] proposed a model to define the tripping time in relation to two exponential terms:

$$t = a \exp(bI) + c \exp(dI). \tag{3}$$

In the above Equations (2) and (3), a_n, a, b, c, and d are the parameters that need to be adjusted/extracted to fit the equations to the proper behavior of the fuse.

Finally, in the works by Conde et al. [30,31] and Costa et al. [32], a comparison between different models to select the best one is carried out.

The fuse trip characteristic curve shown in Figure 1 is characterized by two boundary curves:

- Minimum melting curve (as the lower boundary), which detects the minimum overcurrent causing the link to start melting.
- Full opening curve (as the upper boundary), indicating the complete blowing of the fuse and the circuit opening.

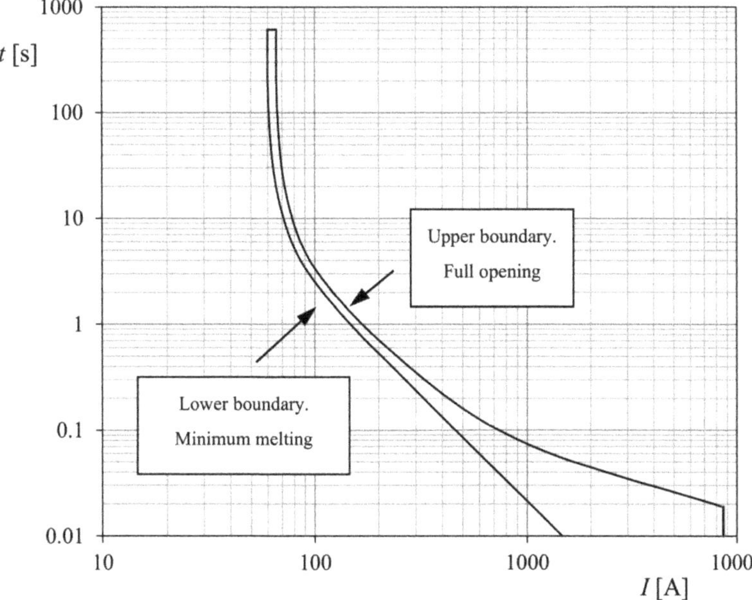

Figure 1. Tripping characteristic curve of an expulsion fuse.

A classic coordination between protective fuses for non-DG rural electrical networks is shown in Figure 2. The sketch in the left side of this figure shows a radial network in which F1 is the fuse which protects the main line and F2 the one protecting the branch line. The criteria for a correct coordination between fuses state the following:

- For faults in the main feeder, the F2 fuse will not detect any fault current, with F1 fuse being the one that will act as main protection.
- For faults in the branch line, F2 will act as the main protection and F1 as a backup protection.

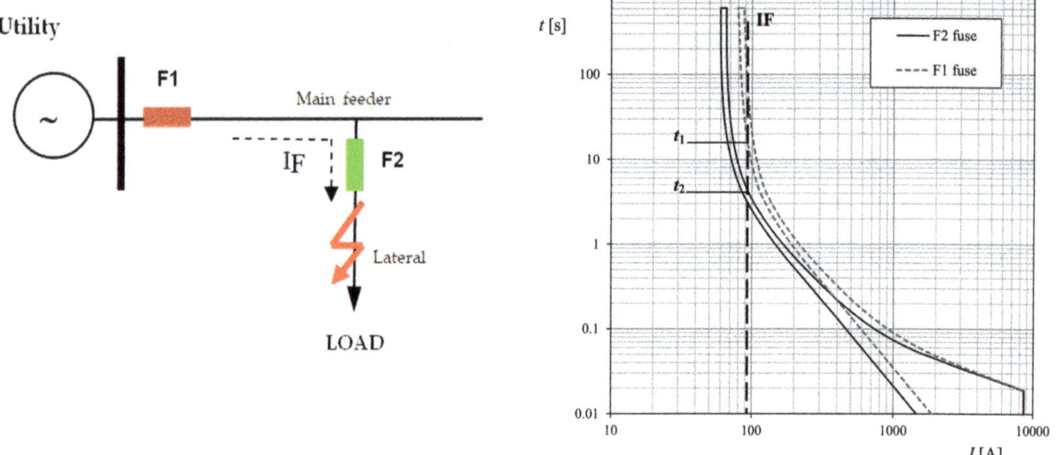

Figure 2. Classic coordination between protective fuses in non-DG radial distribution networks. (**Left**) Sketch of a radial distribution network without the presence of a DG. (**Right**) Operating characteristics curves of protection fuses F1 and F2.

The graph included in Figure 2 shows the tripping characteristics of fuses F1 and F2. At the same fault current, IF, there should be a coordination margin such that the total operating time (upper boundary), t_1, of the fuse acting as main protection, F2, should not exceed 75% of the minimum time, t_2, in which the fuse acting as backup protection, F1, starts blowing (lower boundary) [33].

When the electrical network is working in a permanent regime, power balance is achieved. The generated power, P_G, equals the power demanded by the loads, P_L, plus a power loss, P_P:

$$P_G = P_L + P_P. \tag{4}$$

When DG is installed in the electrical network, the situation is different, as new power sources, P_{DG}, are included:

$$P_G + P_{DG} = P_L + P_P. \tag{5}$$

As a consequence, the values of the currents in the permanent regime may change due to a new distribution of the loads' current flow, causing the activation of the nearest fuse to the DG and compromising the coordination between this fuse and the one located immediately upstream.

Figure 3 shows a simple sketch of an electrical network with DG, and the corresponding curves of fuses F1 (backup protection) and F2 (main protection). In this situation, the power supplied by DG, P_{DG}, might be larger than the one demanded by the load, P_L. This situation is translated into the curves included in the graph from Figure 3: F2 fuse (closest to the DG) acts too close to the blowing starting point of fuse F1. If the power of the DG unit is increased, the coordination margin between the fuses would be altered and would cause both to trip untimely.

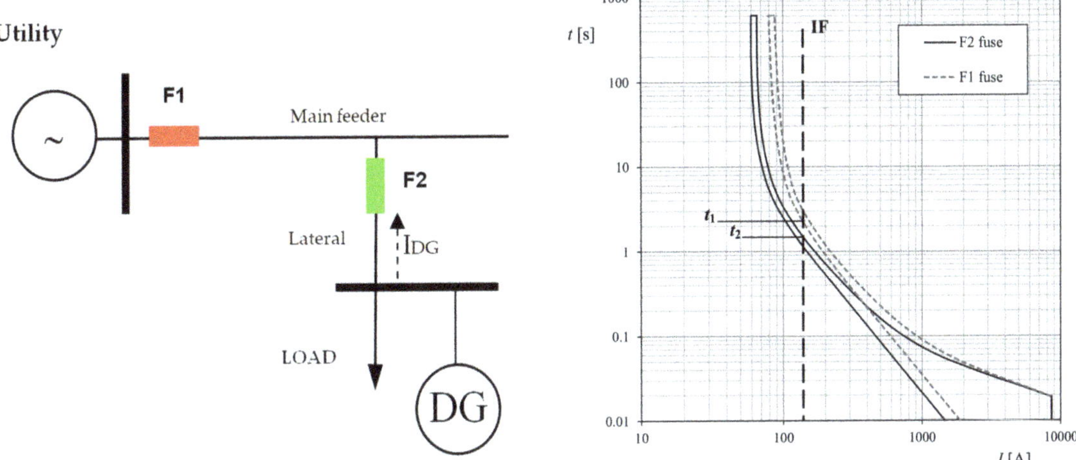

Figure 3. Loss of coordination between fuses in the distribution network with the presence of DG. (**Left**) Sketch of a radial distribution network with DG. (**Right**) Operating characteristics curves of protection fuses F1 and F2.

2.2. Modeling the Performance of Fuses

This article proposes a new method for the selection of protection fuses to avoid untimely tripping, and therefore achieving a more sustainable electrical network while maintaining the functionality of the protections. This procedure is based on a new mathematical modeling of the upper (ES) and lower (EI) boundary tripping curves of the fuses to be coordinated. For the upper boundary curve, two mathematical expressions are proposed, depending on the current, I, in relation to a specified value, I^*:

$$\frac{1}{t_{ES}} = \begin{cases} \sum_{i=0}^{N_1} \beta_i I^{\frac{i}{2}}; I < I^* \\ \sum_{i=0}^{N_2} \beta_i I^{\frac{i}{2}}; I > I^* \end{cases}. \tag{6}$$

For the lower boundary curve, one single equation is proposed:

$$\frac{1}{t_{EI}} = \sum_{i=0}^{N} \beta_i I^{\frac{i}{2}}. \tag{7}$$

In the above equations, t_{ES} and t_{EI} stand for the total operating time corresponding to the upper boundary curve of the fuse acting as the main protection, and the minimum melting time corresponding to the lower boundary curve of the fuse acting as backup protection, respectively. β_i are characterization parameters of both the upper and the lower boundary curve of the fuse acting as backup protection.

Equations (6) and (7), fitted to the upper curve of SMD-50 10E fuse and the lower curve of an SMD-50 15E fuse (both fuses manufactured by S&C Electric Company Chicago, IL, USA) are shown in Figure 4. In the first case, fifth- and second-order polynomials ($N_1 = 5$ and $N_2 = 2$) have been used, whereas in the second case, a fifth-order polynomial is used.

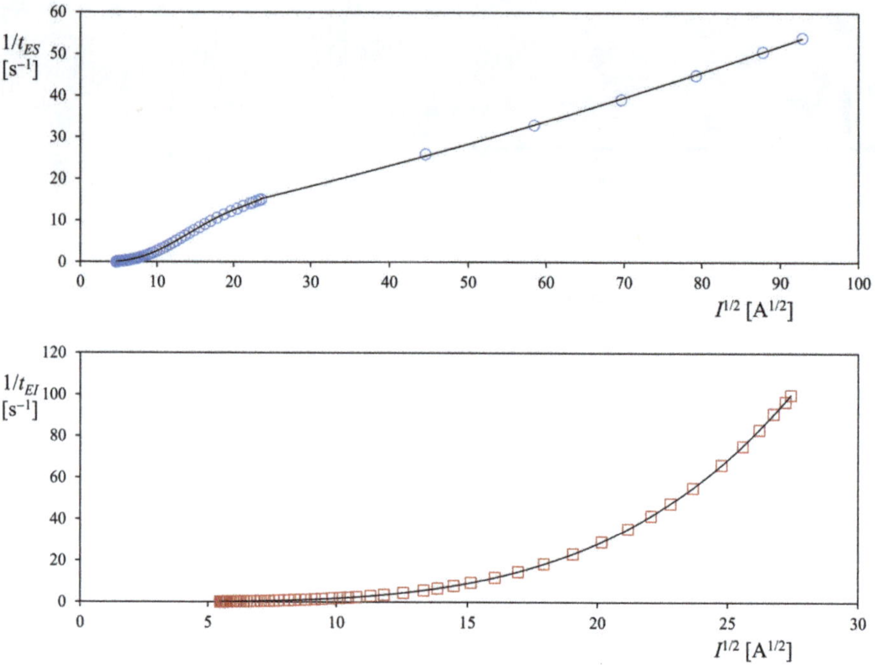

Figure 4. Upper and lower curves that define fuse behavior (Equations (6) and (7)). (Top) Inverse of the total operation time, t_{ES}, in relation to the root of the current for the SMD-50 10E fuse. Equation (6) has been fitted to the manufacturer's data ($I < 558$ A; $N_1 = 5$; $\beta_0 = -1.19851$, $\beta_1 = 7.03361 \times 10^{-1}$, $\beta_2 = -1.91430 \times 10^{-1}$, $\beta_3 = 2.59919 \times 10^{-2}$, $\beta_4 = -1.18321 \times 10^{-3}$, $\beta_5 = 1.79907 \times 10^{-5}$; $I > 558$ A; $N_2 = 2$; $\beta_0 = 5.09321$, $\beta_1 = 3.95777 \times 10^{-1}$, $\beta_2 = 1.40653 \times 10^{-3}$). (Bottom) Inverse of the minimum melting time, t_{EI}, in relation to the root of the current for the SMD-50 15E fuse. Equation (7) has been fitted to the manufacturer's data ($N = 5$; $\beta_0 = 2.93095$, $\beta_1 = 1.17318$, $\beta_2 = -1.84165 \times 10^{-1}$, $\beta_3 = 1.40294 \times 10^{-2}$, $\beta_4 = -3.33170 \times 10^{-4}$, $\beta_5 = 6.99420 \times 10^{-6}$).

Additionally, it should be said that only the left branch of Equation (6) will be needed for the calculations. The translation of the above data to the normal time–current (*t-I*) graphs is included in Figure 5, in which the relative performance of fuses SMD-50 10E and SMD-50 15E can be compared (t_{ES} and t_{EI}, respectively), together with the respective errors of the model, e_{rr}. In Figure 5, it can be appreciated how this model fits the manufacturer's data accurately, with less than 2% error within the normal current interval in which the fuses' performance is analyzed.

For a proper protection coordination there must be a trip delay between two adjacent fuses that overcome the same electrical fault. The coordination criterion for protection fuses can be expressed as:

$$t_{EI} = K(t_{ES} + CTI), \tag{8}$$

where *CTI* is the coordination time interval (200–300 ms) [34,35] and *K* is a proportionality constant.

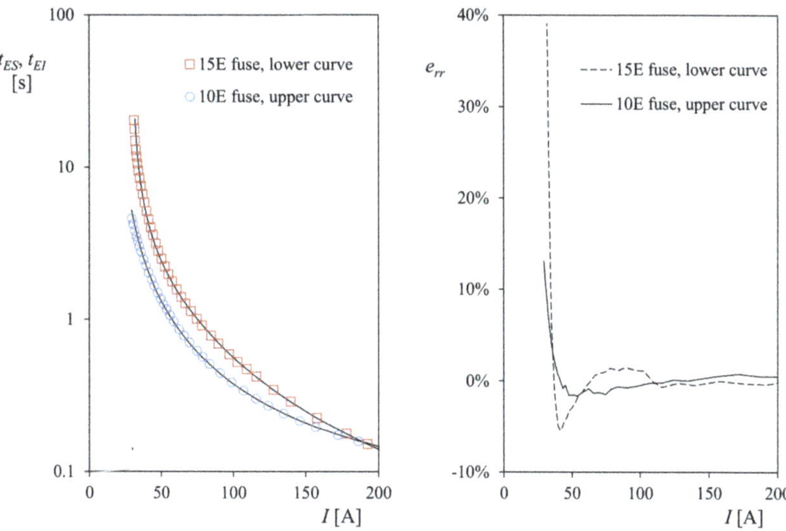

Figure 5. (Left) SMD-50 10E fuse upper curve and SMD-50 15E fuse lower curve from the manufacturer's data. Equations (6) and (7) have been fitted to these data (solid lines). (Right) Error, e_{rr}, of the proposed fuse model with regard to the manufacturer's data.

2.3. Distrtibution Network

In the present work, the IEEE 13-bus test feeder system has been used to analyze the effect of the DG on the protection devices of the distribution network (see Figure 6). The distribution network has been divided into four protection zones:

- Zone 1, connected to the substation through fuse F1
- Zone 2, connected to the substation through fuse F3
- Zone 3, connected to the substation through fuses F2 and F5
- Zone 4, connected to the substation through fuses F2 and F8

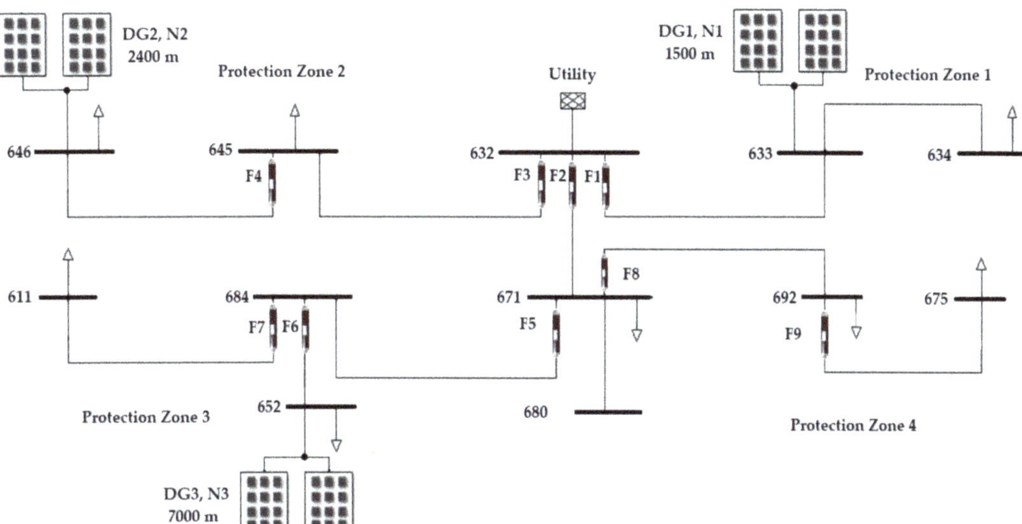

Figure 6. IEEE 13-bus test feeder system used to analyze the performance of the fuses in relation to the DG.

Protection fuses F1, F2, F3 protect the aerial outputs of the main feeders (substation outputs). F4 protects the bus 646 where a DG unit has been installed. F5 and F6 protect the section 671-611, where DG has also been installed, and finally F8 and F9 protect the loads of protection zone 4 where no DG installation exists.

The case studies have been designed for three different locations of the DG in the network (bus/nodes 633, 646, and 652) and different levels of DG power penetration, specifically 25%, 50%, 75%, and 100%. The different case studies have been simulated with DIgSILENT© software from Power Factory [15], as this power system simulation software has a specific library for electrical protection. The locations of the DG in the distribution network are included in Table 1, the distance to the substation being also included in each case.

Table 1. Location (node) and distance of the DG from the substation (SE).

Node	Distance (m)	Description
Node 633 (N1)	1500	Close to SE
Node 646 (N2)	2400	Moderate distance to SE
Node 652 (N3)	7000	Far from SE

A nominal voltage of 20 kV is considered for the analyzed cases, the aerial cable being 47-AL1/8ST1A aluminum steel reinforced (see in Table 2 the main characteristics, according to EN 50182 standard [36]). The total load considered is 3.83 MW, which consists of several three-phase loads distributed according to the information included in Table 3.

Table 2. Characteristics of 47-AL1/8ST1A aluminum steel reinforced cable [36].

Electrical resistance at 20 °C [$\Omega \cdot \mathrm{km}^{-1}$]	0.614
Electrical reactance at 20 °C [$\Omega \cdot \mathrm{km}^{-1}$]	0.41
Max. constant current [kA]	0.202

Table 3. Loads considered in the analyzed network.

Location	Type	Power [MW]
Node 634	Type Y	1.5
Node 611	Type Y	0.056
Node 645	Type Y	0.056
Node 646	Type D	0.5
Node 652	Type Y	1
Node 671	Type D	0.385
Node 675	Type Y	0.281
Node 692	Type D	0.056

When choosing the size of the expulsion fuses, the standards of the supplier [37] and the recommendations indicated by the IEC 60787 standard [38] have been taken into account. In addition, two supplementary criteria were established:

- the nominal fuse current needs to be above the maximum expected generator load current with a sufficient margin (125% load)
- the fuse should not trip with generator connection currents [34]

Each section of the distribution network in which the DG was installed was protected by expulsion fuses chosen to withstand up to 125% of the load current, in order to take into account the possible overloads that could appear in the network. Therefore, the following rules were set regarding the currents:

- $I_{F1} \geq 1.25 \, I_{LOAD\,634}$
- $I_{F2} \geq 1.25 \, (I_{LOAD\,611} + I_{LOAD\,652} + I_{LOAD\,692} + I_{LOAD\,625} + I_{LOAD\,671}) > I_{F5}$
- $I_{F3} \geq 1.25 \, (I_{LOAD\,645} + I_{LOAD\,646}) > I_{F4}$

- $I_{F4} \geq 1.25\, I_{LOAD\,646}$
- $I_{F5} \geq 1.25\, (I_{LOAD\,611} + I_{LOAD\,652}) > I_{F6}$
- $I_{F6} \geq 1.25\, I_{LOAD\,652}$
- $I_{F7} \geq 1.25\, I_{LOAD\,611}$
- $I_{F8} \geq 1.25\, (I_{LOAD\,692} + I_{LOAD\,675}) > I_{F9}$
- $I_{F9} \geq 1.25\, I_{LOAD\,675}$

S&C Electric Company SMD-50 fuses were selected in this work to analyze the coordination. Table 4 includes the nominal currents of these fuses and their technical ID. Additionally, the constants of the model proposed in the present work (Equations (6) and (7)) fitted to the selected fuses are included in Table 5.

Table 4. Nominal currents of the fuses installed in the distribution network.

Fuse	I [A]	Fuse Selected
F1	53.75	S&C SMD-50 30E
F2	62.5	S&C SMD-50 40E
F3	20	S&C SMD-50 13E
F4	17.5	S&C SMD-50 10E
F5	$I_{F2} > 37.5 \geq I_{F6} + I_{F7}$	S&C SMD-50 30E
F6	35	S&C SMD-50 20E
F7	2.5	S&C SMD-50 5E
F8	$I_{F2} > 12.5 \geq I_{F9}$	S&C SMD-50 10E
F9	10	S&C SMD-50 7E

Table 5. Coefficients of Equations (6) and (7) fitted to the upper (upp) and lower (low) tripping time curves, t_{ES} and t_{EI}, of different S&C Electric Company SMD-50 fuses. See also Figures 4 and 5.

Fuse	Curve	β_0 $[\mathrm{s^{-1}}]$	β_1 $[\mathrm{s^{-1} \cdot A^{0.5}}]$	β_2 $[\mathrm{s^{-1} \cdot A^{1}}]$	β_3 $[\mathrm{s^{-1} \cdot A^{1.5}}]$	β_4 $[\mathrm{s^{-1} \cdot A^{2}}]$	β_5 $[\mathrm{s^{-1} \cdot A^{2.5}}]$
5E	upp; $I < 299$ A	-1.87460	1.3974	-4.58631×10^{-1}	7.76704×10^{-2}	-4.78102×10^{-3}	9.97543×10^{-5}
	upp; $I > 299$ A	6.83807	3.61221×10^{-1}	1.58040×10^{-3}	-	-	-
	low	-3.69555	2.55339	-6.63554×10^{-1}	8.13467×10^{-2}	-3.23895×10^{-3}	1.03372×10^{-4}
7E	upp; $I < 430$ A	-1.51111	9.79975×10^{-1}	-2.84227×10^{-1}	4.17970×10^{-2}	-2.15210×10^{-3}	3.72960×10^{-5}
	upp; $I > 430$ A	5.82341	3.86086×10^{-1}	1.42370×10^{-3}	-	-	-
	low	-8.63283×10^{-1}	3.42341×10^{-1}	-4.85652×10^{-2}	3.13876×10^{-3}	6.24309×10^{-4}	9.06465×10^{-7}
10E	upp; $I < 558$ A	-1.19851	7.03361×10^{-1}	-1.91430×10^{-1}	2.59919×10^{-2}	-1.18321×10^{-3}	1.79907×10^{-5}
	upp; $I > 558$ A	5.09321	3.95777×10^{-1}	1.40653×10^{-3}	-	-	-
	low	-2.25987×10^{-1}	-7.04764×10^{-2}	3.90330×10^{-2}	-5.07612×10^{-3}	6.52842×10^{-4}	-4.82039×10^{-6}
13E	upp; $I < 722$ A	-1.78824	8.59426×10^{-1}	-1.82664×10^{-1}	1.98489×10^{-2}	-7.64613×10^{-4}	9.99949×10^{-6}
	upp; $I > 722$ A	3.78076	4.24034×10^{-1}	1.25333×10^{-3}	-	-	-
	low	-5.38095×10^{-1}	5.39468×10^{-2}	1.66285×10^{-2}	-2.91772×10^{-3}	3.92933×10^{-4}	-2.89251×10^{-6}
15E	upp; $I < 826$ A	-1.90389	8.37374×10^{-1}	-1.64212×10^{-1}	1.64897×10^{-2}	-5.90620×10^{-4}	7.21544×10^{-6}
	upp; $I > 826$ A	3.78076	4.24034×10^{-1}	1.25333×10^{-3}	-	-	-
	low	-2.93095	1.17318	-1.84165×10^{-1}	1.40294×10^{-2}	-3.33170×10^{-4}	6.99420×10^{-6}
20E	upp; $I < 1138$	-1.69113	6.44722×10^{-1}	-1.09999×10^{-1}	9.57500×10^{-3}	-2.88718×10^{-4}	2.96029×10^{-6}
	upp; $I > 1138$	6.64811×10^{-1}	4.79303×10^{-1}	9.81018×10^{-4}	-	-	-
	low	-1.55131	4.65904×10^{-1}	-5.60123×10^{-2}	3.39124×10^{-3}	-4.81103×10^{-6}	1.06292×10^{-6}
25E	upp; $I < 1458$	-2.40136	7.94736×10^{-1}	-1.10811×10^{-1}	7.81405×10^{-3}	-2.01186×10^{-4}	1.79140×10^{-6}
	upp; $I > 1458$	-1.97454	5.25978×10^{-1}	7.36144×10^{-4}	-	-	-
	low	-2.86886	8.54336×10^{-01}	-9.67943×10^{-2}	5.21690×10^{-3}	-7.42881×10^{-5}	1.23231×10^{-6}
30E	upp; $I < 1608$ A	-2.18019	6.71422×10^{-1}	-9.07023×10^{-2}	6.20992×10^{-3}	-1.51163×10^{-4}	1.26547×10^{-6}
	upp; $I > 1608$ A	-2.64464	5.27075×10^{-1}	7.90099×10^{-4}	-	-	-
	low	-1.27531	2.68551×10^{-1}	-2.21904×10^{-2}	9.05809×10^{-4}	2.87958×10^{-5}	1.36786×10^{-7}
40E	upp; $I < 2065$ A	-1.56502	4.32637×10^{-1}	-5.62540×10^{-2}	3.66028×10^{-3}	-7.87726×10^{-5}	5.76520×10^{-7}
	upp; $I > 2065$ A	-6.65936	5.92598×10^{-1}	4.68242×10^{-4}	-	-	-
	low	-1.83832×10^{-2}	-1.07938×10^{-1}	1.62356×10^{-2}	-8.70876×10^{-4}	4.91068×10^{-5}	-1.84714×10^{-7}
50E	upp; $I < 2517$ A	-2.63857	6.27277×10^{-1}	-6.44707×10^{-2}	3.34754×10^{-3}	-6.24271×10^{-5}	4.05220×10^{-7}
	upp; $I > 2517$ A	-8.97351	6.01037×10^{-1}	5.73675×10^{-4}	-	-	-
	low	-2.28308	4.41665×10^{-1}	-3.36337×10^{-2}	1.24261×10^{-3}	-2.86096×10^{-6}	1.46350×10^{-7}
65E	upp; $I < 3397$ A	-2.02202	4.32137×10^{-1}	-4.03342×10^{-2}	1.85871×10^{-3}	-2.89685×10^{-5}	1.55962×10^{-7}
	upp; $I > 3397$ A	$-1.68545 \times 10^{+1}$	7.07076×10^{-1}	1.20284×10^{-4}	-	-	-
	low	-6.28233×10^{-1}	5.43519×10^{-2}	-1.10800×10^{-3}	-3.06980×10^{-5}	1.19222×10^{-5}	-1.96977×10^{-8}

3. Cases Studied and Results

Two different groups of cases were analyzed. In the first one, the buses where the DG was installed are considered as PV bus nodes, whereas in the second one, PQ buses are considered. In Tables 6 and 7, the results of the simulations carried out are included. The different cases of DG penetration and distribution (between bus nodes 633, 646, and 652) are indicated in Tables 6 and 7. The information included states either:

- correct performance, that is, no fuse is tripped, indicated by "OK", or
- possible incorrect performance, in which some fuses are tripped.

Table 6. IEEE 13-bus feeder system (Figure 6), simulated considering PV nodes. The status result (either "OK" when no fuse was tripped, or including the unexpected tripped fuses) of the different cases related to DG power penetration level and its distribution among the 3 DG installed (bus nodes 633, 646, and 652) are included in the table.

DG	Case	DG Distribution			Status Result
		Bus 633	Bus 646	Bus 652	
25%	1	-	-	25%	OK
	2	-	25%	-	OK
	3	25%	-	-	OK
50%	4	-	-	50%	OK
	5	-	25%	25%	F3, F4
	6	-	50%	-	F3, F4
	7	25%	-	25%	OK
	8	25%	25%	-	F4
	9	50%	-	-	OK
75%	10	-	-	75%	F6
	11	-	25%	50%	OK
	12	-	50%	25%	F3, F4, F6
	13	-	75%	-	F3, F4
	14	25%	-	50%	OK
	15	25%	25%	25%	F3, F4
	16	25%	50%	-	F3, F4
	17	50%	-	25%	OK
	18	50%	25%	-	OK
	19	75%	-	-	OK
100%	20	-	-	100%	F5, F6
	21	-	25%	75%	F5, F6
	22	-	50%	50%	F4
	23	-	75%	25%	F3, F4, F6
	24	-	100%	-	F4
	25	25%	-	75%	F6
	26	25%	25%	50%	F4
	27	25%	50%	25%	F4
	28	25%	75%	-	F4
	29	50%	-	50%	OK
	30	50%	25%	25%	F4
	31	50%	50%	-	F4
	32	75%	-	25%	OK
	33	75%	25%	-	OK
	34	100%	-	-	F1

Table 7. IEEE 13-bus feeder system (Figure 6), simulated considering PQ nodes. The status result (either "OK" when no fuse was tripped, or including the unexpected tripped fuses) of the different cases related to DG power penetration level and its distribution among the 3 DG installed (bus nodes 633, 646, and 652) are included in the table.

DG	Case	DG Distribution			Result
		Bus 633	**Bus 646**	**Bus 652**	
25%	1	-	-	25%	OK
	2	-	25%	-	OK
	3	25%	-	-	OK
50%	4	-	-	50%	OK
	5	-	25%	25%	OK (*)
	6	-	50%	-	F3, F4
	7	25%	-	25%	OK
	8	25%	25%	-	OK (*)
	9	50%	-	-	OK
75%	10	-	-	75%	F6
	11	-	25%	50%	OK
	12	-	50%	25%	F3, F4
	13	-	75%	-	F3, F4
	14	25%	-	50%	OK
	15	25%	25%	25%	OK (*)
	16	25%	50%	-	F3, F4
	17	50%	-	25%	OK
	18	50%	25%	-	OK
	19	75%	-	-	OK
100%	20	-	-	100%	F5, F6
	21	-	25%	75%	F6
	22	-	50%	50%	F4
	23	-	75%	25%	F3, F4
	24	-	100%	-	F3, F4
	25	25%	-	75%	F6
	26	25%	25%	50%	OK (*)
	27	25%	50%	25%	F3, F4
	28	25%	75%	-	OK
	29	50%	-	50%	OK
	30	50%	25%	25%	OK
	31	50%	50%	-	F3, F4
	32	75%	-	25%	OK
	33	75%	25%	-	OK
	34	100%	-	-	F1

(*) Fuse is tripped if PV nodes are considered (see Table 6).

The cases of the first group (DG installed in PV bus) in which two consecutive fuses were tripped are analyzed in Table 8. In this table, the correct coordination between fuses is evaluated. The coordination time interval (Equation (8)) is derived from the tripping time of the fuses taking into account a proportionality constant $K = 1$. In all cases, CTI is larger than 0.3 s, with the exception of case 23 (100% DG penetration, distributed among bus 646, 75% penetration, and node 652, 25% penetration). In this case $CTI = 0.088$ s < 0.3 s, therefore a lack of coordination between fuses is observed. The cases corresponding to the second group (DG installed in PQ nodes) in which two consecutive fuses are tripped, are analyzed in Table 9. Among these cases, case 24 (100% DG penetration, distributed at node 646) resulted as Not OK (as $CTI = 0.07$, lower than the limit value). Finally, it should be emphasized that taking into account the whole set of simulations, DG considered as PV nodes had a larger amount of cases (59%) in which at least one fuse was tripped, whereas DG considered as PQ obtained a lower value of fuse-tripping cases (41%). This is caused by a larger current supply from the DG in the case of PV nodes.

Table 8. IEEE 13-bus feeder system (Figure 6), simulated considering PV nodes. Fuse coordination results (tripping times) and compliance in cases where two consecutive fuses tripped (see Table 6).

Case	Zone 1	Zone 2		Zone 3			CTI [s]	Status Result
	t_{F1} [s]	t_{F3} [s]	t_{F4} [s]	t_{F2} [s]	t_{F5} [s]	t_{F6} [s]		$CTI > 0.3$ s
5	-	7.84	3.48	-	-	-	4.36	OK
6	-	3.16	1.61	-	-	-	1.55	OK
12	-	1.37	0.75	-	-	3.11	0.62	OK
13	-	0.94	0.52	-	-	-	0.42	OK
15	-	5.01	2.55	-	-	-	2.46	OK
16	-	1.16	0.65	-	-	-	0.51	OK
20	-	-	-	-	4.16	1.43	2.73	OK
21	-	5.45	2.73	-	29.58	2.84	2.72	OK
23	-	0.56	0.47	-	-	6.16	0.09	Not OK

Table 9. IEEE 13-bus feeder system (Figure 6), simulated considering PQ nodes. Fuse coordination results (tripping times) and compliance in cases where two consecutive fuses tripped (see Table 7).

Case	Zone 1	Zone 2		Zone 3			CTI [s]	Status Result
	t_{F1} [s]	t_{F3} [s]	t_{F4} [s]	t_{F2} [s]	t_{F5} [s]	t_{F6} [s]		$CTI > 0.3$ s
6	-	3.27	1.67	-	-	-	1.60	OK
12	-	3.29	1.67	-	-	-	1.62	OK
13	-	1.00	0.55	-	-	-	0.45	OK
16	-	3.28	1.67	-	-	-	1.67	OK
20	-			-	6.24	1.78	4.46	OK
23	-	1.00	0.55	-	-	-	0.45	OK
24	-	0.49	0.42	-	-	-	0.07	Not OK
27	-	3.30	1.68	-	-	-	1.62	OK
28	-	1.00	0.55	-	-	-	0.45	OK
31	-	3.29	1.68	-	-	-	1.61	OK

In case 23 of the IEEE 13-bus feeder system simulation considering DG installed in PV bus, fuses F3 and F4 tripped for $I = 87.2$ A current with a lack of coordination (that is, the time distance corresponding to the upper boundary curve of fuse F4, t_{ES}, and the lower boundary curve of fuse F3, t_{EI}, is smaller than the coordination time interval, $CTI = 0.3$ s), that should be avoided by replacing one of the fuses. In Figure 7, the upper curve of fuse F4 (SMD-50 E10) and the lower curve of fuse F3 (SMD-50 E13) are plotted. It can be observed that the current intersects the fuse F3 curve at a point (indicated by an open square) located below the curve corresponding to fuse F4 plus a $CTI = 0.3$ s (which imply a lack of coordination). If fuse F3 is replaced by the next one in the family, characterized by a larger nominal current (SMD-50 E15), the situation is not solved, as at $I = 87.2$ A current the lower curve of the fuse has a tripping time lower than the one stated by the upper curve from fuse F4 plus a $CTI = 0.3$ s (also indicated by an open square in Figure 7). The problem is finally solved by selecting an SMD-50 E20 for the F3 fuse. See in Figure 7 how the tripping time of its lower curve at $I = 87.2$ A (indicated by an open square) detaches more than 0.3 s in relation to the SMD-50 E10 upper curve.

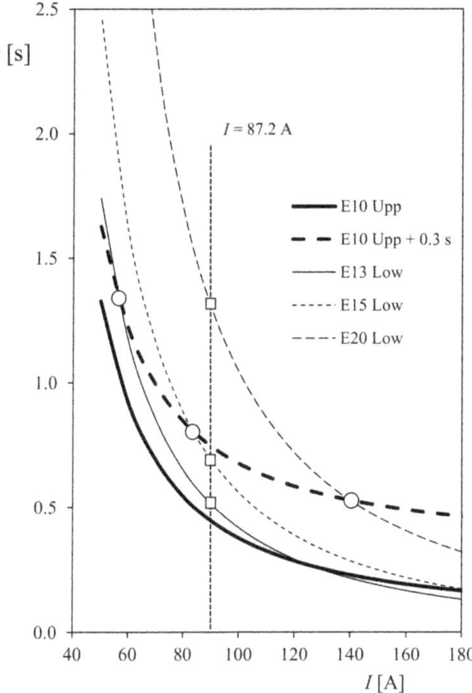

Figure 7. Selection of fuses to avoid the uncorrected coordination detected in case 23 of the calculations considering PV nodes where the photovoltaic DG are installed. The upper tripping time curve, t_{ES}, of SMD-50 E10 fuse selected for F4 fuse location (see Figure 6) is plotted, together with the lower tripping time curves, t_{EI}, corresponding to SMD-50 E13, SMD-50 E15, and SMD-50 E20 fuses. The points of these curves corresponding to current I = 87.2 A obtained from the simulation of case 23 are indicated (open squares). The points corresponding to a proper coordination between fuse SMD-50 E10 as F4 with SMD-50 E13, SMD-50 E15, and SMD-50 E20 fuses as F3 are also indicated (open circles).

Besides, it is possible to work alternatively with the points of intersection between the lower curves of the selected fuses and the upper curve of fuse F4 (SMD-50 E10) plus a CTI = 0.3 s (open circles in Figure 7). From Equations (6)–(8), it is possible to derive the following equation:

$$\frac{1}{\left.\sum\limits_{i=0}^{N} \beta_i I^{\frac{i}{2}}\right|_{lower}} = \frac{1}{\left.\sum\limits_{i=0}^{N} \beta_i I^{\frac{i}{2}}\right|_{upper}} + 0.3, \tag{9}$$

that includes two families of polynomials, the one corresponding to the upper curve of fuse F4, and the one corresponding to the lower value of the considered fuse for F3. From the above equation, it is possible to derive the value of the current which implies a correct coordination between fuses. The following current values were calculated with the data from Table 5: I = 56 A, I = 83.54 A, and I = 140 A, for the corresponding SMD-50 E13, SMD-50 E15, and SMD-50 E20 fuses (these values are the ones indicated in Figure 7 with open circles). As the first two values of the current are below the tripping current I = 87.2 A, there is a lack of coordination. In the latter case, the calculated value is larger than the tripping current. Therefore, for this current, the lower curve of SMD-50 E20 will be detached from the SMD-50 E10 upper curve a larger value than the one stated by the CTI (=0.3 s).

Going one step further, the fuse selected for F3 initially might be preserved and fuse F4 replaced by another one with smaller nominal current (if it is possible, bearing in mind the required nominal currents). The situation is very similar to the one described above; see Figure 8. In this case, fuses SMD-50 E10 and SMD-50 E7 cannot produce a proper coordination, the values of their upper curves at $I = 87.2$ A being detached from the SMD-50 E13 lower curve less than the *CTI* (points indicated with open squares). Nevertheless, fuse SMD-50 E5 selected as F4 fuse produces a proper coordination. Additionally, if Equations (6)–(8) are considered:

$$\left. \frac{1}{\sum\limits_{i=0}^{N} \beta_i I^{\frac{i}{2}}} \right|_{lower} - 0.3 = \left. \frac{1}{\sum\limits_{i=0}^{N} \beta_i I^{\frac{i}{2}}} \right|_{upper}, \tag{10}$$

where the polynomial corresponding to the lower curve of fuse F3 (fuse SMD-50 E13, in this case) is constant, the polynomial corresponding to the upper curve being selected from the studied fuse at F4 in order to calculate the current that produces a proper coordination with fuse F3. These current values, $I = 56$ A (fuse SMD-50 E10), $I = 79.1$ A (fuse SMD-50 E7), and $I = 93.5$ A (fuse SMD-50 E5), are indicated in Figure 8 with open circles. From these values, it is possible to state that only selecting fuse SMD-50 E5 for fuse F4 will ensure a proper coordination. It should be underlined that the selection of fuses is also conditioned by their nominal current values.

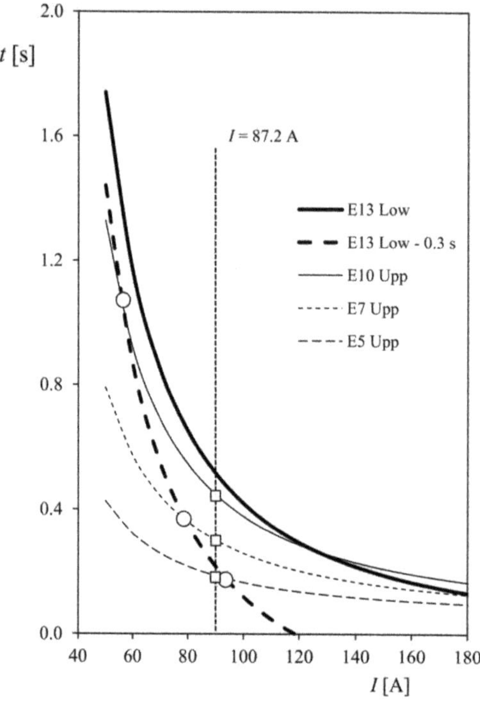

Figure 8. Selection of fuses to avoid the uncorrected coordination detected in case 23 of the calculations considering PV nodes where the photovoltaic DG are installed. The lower tripping time curve, t_{EI}, of SMD-50 E13 fuse selected for F3 fuse location (see Figure 6) is plotted, together with the upper tripping time curves, t_{ES}, corresponding to SMD-50 E10, SMD-50 E7, and SMD-50 E5 fuses. The points of these curves corresponding to current $I = 87.2$ A obtained from the simulation of case 23 are indicated (open squares). The points corresponding to a proper coordination between fuse SMD-50 E13 as F3 with SMD-50 E10, SMD-50 E7, and SMD-50 E5 fuses as F4 are also indicated (open circles).

The analysis corresponding to the other case with improper coordination detected between fuses (case 24 of simulations carried out considering DG installed in PQ nodes) is very similar to the one analyzed. In this case, the current at the line protected by fuses F3 and F4 is I = 93.2 A, which is very close to the mentioned current that indicates a proper coordination between fuse SMD-50 E5 at F4 and fuse SMD-50 E13 at F3. Nevertheless, the result is still valid.

4. Conclusions

The aim of this work is to analyze the fuse coordination in rural (or not too large) power networks with presence of photovoltaic DG. The performance of the fuses in the IEEE 13-bus feeder system with DG was studied with DIgSILENT©, and considering different power penetration levels and power distribution between the three bus nodes where the DG sources are installed (at close, moderate, and large distance from the substation). Besides, two options were considered in relation to these bus nodes, PV and PQ nodes, in order to study the widest scope of possibilities. The most relevant results of this work are:

- The effect of photovoltaic DG on fuse coordination in a power network similar to the ones present in small rural villages has been checked. The results indicate that an improper coordination of fuses (correctly selected in the case of no DG presence) might be produced.
- A new and easy way to simulate the performance of fuses has been described.
- Thanks to the aforementioned new fuse modeling, criteria to select new fuses in power networks that can ensure a proper coordination between them have been described.
- The present study can be extrapolated to networks with DG consisting of other kinds of sources (different from the photovoltaic ones), as the main control parameter is the current supplied to the network.

Author Contributions: Conceptualization, D.A.-G. and S.P.; methodology, D.A.-G., E.M.G.d.T., M.I.M.-L. and S.P.; software, D.A.-G.; validation, D.A.-G., D.A.-C., E.M.G.d.T. and M.I.M.-L.; formal analysis, D.A.-G.; investigation, D.A.-G., D.A.-C., E.M.G.d.T., M.I.M.-L. and S.P.; data curation, D.A.-G.; writing—original draft preparation, D.A.-C., D.A.-G. and S.P.; writing—review and editing, D.A.-C., D.A.-G. and S.P.; supervision, D.A.-G. and S.P. All authors have read and agreed to the published version of the manuscript.

Funding: This research received no external funding.

Institutional Review Board Statement: Not applicable.

Informed Consent Statement: Not applicable.

Acknowledgments: The authors are indebted to the head of the IDR/UPM Institute, Ángel Sanz-Andrés, for his support.

Conflicts of Interest: The authors declare no conflict of interest.

References

1. Paithankar, Y.G.; Bhide, S.R. *Fundamentals of Power System Protection*; PHI Learning Pvt. Ltd.: Delhi, India, 2010.
2. Barrenetxea, A.I. *Protecciones de Sistemas de Potencia*; Universidad del País Vasco, Servicio Editoria/Euskal Herriko Unibertsitatea, Argitalpen Zerbitzua: Biscay, Spain, 1997.
3. Dugan, R.C.; McDermott, T.E. Operating conflicts for distributed generation on distribution systems. In Proceedings of the Papers—Rural Electric Power Conference, Little Rock, AR, USA, 29 April–1 May 2001.
4. Salman, S.K.; Rida, I.M. Investigating the impact of embedded generation on relay settings of utilities' electrical feeders. *IEEE Trans. Power Deliv.* **2001**, *16*, 246–251. [CrossRef]
5. Bastião, F.; Cruz, P.; Fiteiro, R. Impact of distributed generation on distribution networks. In Proceedings of the 2008 5th International Conference on the European Electricity Market, EEM, Lisboa, Portugal, 28–30 May 2008.
6. Barker, P.P.; De Mello, R.W. Determining the impact of distributed generation on power systems: Part 1—Radial distribution systems. In Proceedings of the IEEE Power Engineering Society Transmission and Distribution Conference, Seattle, WA, USA, 16–20 July 2000; Volume 3, pp. 1645–1656.
7. Dadkhah, M.; Fani, B.; Heydarian-Forushani, E.; Mohtaj, M. An off-line algorithm for fuse-recloser coordination in distribution networks with photovoltaic resources. *Int. Trans. Electr. Energy Syst.* **2020**, *30*, 1–16. [CrossRef]

8. Deng, W.; Pei, W.; Qi, Z. Impact and improvement of distributed generation on voltage quality in micro-grid. In Proceedings of the 3rd International Conference on Deregulation and Restructuring and Power Technologies, DRPT 2008, Nanjing, China, 6–9 April 2008; pp. 1737–1741.
9. Wang, S. Distributed generation and its effect on distribution network system. In Proceedings of the IET Conference Publications, Prague, Czech Republic, 8–11 June 2009.
10. Naiem, A.F.; Hegazy, Y.; Abdelaziz, A.Y.; Elsharkawy, M.A. A classification technique for recloser-fuse coordination in distribution systems with distributed generation. *IEEE Trans. Power Deliv.* **2012**, *27*, 176–185. [CrossRef]
11. Hadjsaid, N.; Canard, J.; Dumas, F. Dispersed generation impact on distribution networks. *IEEE Comput. Appl. Power* **1999**, *12*, 22–28. [CrossRef]
12. Girgis, A.; Brahma, S. Effect of distributed generation on protective device coordination in distribution system. In Proceedings of the LESCOPE 2001—2001 Large Engineering Systems Conference on Power Engineering: Powering Beyond 2001, Halifax, NS, Canada, 11–13 July 2001; pp. 115–119.
13. Brahma, S.M.; Girgis, A.A. Development of Adaptive Protection Scheme for Distribution Systems with High Penetration of Distributed Generation. *IEEE Trans. Power Deliv.* **2004**, *19*, 56–63. [CrossRef]
14. Chaitusaney, S.; Yokoyama, A. Prevention of reliability degradation from recloser-fuse miscoordination due to distributed generation. *IEEE Trans. Power Deliv.* **2008**, *23*, 2545–2554. [CrossRef]
15. Boonyapakdee, N.; Konghirun, M.; Sangswang, A. Separated phase-current controls using inverter-based DGs to mitigate effects of fault current contribution from synchronous DGs on recloser-fuse. *Appl. Sci.* **2019**, *9*, 4311. [CrossRef]
16. Razavi, S.E.; Rahimi, E.; Javadi, M.S.; Nezhad, A.E.; Lotfi, M.; Shafie-khah, M.; Catalão, J.P.S. Impact of distributed generation on protection and voltage regulation of distribution systems: A review. *Renew. Sustain. Energy Rev.* **2019**, *105*, 157–167. [CrossRef]
17. Bayati, N.; Baghaee, H.R.; Hajizadeh, A.; Soltani, M. A Fuse Saving Scheme for DC Microgrids with High Penetration of Renewable Energy Resources. *IEEE Access* **2020**, *8*, 137407–137417. [CrossRef]
18. Alam, M.N.; Das, B.; Pant, V. Protection scheme for reconfigurable radial distribution networks in presence of distributed generation. *Electr. Power Syst. Res.* **2020**, 106973. [CrossRef]
19. Santoso, S.; Short, T.A. Identification of fuse and recloser operations in a radial distribution system. *IEEE Trans. Power Deliv.* **2007**, *22*, 2370–2377. [CrossRef]
20. Yazdanpanahi, H.; Li, Y.W.; Xu, W. A new control strategy to mitigate the impact of inverter-based DGs on protection system. *IEEE Trans. Smart Grid* **2012**, *3*, 1427–1436. [CrossRef]
21. Hussain, B.; Sharkh, S.M.; Hussain, S.; Abusara, M.A. An adaptive relaying scheme for fuse saving in distribution networks with distributed generation. *IEEE Trans. Power Deliv.* **2013**, *28*, 669–677. [CrossRef]
22. Nojavan, M.; Seyedi, H.; Mahari, A.; Zare, K. Optimization of Fuse-Recloser Coordination and Dispersed Generation Capacity in Distribution Systems. *Majlesi J. Electr. Eng.* **2014**, *8*, 15–24.
23. Yousaf, M.; Mahmood, T. Protection coordination for a distribution system in the presence of distributed generation. *Turkish J. Electr. Eng. Comput. Sci.* **2017**, *25*, 408–421. [CrossRef]
24. Plesca, A. Temperature distribution of HBC fuses with asymmetric electric current ratios through fuselinks. *Energies* **2018**, *11*, 1990. [CrossRef]
25. Chaitusaney, S.; Yokoyama, A. Reliability analysis of distribution system with distributed generation considering loss of protection coordination. In Proceedings of the 2006 9th International Conference on Probabilistic Methods Applied to Power Systems, PMAPS, Stockholm, Sweden, 11–15 June 2006.
26. Braga, G.; Zanin Bertoletti, A.; Peres de Morais, A.; Ghendy Cardoso, J. Curve Fitting Analysis of Expulsion Fuse Links through the Cross-Validation Technique. In Proceedings of the 2018 IEEE PES Transmission & Distribution Conference and Exhibition—Latin America (T&D-LA), Lima, Peru, 18–21 September 2018; pp. 1–5.
27. Fazanehrafat, A.; Javadian, S.A.M.; Bathaee, S.M.T.; Haghifam, M.R. Maintaining the recloser-fuse coordination in distribution systems in presence of DG by determining DG's size. In Proceedings of the IET 9th International Conference on Developments in Power Systems Protection (DPSP 2008), Glasgow, UK, 17–20 March 2008; pp. 132–137.
28. Tang, Y.; Ayyanar, R. Verification of Protective Device Coordination in Distribution Systems with Photovoltaic Generation. In Proceedings of the 2014 IEEE 40th Photovoltaic Specialist Conference (PVSC), Denver, CO, USA, 8–13 June 2014; pp. 2100–2105. [CrossRef]
29. Abdel-ghany, H.A.; Azmy, A.M.; Elkalashy, N.I.; Rashad, E.M. Optimizing DG penetration in distribution networks concerning protection schemes and technical impact. *Electr. Power Syst. Res.* **2015**, *128*, 113–122. [CrossRef]
30. Conde Enríquez, A.; Vázquez-Martínez, E. Enhanced time overcurrent coordination. *Electr. Power Energy Syst.* **2006**, *76*, 457–465. [CrossRef]
31. Conde Enríquez, A.; Vázquez-Martínez, E.; Altuve-Ferrer, H.J. Time overcurrent adaptive relay. *Electr. Power Energy Syst.* **2003**, *25*, 841–847. [CrossRef]
32. Costa, G.B.; Marchesan, A.C.; Morais, A.P.; Cardoso, G.; Gallas, M. Curve Fitting Analysis of Time-Current Characteristic of Expulsion Fuse Links. In Proceedings of the 2017 IEEE International Conference on Environment and Electrical Engineering and 2017 IEEE Industrial and Commercial Power Systems Europe (EEEIC/I&CPS Europe), Milan, Italy, 6–9 June 2017; pp. 1–6.
33. Gers, J.M.; Holmes, E.H. *Protection of Distribution Networks*; The Institution of Electrical Engineers: London, UK, 2004.
34. Anderson, P. *Power System Protection*; Wiley: Hoboken, NJ, USA, 1998.

35. Viawan, F.A.; Karlsson, D.; Sannino, A.; Daalder, J. Protection scheme for meshed distribution systems with high penetration of distributed generation. In Proceedings of the 2006 Power Systems Conference: Advanced Metering, Protection, Control, Communication, and Distributed Resources, Clemson, SC, USA, 14–17 March 2006; pp. 99–104.
36. UNE-EN 50182:2002/AC:2013: Conductores para Líneas Eléctricas Aéreas. Conductores de Alambres Redondos Cableados en Capas Concéntricas. Available online: https://www.une.org/encuentra-tu-norma/busca-tu-norma/norma/?c=N0051959 (accessed on 24 March 2021).
37. Iberdrola MT 2.13.40: Criterios de Selección y Adaptación del Calibre de los Fusibles de MT para Centros de Transformación. 2003. Available online: https://docplayer.es/40982302-Procedimiento-de-seleccion-y-adaptacion-del-calibre-de-los-fusibles-de-mt-para-centros-de-transformacion.html (accessed on 24 March 2021).
38. IEC 60783. Application Guide for the Selection of Fuse-Links of High-Voltage Fuses for Transformer Circuit Applications. 1983. Available online: https://standards.globalspec.com/std/113389/IEC%2060787 (accessed on 24 March 2021).

 *agronomy*

Article

Micro-Grid Solar Photovoltaic Systems for Rural Development and Sustainable Agriculture in Palestine

Imad Ibrik

Energy Research Centre, Electrical Department, An-Najah National University, Nablus,
P.O. Box 721 West Bank, Palestine; iibrik@najah.edu; Tel.: +970-599275293

Received: 30 August 2020; Accepted: 24 September 2020; Published: 26 September 2020

Abstract: The objective of this paper is to study the impact of using micro-grid solar photovoltaic (PV) systems in rural areas in the West Bank, Palestine. These systems may have the potential to provide rural electrification and encourage rural development, as PV panels are now becoming more financially attractive due to their falling costs. The implementation of solar PV systems in such areas improves social and communal services, water supply and agriculture, as well as other productive activities. It may also convert these communities into more environmentally sustainable ones. The present paper details two case studies from Palestine and shows the inter-relation between energy, water and food in rural areas to demonstrate how the availability of sustainable energy can ensure water availability, improve agricultural productivity and increase food security. Further, the paper attempts to evaluate the technical and economic impacts of the application of nexus approaches to Palestine's rural areas. The results of this study are for a real implemented project and predict the long-term success of small, sustainable energy projects in developing rural areas in Palestine.

Keywords: sustainable agriculture; nexus approach; rural development; solar micro-grid; techno-economic impact

1. Introduction

Palestine still has a number of remote small communities without access to electricity. It is unlikely for these communities to be connected to the local power grid in the near future as a result of political conflicts and financial issues. The unavailability as well as the lack of sufficient electricity is still one of the main issues hindering socio-economic development in Palestine, especially in its rural areas. The electricity is typically used for potable water pumping, irrigation, lighting and cooking (Imad, 2019) [1].

In some remote areas located in the Palestinian territories, diesel generators are still used to power homes and pump water for a limited period of time during a day. Therefore, a solar photovoltaic (PV) powered irrigation system can be a practical choice for irrigating by utilizing solar PV systems. Such a system can be employed as an alternative so as to provide isolated villages and localities with energy, especially given that Palestine has a daily mean of 5.6 kWh/m^2 of solar radiation and 3000 sunshine hours per year (Mason, 2009) [2], that is to say the region is well-suited to PV installations, (Juaidi et al., 2016) [3]. At the same time, Palestine suffers from scarcity of water and arable land. As a result, the Palestinian government provides assistance with PV schemes to encourage rural farmers to install solar PV pump systems.

Generally, Palestine has a Mediterranean climate characterized by long, hot, dry summers and short, cool, rainy winters, Table 1 below shows the maximum, minimum and mean temperatures, annual rainfall and number of cloudy days in Palestine.

Table 1. Climate in Palestine.

Temperature	Maximum (30 °C), Minimum (10 °C), Average (25.5 °C)
Annual rainfall	450 and 500 mm/year
Number of cloudy days	Partly cloudy (156 days/year), Totally cloudy (16)

This paper describes how a micro grid solar PV system with lead-acid storage batteries may be utilized for rural electrification and water pumping. Two PV system installation processes have been completed, in both Al-Birin and Dir Ammar small village (hamlet) communities, in order to provide electricity access and pump water. In this paper, a solar PV system design for electrification and irrigation is presented, along with the techno-economic feasibility of substituting the existing diesel engines for solar photovoltaic (PV) systems. Solar PV systems were found to be more economic in comparison with diesel use in rural, urban and remote regions in Palestine. The investment payback for solar PV systems rather than diesel was estimated at 3.5 years.

Therefore, the main goal of this paper is to illustrate the real feasibility of using micro-grid solar PV systems instead of diesel generators in different areas to promote rural development and sustainable agriculture in Palestine by drawing on the performance assessment results. The monthly amount of the energy generated from solar PV systems was recorded by data loggers and analyzed against the total solar irradiation measured by a local weather station.

The Energy Research Center at An-Najah National University designed and installed two PV irrigation systems for remote Palestinian communities in 2017. This paper summarizes the design and documents the systems' performance over their first year of operation.

2. Literature Review

Existing research literature has indicated that more than 1.5 billion people worldwide are living in rural areas in developing countries without access to electricity. Many countries seek to improve the quality of life of their citizens and increase the economic well-being of the families who live in rural areas, even though they are relatively isolated and live in small families, which are few in number (Feron, 2016) [4].

Over the past years, health and education have mainly been the focal points of social development (Rowley, 1996) [5] and these sectors have been acknowledged, along with others, such as tourism, recreation and decentralized manufacturing. Nowadays, one of the main sectors which is perceived as the core of rural development is agriculture, i.e., food security, since it can be considered as the most important sector for developing rural areas in the world. In fact, many communities in Palestine have serious problems in terms of the scarcity of water and energy (Imad, 2019; Rehan, 2019) [1,6].

Many researchers have investigated the sustainable development of rural communities, and it has been shown that there is a link between the enhancement of energy and that of water and food supplies. Querikiol (2018) [7] evaluated the performance of a 1.5 kW solar PV system in an agricultural farm located in Camotes Island, mainly for agricultural water use; it was found that around three cubic meters of water per day would be necessary for land irrigation. Additionally, it was concluded that the capacity of the required water pump in order to provide the required irrigation would be 360 W. It was pointed out that it had a very good overall performance; this would prove the potential success of other applications of solar systems operationalized for agricultural purposes.

Santos et al. (2018) [8] proposed a framework for designing a micro-grid system after examining technical, economic as well as social issues so as to determine the optimum required system.

As for research, Chel and Kaushik (2011) [9] analyzed the economic impact, along with the environmental impact, of using solar pumps so as to attain sustainable agriculture. The author of [8] pointed out the role that solar energy plays in farming, namely strengthening all agronomic parameters with regard to ecological efficiency, environment and social impacts, in addition to feasibility. Al-Saidi and Lahham (2019) [10] evaluated the nexus approach, which has been adopted and whose projects have been implemented in the Azraq Basin, Jordan. Through adopting a nexus perspective, the authors of [8] assessed the feasibility and the requested incentives in order to encourage farmers to use standalone solar PV systems. Furthermore, Kyriakarakos et al. (2020) [11] discussed high-cost rural electrification projects by examining a number of methods which could be followed to cover the costs required to implement them, including increasing the cost of produce and plant products in addition to subsidizing the cost of solar PV systems. This approach has been implemented in Rwanda and has led to subsidizing the local agricultural cooperatives and promoting electrification activities in rural areas.

This paper evaluated the performance of an installed 6.2 kW, 9.6 kW off-grid micro-grid solar PV system in terms of its ability to meet the irrigation and other operational requirements of a 4-hectare and 5.5-hectar plantations located in Dir Ammar and Al-Birin areas, respectively. The agricultural lands vary between flat and wavy lands for the cultivation of various types of grains and vegetables, sloping lands for the cultivation of fruit trees, and steep, rugged lands in which forests and natural herbs grow suitable for grazing. Olives are one of the most important agricultural crops in these areas, as they occupy the largest cultivated area and almost surround the town in all directions.

The main aim of this study is, therefore, to contribute to the evaluation of the potential impact of implementing solar PV systems on sustainable agriculture and rural development in Palestine, especially concerning the possibility of income-generating activities. It is important to identify the potential contribution of solar PV, as a replacement of diesel generators, to ensure rural development and gain further income and political commitment because such solar projects may expand to other areas and may help ensure solar PV is designed appropriately, under real-life environmental conditions.

3. Case Studies: Dir Ammar and Al-Birin Small Villages (Hamlets) in the West Bank, Palestine

Both Dir Ammar and Al-Birin small villages (hamlets) are located in Palestine and face relatively similar circumstances in terms of their access to electricity. On the one hand, Dir Ammar is a town located in Ramallah Governorate, 20 km northwest of the city of Ramallah in the north of the West Bank, located at latitude 31°58′00″ N and longitude 35°06′07″ E.A. The community in the above-mentioned hamlet suffers from lack of supplies and relies on diesel generators for household electrification and land irrigation. On the other hand, Al-Birin hamlet is in the southeast of the Hebron District. At a distance of 10 km, the city of Hebron is the closest to this community; it is located at latitude 31.489668° and longitude 35.147839°. Similar to Dir Ammar, this community also depends on a diesel generator for generating electric power.

Through the assessment of non-electrified villages in Palestine in 2017, we found that Dir Ammar and Al-Birin communities are two suitable villages for the implementation of micro-grid solar PV systems. The villages are located near Israeli settlements; thus, the process of supplying them with the conventional power supply from the grid proved to be challenging for implementation. Funded by the Spanish Agency for International Development Cooperation (AECID), micro-grid centralized solar PV systems were installed in 2018 as rural development projects in Palestine. The present paper examines the socio-techno-economic impact of these projects under the circumstances (Ibrik, 2016) [12].

The number of the inhabitants of Dir Ammar and Al-Birin does not exceed 180 individuals who live in 24 houses. Most of those who live in the aforementioned communities mainly work as farmers and cattle breeders, whereas some are construction workers. The location of these communities are in area C, where Israel does not allow Palestinian to expand the electrical network to this area. This encouraged us to select these remote areas to be a model for a solar electrification and water irrigation in Palestine. Local wells supply water to the villages for the most part. Old generators were only used for 4–5 h per day due to their high fuel prices and high-level consumption; the cost of 1 kWh electricity production was around $0.5. Table 2 shows the daily consumption allocated for these communities. The cost of diesel/liter is around 1.5 $/L, because the diesel generators are very old, and the diesel consumption is around 0.3 L/kWh. The overall efficiency of the existing generators is around 32%.

Table 2. Total daily consumption/family.

Application	Quantity	Power (Watt)	h/Day	W.h/Day
PL lamp	2	13	5	130
TV	1	100	3	300
Mobile charge	1	10	1	10
Small refrigerator	1	200	5	1000
High efficiency washing machine	1	180	2	360
Total				1800

In winter, it is not necessary to use energy for irrigation. For some days in winter and cloudy weather, the output of the PV system is very low. In summer, there is more output energy and, at the same time, more need for drinking and irrigation.

Each house is fitted with an energy dispenser and meter, which limits the amount of energy available for each user in accordance with their predetermined needs and the contracted tariff. In order to avoid flattening of batteries, the diesel generator works for a few hours in winter to fill the batteries.

Instead, solar PV systems are now installed for electrification and for water pumping in these communities as such systems, rather than diesel generators, are now deemed to offer the best solution and feasible method for irrigation in Palestinian rural communities. Table 3 compares the use of diesel generators for irrigation as opposed to solar PV systems, taking Dir Ammar as a case study. It can be noticed that using solar PV is more sustainable, eco-friendly and encourages the socio-economic development of these communities, and it can help to solve the energy crisis which Palestinian farmers face.

Table 3. Irrigation electrical component parts and characteristics (Madziga et al., 2018) [13].

Solar PV	DC
Capital cost	3000 $/kW
Replacement cost	2000 $/kW
Q&M cost	$0
Efficiency	16%
Life time	20
Tracking system	No
Co_2 pollution	0
Diesel Engine	**AC**
Capital cost	700 $/kW
Replacement cost	450 $/kW
Q&M cost	0.088 4/h
Life time	18,000 h
Co_2 pollution	2830 kg·CO_2/year
Battery	**DC**
Technology	Lead Acid
Capacity	7.6 kWh
Nominal capacity	1800
Voltage	2 V
Min. state of charge	35%
Capital cost	320 $
Replacement cost	100 $
Q&M cost	50 $/year

Table 3. *Cont.*

Solar PV	DC
Efficiency	85%
Life time	10 year
Converter	AC/DC/AC
Capacity	7 kW
Capital cost	300 $/kW
Replacement cost	300 $/kW
Q&M cost	0
Efficiency	92%
Life time	10 year

In this study, the micro-grid in each community was built, and it consists of an over-head line (3 × 6 mm^2 PVC), 11 poles and cables (3 × 4) mm^2 for the connection users in each location. The performance analysis and a feasibility study of deploying solar photovoltaic systems for water pumping and for electrification of rural areas in Palestine are presented based on real input data from both implemented projects.

4. System Design

4.1. Determining Solar Irradiation

The solar energy data were collected from weather stations near the hamlets Dir Ammar and Al-Birin. The average recorded data indicated that the solar radiation rate was 5.5 kWh/m^2-day, and the maximum solar radiation almost reached 8.2 kWh/m^2-day in July, while the minimum was about 2.8 kWh/m^2-day in December. Figure 1 shows the average monthly irradiation for both sites.

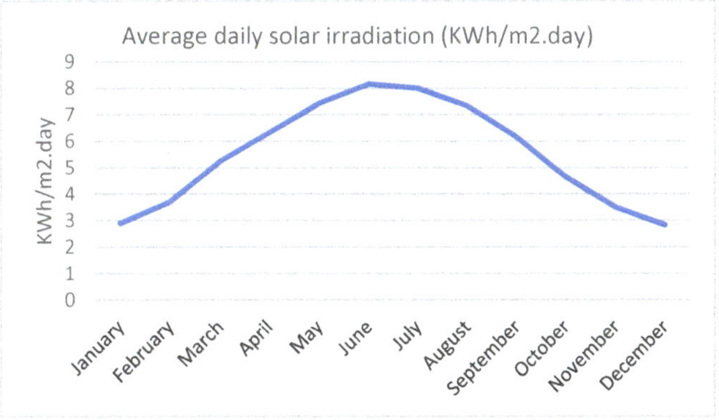

Figure 1. Dir Ammar hamlet.

4.2. Elements of System Design

4.2.1. Electrical Load

The main loads in each village reflect the inhabitants' daily power consumption and water pumping as well as their electricity demand. These figures have been obtained using a questionnaire.

Dir Ammar's deep water well specifications are listed below:

- Total dynamic head = 20 m

- Daily water consumption required: 60 m^3/day.
- Diesel consumption = 3 L/day, needed monthly diesel = 90 (L/month)

The specifications of deep water wells in Al-Birin are as follows:

- Total dynamic head = 30 m
- Daily water consumption required: 80 m^3/day.
- Diesel consumption = 6.5 L/day, needed monthly diesel = 200 (L/month)

The dynamic head is the total equivalent height that a fluid is to be pumped. The hydraulic energy (HE) can be calculated as in (1) in Amjath et al. (2019) [13].

$$HE\left(\frac{kWh}{day}\right) = 0.002725 \times Q \times TDH \tag{1}$$

where Q is the water pumping rate (m^3/day) and TDH is the total dynamic head (m).

The electrical energy required for water pumping is calculated as in (2).

$$Electrical\ Energy\ = \frac{HE}{\eta s} \tag{2}$$

where ηs is the efficiency of the system components.

The calculated total daily loads including water pumping for these communities are indicated in Table 4.

Table 4. Total daily consumption in communities.

Community	Water Pumping Consumption kWh/Day	Household Consumption kWh/Day-Family	No. Houses	Total Consumption kWh/Day
Dir Ammar	5.5	1.8	10	23.5
Al-Birin	11	1.8	14	36.2

Table 2 shows that by deploying high efficiency pumps and appliances to carry out these projects, the total consumption may be around 23.5 kWh/day in Dir Ammar but 36.2 kWh/day in Al-Birin.

4.2.2. Sizing Solar PV Systems

To determine the capacity of the required solar PV system to supply the average daily load consumptions of these communities, Equation (3) was used (Imad, 2019) [1].

$$P_{PV-system}\ = \frac{E_{con}}{\left(\eta_{inv} \times \eta_{bat} \times P.S.H\right)} \times S_{fac} \tag{3}$$

Ecos: average daily consumption in kWh/day
P.S.H: peak sunshine hours (5.5 h) (Imad, 2019) [1]
ηinv: inverter efficiency (97%)
ηbat: battery efficiency (85%)
Sfac: factor of safety (1.2)

The total power of Dir Ammar's PV system, PV-system = 6.2 kWp.
The total power of Al-Birin's PV system, PV-system = 9.6 kWp
In both projects, a solar PV module capacity of 395 W was installed, so the number of modules in the system was determined as in (4).

$$N = \frac{P_{PV\ sys}}{P_{module}} \tag{4}$$

The number of solar PV system modules in Dir Ammar is 18; however, there are 24 modules in Al-Birin.

4.2.3. Determining DC System Voltage

The selected DC system voltage of the micro-grid solar PV system equals 48 Vdc. The number of series modules, Ns, can then be calculated as in (5).

$$N_s = \frac{V_{d.c(PV\ sys)}}{V_{module}}$$ (5)

$= 48/18.1 = 3$ module in series

The PV system in Dir Ammar is composed of two arrays; each consists of 3×3 PV modules. Similarly, Al-Birin's is composed of two solar arrays; however, each array has 3×4 PV modules, which are installed on galvanized steel supports, a south-facing, horizontally-oriented surface at a tilt angle of 30 degrees for optimum performance throughout the years.

4.2.4. Selection of Battery Bank

In micro-grid solar PV, the battery constitutes an important part of system. The needs of the community are met whether at night or on cloudy days, which requires a high number of charge-discharge cycles, by selecting the most appropriate battery.

The battery is selected in ampere hours, as in Equation (6).

$$C_{Ah} = \frac{N_a \times E_{con}}{V_{bat} \times DOD \times \eta_{in} \times \eta_{bat}}$$ (6)

Na: autonomy days (1.5–3 days)
Vbat: system battery voltage =, 48 V
DOD: depth of discharge (0.35)
For Dir Ammar, total CAh = 1700 Ah, and for Al-Birin, total CAh = 2600 Ah.

For limitation of budget, we selected 1800 Ah for both sites, the characteristics of the lead-acid batteries deployed in both sites are mentioned in Table 5. The storage system is composed of 24 deep cycle batteries. Each element is a 2 V battery with a capacity of 1800 Ah (C10), connected in a series.

Table 5. Characteristics of used batteries.

Type of Battery	AGM Block 2 V
Number of batteries	24
Capacity of battery (C10)	1800 Ah
Autonomy days	2

The total available energy in batteries can be calculated as in (7).

$$C_{Wh} = C_{Ah} \times V_{Bat}$$ (7)

C_{Wh} = 1800 Ah × 48 V = 86.4 kWh for Dir Ammar and C_{Wh} = 86.4 kWh for Al-Birin

4.2.5. Selection of Charge Controller

For controlling the charge-discharge cycles, the selected charge controller prevents issues related to overcharging as well as deep discharging, the selected controllers are using a maximum power point tracker (MPPT) in order to maximize the solar PV power.

The capacity of the charge controller is selected based on the PV maximum current, taking safety, as a factor, into account as shown in (8).

$$I_c = I_{PV} \times S_{factor} \tag{8}$$

S_{factor}: safety factor (1.25).
Ic = 73.5 A for Dir Ammar, and Ic = 98 for AL-Birin

Therefore, a 100 A MPPT, STUDER VarioTrack VT-100 was selected as the installed PV charge controller; the maximum efficiency reached 97.5%.

4.2.6. Selection of Inverter

We selected single phase inverters to supply single-phase water pumps used in the projects whose system output voltage was 230 V/50 Hz each, as follows (Imad, 2019) [1]:
Rating inverter = PV rating = 395 Wp × 18 = 7110 W for Dir Ammar and 9480 W for Al-Birin.
The inverter was used to convert DC power of PV arrays to AC power; the voltages and currents were suitable for operating domestic appliances in a consumer's household and driving the pump electric motor.

The input energy of inverter = PV rating × P.S.H.
= 7110 × 5.5 = 39.1 kWh/day
The total energy output of inverter = total energy input of inverter × ηinv
= 39.1× 0.95 = 37.1 kWh for Dir Ammar and 49.5 kWh for Al-Birin.

We selected 48 Vdc operating voltage inverters whose output voltage was 230 V/50 Hz, with a capacity of 10 kW in Dir Ammar and 12 kW in Al-Birin.
A comparison between different scenarios regarding solar capacity and storage system were studied and the results are shown in Tables 6 and 7.

Table 6. Different scenarios comparison for optimal design—Dir Ammar.

N	Capacity of Solar PV (kWp)	Storage System	Micro-Grid System Cost ($)
1	5	24 * 2500 Ah	21,340
2	6.2	24 * 1800 Ah	19,000
3	8	24 * 1200 Ah	19,150

Table 7. Different scenarios comparison for optimal design—Al-Birin.

N	Capacity of Solar PV (kWp)	Storage System	Micro-Grid System Cost
1	7	24 * 3500 Ah	31,760
2	10	24 * 2600 Ah	27,200
3	14	24 * 1800 Ah	28,350
4	10	24 * 1800 Ah	25,600

We selected the optimum scenarios which would give minimum cost for each community.

4.2.7. Electricity Installation

Each house is fitted with an energy dispenser, Figure 2, and meter, which limits the amount of energy available for each user in accordance with their predetermined needs and the contracted tariff.
Users are trained to make efficient and rational use of the energy in the household. The corrective (vs. preventive) maintenance technician is a contracted professional in charge of repairing potential

failures of the system and of keeping it in optimal condition. He is trained on the functioning of the micro-grid with adequate workshops and technical materials and manuals. The users in both communities are enjoying a 24-h quality electricity service.

As to the economic structure of the service, its goal is to guarantee the economic sustainability of the project. Thus, the fees paid by users stay in the community and are kept by the operator (a special bank account is created for the project in each community). The payment of fees is aimed at covering the expenses of operation and maintenance (O&M) of the system, that is, the cost of components, diesel for the generator to operate for a few hours in winter time, transport, spare parts, etc. This economic sustainability is highly dependent on the users' capacity and willingness to pay the fees.

Figure 2. Electricity dispenser meter.

One very essential element for long-term sustainability we are dealing with is the feeling of appropriation of the project by the community. Our experience has shown that the more informed the people are, the more involved. Thus, regular informative meetings are very necessary. Other key elements here are the local management of the service, which gives the community a sense of sovereignty (that is, they are not dependent on the electricity company or of Israeli supply); good training (on the rational use of energy, the individual energy control, . . .), which makes them feel more confident in the use of the service; and finally the empowerment of women, who at the end of the day are the ones managing the electricity use in the households.

4.3. Energy Generated by Micro-Grid Solar PV

The hourly data was collected using a data logger and recorded in the monitoring system each hour. The total annual output energy can be calculated as in (9).

$$E_{year} = \Sigma(E_{n1} = E_{n2}\ldots\ldots\ldots + E_{n12}) \tag{9}$$

where n1 = January ... n12 = December

E_n: monthly output energy (kWh)
E_{year}: yearly output energy (kWh)

As illustrated in Figures 3 and 4, the power output and the actual monthly consumption are compared with an estimated solar panel output using "PV sys" Software program.

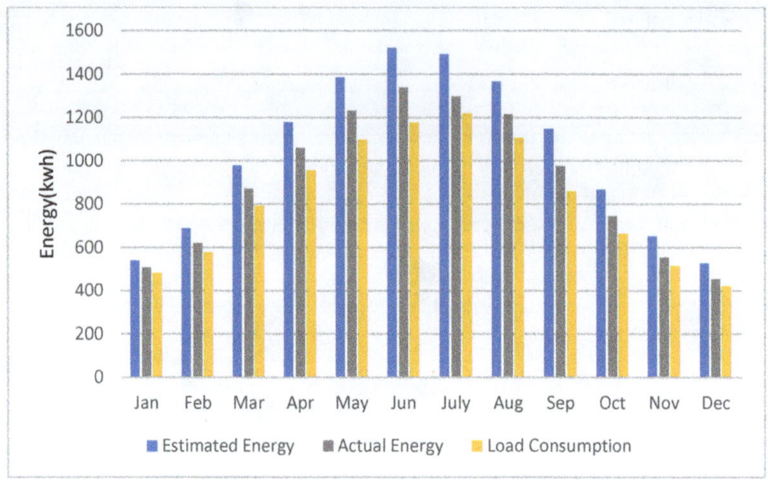

Figure 3. The actual and estimated energy consumption in Dir Ammar community in 2018.

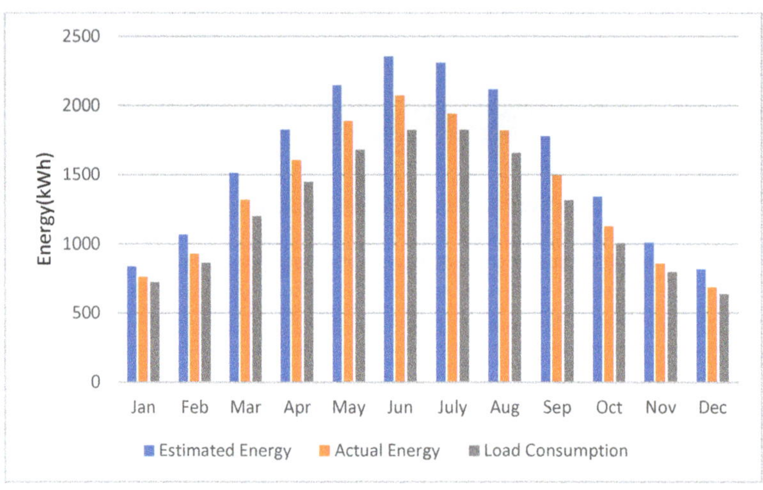

Figure 4. The actual and estimated energy consumption in Al-Birin community in 2018.

Estimated output is the output energy expected to be produced according to average solar radiation in the West Bank (Palestine), while the actual output is the real output based on real solar radiation. The solar PV system is normally expected to produce the most ideal output "optimum" energy, but in reality the output power will be dependent on real solar irradiance and temperature and on the electrical losses in the system; therefore, the actual energy output is usually less than the estimated.

In Dir Ammar the maximum output was 1337.2 kWh in June 2018, yet in December, it reached 452.6 kWh. As for the Al-Birin system, the maximum output energy was 2070.6 kWh in June 2018 and 684.6 kWh in December.

5. Performance Analysis of Rural Electrification Using Micro-Grid Systems in Palestine

5.1. Performance Ratio (PR) of Solar PV Systems

The PR in Equation (10) can be calculated by dividing the actual output generated energy by the estimated output energy as in (10); the suitable range of this factor is 68–90% (Ghouari, 2016; Otaibi, 2015) [14,15]. The calculated PR for both systems is shown in Figure 5.

$$PR = \frac{E_{actual}}{E_{estimated}} \tag{10}$$

For Dir Ammar, the yearly average PR is 88%, and for Al-Birin, the yearly average PR is 86%.

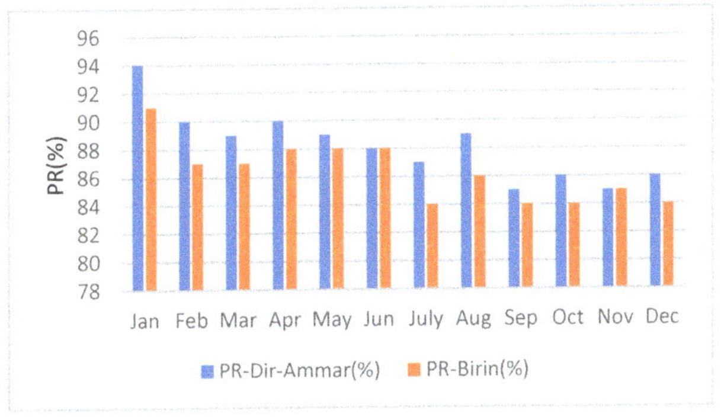

Figure 5. The performance ratio (PR) of both systems.

5.2. Capacity Utilization Factor (CUF)

The CUF is a factor which measures the actual energy output against the nominal energy output of the installed power at a specific period, as in (11) (Amjath, 2019; Ibrik and Hashaika, 2019) [16,17].

$$CUF = \frac{E_{real}}{h \times P_{ins}} \times 100\% \tag{11}$$

P_{ins}: rated capacity of solar power plant (Wp).
h: hours in specific period "during year/month/day".

The CUF has been calculated for both systems, and the results are shown in Figure 6.

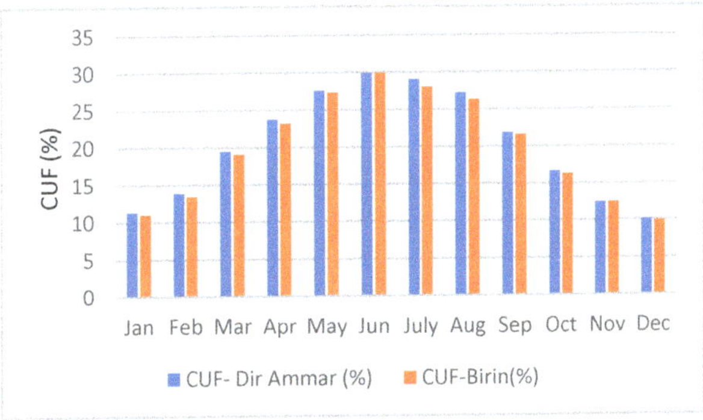

Figure 6. The capacity utilization figure (CUF) of Dir Ammar and Al-Birin solar PV power systems.

Figure 6 indicates that the annual average CUF is 20.27%; the maximum value is 29.9% in June, and the minimum is 10.14%, in December.

6. Social and Economic Impact of Solar PV Systems in Dir Ammar and Al-Birin Communities

6.1. Social Impact

Based on real-time data collection from both communities, Dir Ammar and Al-Birin, it was concluded that rural electrification had changed the status of these small villages and had created economic, socio-cultural and demographic impacts on the daily lives of both communities.

Energy access has considerable, sustained impacts on poverty as it helps to reduce it. Moreover, other significant aspects may be influenced by rural electrification, including health, education and childcare as well as female employment (Mondal, 2011; Khan, 2014) [18,19]. It was found that the availability of energy provided access to potable water. In fact, not only does safe drinking water play an essential role in maintaining an individual's health, but it also has a major impact on agricultural development. Furthermore, having energy access stimulates agro-food industries since micro-grid solar PV systems are likely to provide electricity services to these remote communities as well as rural areas, which still deploy diesel engines. Solar PV may also improve healthcare quality, agriculture and the availability of electricity and water supplies in these poor areas (Mala, 2009; Epstein, 2016) [20,21].

Using solar pump systems provides a reliable, sustainable energy source for irrigation. In terms of the economic impacts which these systems have, the farmers cut back on diesel consumption and save money. The agricultural output increases and farmers' incomes as well as increases, now they can enhancing crop productivity and being able to perform multiple cropping cycles during the year, which boosts their income, enhances their resilience, improves food security and contributes to cutting poverty in the communities (Hirmer, 2014; Müggenburg, 2012) [22,23].

These implemented projects proved to be successful as the beneficiaries reported their satisfaction. While the two rural areas' inhabitants had no access to electricity and, instead, had to use primitive tools, such as candles, before carrying out the projects, they now feel safer and closer to one another as they can participate in other nighttime activities, such as spending quality time with their families, which helps strengthen their ties.

Moreover, the time spent by the women involved in dairy production or doing house chores was reduced, allowing them to pay visits to one another and take better care of their children, so 63% of women reported the positive effects of the PV system on their lives.

The project affected the behavior of children positively. Twenty percent of the children who live in the villages under the study pointed out that studying at night was no longer impossible. Prior to the implementation of the projects, those children were without light at night; keeping in mind that 50% of the children did not attend school because they were either in their pre-school years or females who were not allowed to go to school because of the fear for their safety due to the dangers imposed by settlers, 20% could be considered as an acceptable percentage.

The houses in each community are close to each other as clear in Figures 7 and 8.

Figure 7. Houses in Dir Ammar.

Figure 8. Houses in Al-Birin.

Figure 9 indicates that through implementing the aforementioned solar PV projects in both communities, both villages now have access to electricity round the clock. In the same way, the generated energy has also improved access to water sources; water is stored in water tanks and later used for drinking (see Figure 10) and for solar-powered irrigation pumping, as shown in Figure 11. These sources have enabled poor farmers to improve their agricultural productivity as well as their vegetable production and cropping intensity at low costs while providing a cleaner and greener alternative for irrigation, as demonstrated in Figure 12. The water pump fills the water storage tanks during sunshine hours and the farmers irrigate the crop fields mainly at night and according to a specific schedule. Farmers can now cultivate more crops annually which will boost their incomes, enhance crop resilience, improve food security and alleviate poverty.

Figure 9. Installed solar PV system in Al-Birin.

Figure 10. Water tank for drinking and irrigation.

Figure 11. Water borehole in Al-Birin.

Figure 12. Increasing food production in community.

6.2. Economic Impact

The economic impact of the proposed micro-grid solar PV systems can be determined using different methods, all of which depend on the life cycle costing (LCC) (Barringer, 1995; Fuller, 1996) [24,25].

The proposed LCC consists of investment, replacement and operation costs, in addition to the cost of owning it over its lifetime.

The breakdown cost for each system is shown in Table 8.

Table 8. Breakdown cost in each community.

N	Item	Cost ($)—Dir Ammar	Cost ($)—Birin
1	Solar PV	2170	3500
2	Steal Structure	2000	2500
3	Batteries	7500	7500
4	Charge Controller	1500	2000
5	Inverter	2100	3000
6	Installation	2500	3000
7	Electrical Grid and Installation of Users	1230	4100
	Total	19,000	25,600

The replacement cost in year 10 will include the cost of replacement batteries, charge controller, and inverter; the estimated cost can be considered as the following:

- For Dir Ammar: 13,000 $ including transportation and installation.
- For Al-Birin: 15,000 $ including transportation and installation.

The rate of return (ROR) and simple pay back period (SPBP) methods were used in this evaluation.

6.2.1. Rate of Return Method

For feasibility evaluation, the rate of return (ROR) method can be employed so as to assess if a project or investment is economically justified (Yoomak, 2019; Firouzjah, 2018) [26,27]. The ROR method is expressed as the rate of interest earned on the unrecovered balance of the capital cost. The process of calculating the ROR value does not resemble that of calculating the present worth (PW) or annual worth (AW) for a series of income and outcome cash flows. The PW technique through the LCC of cash flow was adopted in the present study, as shown in Figure 12, with regard to the Dir Ammar project, and Figure 13, with regard to Al-Birin's. The LCC includes the initial cost of the project, operation and maintenance (O & M) costs, battery replacement costs, scrap value and saving 'revenue' from annually produced energy. In the savings calculation, the output energy cost equaled 0.5 $/kWh (Imad, 2019) [28], as it replaced the cost of using diesel. The annual saving of produced energy is determined as in (12).

$$Annual\ saving = \text{Total Annual output Energy (kWh)} \times \text{Energy Cost (\$/kWh)} \tag{12}$$

Annual saving for Dir Ammar = 10,861 × 0.5 = 5430.25 $ and for Al-Birin = 16,478 × 0.5 = 8238.5 $.

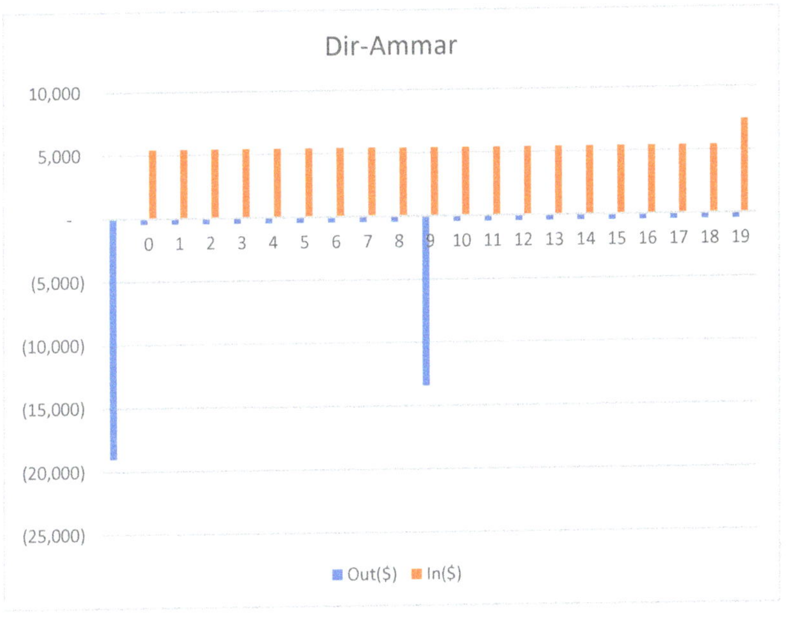

Figure 13. Lifetime cash flow—Dir Ammar.

The cash flow that is shown in Figures 13 and 14, the investment cost, battery replacement and O & M costs are considered as outcomes while the annual savings and the scrap value are considered as incomes.

PW (output) = investment cost + (P/A, i, 20) of operation cost + (P/F, i, 10) of battery replacement cost.
P/F,i,10 = finds the equivalent present value from future value at i% interest for 10 years.
P/A,i,20 = finds the equivalent present value from given annual value at i% interest for 20 years.

Where the interest rate i = 10%

PW (input) = (P/A, i, 20) of Energy savings + (P/F, i, 20) of scrap value.

For Dir Ammar:

PW (output) = 19,000 \$ + 420 \$ *(P/A, i, 20) + 13,000 * (P/F, i, 10)

PW (input) = 5430 * (P/A, i, 20) + 2000 * (P/F, i, 20).

For Al-Birin:

PW (output) = 25,600 \$ + 520 \$ * (P/A, i, 20) + 15,000 * (P/F, i, 10)

PW (input) = 8238.5 * (P/A, i, 20) + 3000 * (P/F, i, 20).

Using an Excel sheet, the ROR for Dir Ammar equaled 23.28%, yet for Al-Birin, it was 27.83%. This suggests that these projects will return 25.55% of their initial costs annually; in other words, these projects are feasible.

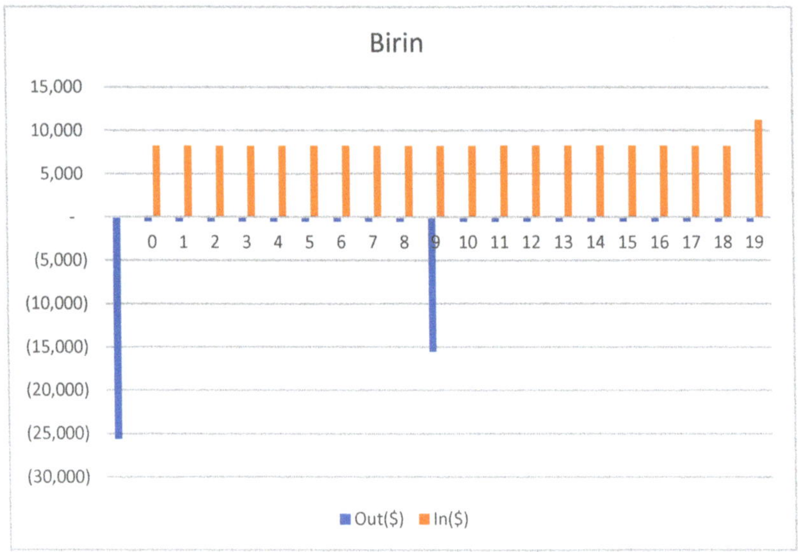

Figure 14. Lifetime cash flow—Al-Birin.

6.2.2. Payback Period (SPBP)

The SPBP method is used also to determine the project feasibility, and it can be calculated using Equation (13) (Berk, 2012; Brealey, 2012) [29,30].

$$S.P.B.P = \frac{Investment}{Saving} \tag{13}$$

S.P.B.P for Dir Ammar = 3.5 years, and for Al-Birin the S.P.B.P = 3.1 Years

The SPBP for both projects is around 3.5 years. Alternatively stated, all the expenses are to be recovered by the first 3.5 years, whereas the rest, which add up to 16.5 years, are to be considered as a profit, which proves the feasibility of implementing these projects.

7. Environmental Impact

The replacement of diesel generators with solar PV systems has a significant environmental impact especially on the atmosphere due to the entailed reduction of combustion processes. The amount of CO_2 emitted per kilowatt-hour (kWh) depends on the method of generation, diesel, nuclear, coal, gas … and so forth. The estimated annual reduction of CO_2 emissions in Dir Ammar is about 2830 kg CO_2 and Al-Birin villages about 6288 kg CO_2.

8. Conclusions

This study presented a design of a micro-grid solar PV system for electrification and irrigation systems in two rural communities (Dir Ammar and Al-Birin hamlets) in Palestine since this technology is reliable and feasible for irrigation of agriculture crops. The solar PV systems minimize the dependence on diesel as well as conventional electricity sources, which may help solve the problems related to the lack of energy supply in Palestine. This study points out that the total cost of installing solar PV systems, including fixed, running and replacement costs are lower than those of diesel engines.

The implementation of solar micro-grid systems in rural areas suggests a diversity of approaches that address many objectives, such as rural electrification, solar PV dissemination, water availability and increasing agricultural productivity. The implemented projects in the discussed two communities, Dir Ammar and Al-Birin, are integrated into the processes of establishing more direct correlations between producing energy, availability of water and agricultural activities, not to mention increasing the created opportunities with respect to energy, water and food securities.

The performance analysis of micro-grid solar PV systems for electrification and irrigation of land for small communities in Palestine shows very good results. The installation of an electricity dispenser and training for the community on load management and using water for irrigation at night are key factors for no black outs and keeping batteries in good condition. The degree of satisfaction within the community is high, and the social development and utilizing of PV systems is more economically feasible for electrification and irrigation of remote villages of geographic, climate and load conditions similar to Dir Ammar and Al-Birin in Palestine. In addition, micro-grid solar PV systems do not pollute the environment, unlike the use of diesel generators.

Funding: This research received no external funding.

Acknowledgments: This research was sponsored by the Med-EcoSuRe project (www.enicbcmed.eu/projects/med-ecosure), which receives funding from the European Union ENI CBC Mediterranean Sea Basin Programme (2014–2020) (grant agreement no. 26/1233).

Conflicts of Interest: The authors declares no conflict of interest.

References

1. Ibrik, I.H. An overview of electrification rural areas in Palestine by using micro-grid solar energy. *Cogent Eng.* **2019**, *6*, 1638574. [CrossRef]
2. Mason, M.; Mor, A. Chapter 5: Energy Profile and the Potential of Renewable Energy Sources in Palestine. In *Renewable Energy in the Middle East Enhancing Security through Regional Cooperation*; Springer: Dordrecht, The Netherlands, 2009; pp. 71–89.
3. Juaidi, A.; Montoya, F.G.; Ibrik, I.H.; Manzano-Agugliaro, F. An overview of renewable energy potential in Palestine. *Renew. Sustain. Energy Rev.* **2016**, *65*, 943–960. [CrossRef]
4. Feron, S. Sustainability of Off-Grid Photovoltaic Systems for Rural Electrification in Developing Countries: A Review. *Sustainability* **2016**, *8*, 1326. [CrossRef]
5. Rowley, T.D. *Rural Development Research: A Foundation for Policy*; Greenwood Press: Westport, CT, USA, 1996.
6. Rehan, A.; Sahito, A.R.; Maheshwari, M.K.; Shafi, A. Assessing the Energy & Environmental Benefits and Socio-Economic Impacts of Solar Microgrid in Rural Areas in Sindh. *Int. J. Mod. Res. Eng. Manag.* **2019**, *2*, 13–29.
7. Querikiol, E.M.; Taboada, E.B. Performance Evaluation of a Micro Off-Grid Solar Energy Generator for Islandic Agricultural Farm Operations Using HOMER. *J. Renew. Energy* **2018**, 1–9. [CrossRef]
8. Santos, A.; Ma, Z.; Olsen, C.; Jørgensen, B. Framework for Microgrid Design Using Social, Economic, and Technical Analysis. *Energies* **2018**, *11*, 2832. [CrossRef]
9. Chel, A.; Kaushik, G. Renewable energy for sustainable agriculture. *Agron. Sustain. Dev.* **2011**, *31*, 91–118. [CrossRef]
10. Al-Saidi, M.; Lahham, N. Solar energy farming as a development innovation for vulnerable water basins. *Dev. Pract.* **2019**, 1–16. [CrossRef]

11. Kyriakarakos, G.; Balafoutis, A.T.; Bochtis, D. Proposing a Paradigm Shift in Rural Electrification Investments in Sub-Saharan Africa through Agriculture. *Sustainability* **2020**, *12*, 3096. [CrossRef]

12. Ibrik, I. Technical and social innovation through electrification small communities in Palestine by multi-user solar PV mini grids—Case study—Birin community. In Proceedings of the Solar Technologies & Hybrid Mini Grids to Improve Energy Access Conference, Bad Hersfeld, Germany, 21–23 September 2016.

13. Madziga, M.; Rahil, A.; Mansoor, R. Comparison between Three Off-Grid Hybrid Systems (Solar Photovoltaic, Diesel Generator and Battery Storage System) for Electrification for Gwakwani Village, South Africa. *Environments* **2018**, *5*, 57. [CrossRef]

14. Ghouari, A.; Hamouda, C.; Chaghi, A.; Chahdi, M. Data Monitoring and Performance Analysis of a 1.6 kWp Grid Connected PV System in Algeria. *Int. J. Renew. Energy Res.* **2016**, *6*, 34–42.

15. Al-Otaibi, A.; Al-Qattan, A.; Fairouz, F.; Al-Mulla, A. Performance evaluation of photovoltaic systems on Kuwaiti schools' rooftop. *Energy Convers. Manag.* **2015**, *95*, 110–119. [CrossRef]

16. Amjath-Babu, T.S.; Sharma, B.; Brouwer, R.; Rasul, G.; Wahid, S.M.; Neupane, N.; Sieber, S. Integrated modelling of the impacts of hydropower projects on the water-food-energy nexus in a transboundary Himalayan river basin. *Appl. Energy* **2019**, *239*, 494–503. [CrossRef]

17. Ibrik, I.; Hashaika, F. Techno-economic impact of grid-connected rooftop solar photovoltaic system for schools in Palestine: A case study of three schools. *Int. J. Energy Econ. Policy* **2019**, *9*, 291–300. [CrossRef]

18. Mondal, A.H.; Klein, D. Impacts of solar home systems on social development in rural Bangladesh. *Energy Sustain. Dev.* **2011**, *15*, 17–20. [CrossRef]

19. Khan, S.A.; Azad, A.K. Social Impact of Solar Home System in Rural Bangladesh: A Case Study of Rural Zone. *Energy Environ.* **2014**, *1*, 5–22. [CrossRef]

20. Mala, K.; Schlapfer, A.; Pryor, T. Better or worse? The role of solar photovoltaic (PV) systems in sustainable development: Case studies of remote atoll communities in Kiribati. *Renew. Energy* **2009**, *34*, 358–361. [CrossRef]

21. Epstein, M.B.; Bates, M.N.; Arora, N.K.; Balakrishnan, K.; Jack, D.W.; Smith, K.R. Household fuels, low birth weight, and neonatal death in India: The separate impacts of biomass, kerosene, and coal. *Int. J. Hyg. Environ. Health* **2013**, *216*, 523–532. [CrossRef]

22. Hirmer, S.; Cruickshank, H. The user-value of rural electrification: An analysis and adoption of existing models and theories. *Renew. Sustain. Energy Rev.* **2014**, *34*, 145–154. [CrossRef]

23. Müggenburg, H.; Tillmans, A.; Schweizer-Ries, P.; Raabe, T.; Adelmann, P. Social acceptance of PicoPV systems as a means of rural electrification—A socio-technical case study in Ethiopia. *Energy Sustain. Dev.* **2012**, *16*, 90–97. [CrossRef]

24. Barringer, H.P.; Weber, D.P.; Westside, M.H. Life-cycle cost tutorials. In *Fourth International Conference on Process Plant Reliability*; Gulf Publishing Company: Houston, TX, USA, 1995.

25. Fuller, S.; Petersen, S. *Life-Cycle Costing Manual for the Federal Energy Management Program*; NIST Handbook 135; NIST: Gaithersburg, MD, USA, 1996.

26. Yoomak, S.; Patcharoen, T.; Ngaopitakkul, A. Performance and Economic Evaluation of Solar Rooftop Systems in Different Regions of Thailand. *Sustainability* **2019**, *11*, 6647. [CrossRef]

27. Firouzjah, K.G. Assessment of small-scale solar PV systems in Iran: Regions priority, potentials and financial feasibility. In *Renewable and Sustainable Energy Reviews*; Elsevier: Amsterdam, The Netherlands, 2018; Volume 94(C), pp. 267–274.

28. Ibrik, I. Multi-User Solar Hybrid Micro-Grid Technologies can Overcome Energy Poverty in Palestinian Villages. *J. Fundam. Renew. Energy Appl.* **2019**, *9*, 1.

29. Jonathan, B.; DeMarzo, P.; Harford, J. *Fundamentals of Corporate Finance*, 2nd ed.; Prentice Hall: Boston, MA, USA, 2012.

30. Brealey, R.A.; Myers, S.C.; Marcus, A.J. *Fundamental of Corporate Finance*, 7th ed.; McGraw-Hill, Inc.: New York, NY, USA, 2012.

Article

Design and Analysis of Photovoltaic Powered Battery-Operated Computer Vision-Based Multi-Purpose Smart Farming Robot

Aneesh A. Chand [1,*], Kushal A. Prasad [1], Ellen Mar [1], Sanaila Dakai [1], Kabir A. Mamun [1], F. R. Islam [2], Utkal Mehta [1] and Nallapaneni Manoj Kumar [3,*]

1 School of Engineering and Physics, The University of the South Pacific, Suva, Fiji; kushalaniketp@gmail.com (K.A.P.); s11086154@student.usp.ac.fj (E.M.); s11086783@student.usp.ac.fj (S.D.); kabir.mamun@usp.ac.fj (K.A.M.); utkal.mehta@usp.ac.fj (U.M.)

2 School of Science and Engineering, University of Sunshine Coast, Sippy Downs, QLD 4556, Australia; fislam@usc.edu.au

3 School of Energy and Environment, City University of Hong Kong, Kowloon, Hong Kong

* Correspondence: aneeshamitesh@gmail.com (A.A.C.); mnallapan2-c@my.cityu.edu.hk (N.M.K.)

check for updates

Citation: Chand, A.A.; Prasad, K.A.; Mar, E.; Dakai, S.; Mamun, K.A.; Islam, F.R.; Mehta, U.; Kumar, N.M. Design and Analysis of Photovoltaic Powered Battery-Operated Computer Vision-Based Multi-Purpose Smart Farming Robot. *Agronomy* **2021**, *11*, 530. https://doi.org/10.3390/agronomy11030530

Academic Editors: Miguel-Ángel Muñoz-García and Luis Hernández-Callejo

Received: 14 January 2021
Accepted: 7 March 2021
Published: 11 March 2021

Publisher's Note: MDPI stays neutral with regard to jurisdictional claims in published maps and institutional affiliations.

Abstract: Farm machinery like water sprinklers (WS) and pesticide sprayers (PS) are becoming quite popular in the agricultural sector. The WS and PS are two distinct types of machinery, mostly powered using conventional energy sources. In recent times, the battery and solar-powered WS and PS have also emerged. With the current WS and PS, the main drawback is the lack of intelligence on water and pesticide use decisions and autonomous control. This paper proposes a novel multi-purpose smart farming robot (MpSFR) that handles both water sprinkling and pesticide spraying. The MpSFR is a photovoltaic (PV) powered battery-operated internet of things (IoT) and computer vision (CV) based robot that helps in automating the watering and spraying process. Firstly, the PV-powered battery-operated autonomous MpSFR equipped with a storage tank for water and pesticide drove with a programmed pumping device is engineered. The sprinkling and spraying mechanisms are made fully automatic with a programmed pattern that utilizes IoT sensors and CV to continuously monitor the soil moisture and the plant's health based on pests. Two servo motors accomplish the horizontal and vertical orientation of the spraying nozzle. We provided an option to remotely switch the sprayer to spray either water or pesticide using an infrared device, i.e., within a 5-m range. Secondly, the operation of the developed MpSFR is experimentally verified in the test farm. The field test's observed results include the solar power profile, battery charging, and discharging conditions. The results show that the MpSFR operates effectively, and decisions on water use and pesticide are automated.

Keywords: solar photovoltaics; agricultural robots; Agri 4.0; battery-based farm machinery; mobile irrigation system; smart water sprinklers; smart pesticide sprayers; multi-purpose farming robots

1. Introduction

Farming is one of the primary activities in agricultural sectors that demand freshwater use worldwide [1]. Presently, the farming community uses more than 70% of the freshwater that is withdrawn from water resources (e.g., well, boreholes, lakes, rivers, etc.) to produce goods for the fast-growing food demand [2]. On the other side, freshwater scarcity is increasing from the water resources mentioned above due to weather patterns changes and uneven rainfall. At the same time, water demands are growing mostly driven by industrial activity. The increasing water demand is emerging as a more immediate threat to society and other interdependent sectors, such as food and agriculture [3–5]. Global estimates and projections are quite uncertain in water use, especially the relationship between water use and crop productions. However, the predictions on food use suggest that the demand for food would rise by 60% by 2050 [6], and this increase will require more arable land with sufficient water resources. The population, urbanization, clean energy, and climate change are the primary concerns concerning food security in the modern world. Globally, food security is becoming a prime concern for government agencies and health organizers. Hence, significant attention is essential in the agricultural product supply to stay afloat with a large population to be fed. Many countries use more than 90% of water withdrawals for agriculture, especially in South Asia, Africa, and Latin America. The freshwater withdrawals in Sudan record highest, i.e., 96%, whereas Germany and the Netherlands record less than 1% [6]. Hence, it is clear that there exists an uneven distribution of water withdrawals for the agricultural sector across the globe, which results in food security and water supply issues. In parallel with this, various techniques are being used to solve food security and water supply issues for multiple uses and ensure an environmental balance [7–9]. Some studies in the literature suggested creating water demand management measures [10,11].

Smart irrigation could be one feasible solution in the context of efficient water management. Assessing the overall water availability, preparing coping mechanisms for drought times, improving irrigation scheduling tools, and combining the technology and management techniques should be considered when upgrading the irrigation systems [11]. Smart irrigation management helps to detect when to irrigate, the amount of water required, and the irrigation frequency, based on the monitoring of crop evapotranspiration and soil moisture situations [12]. The cheapest way to irrigate large farmland is to lay down water pipes connected to sprinklers and pump water from a nearby water source. In the case of rugged and uneven terrain, setting up irrigation pipes may be difficult, and a more efficient pump could be required if there are significant elevation changes. In situations like this, there are apparent disadvantages involved with the use of power and maintenance.

Apart from the irrigation system, pest control is another major issue. There are massive developments in the pest control field, but capital cost, size, and complex nature made it less popular. More advanced machinery can perform the pest controlling task with excellent efficiency, but it often comes with a high price. Overall, in combatting the pest, chemicals are often sprayed over the farm area, assuming they reach each plant on the farm. Different types of plants have different lifespans at some stages; a certain amount of fertilizers or chemicals are required for maximized production. To achieve this, the chemical used must reach the plant correctly and efficiently.

Hence, there is a great potential for mobile robotic systems that can spray the chemicals and other types of fertilizers directly to the plant to satisfy the current challenges. Massive research is currently undertaken on individual systems that serve water sprinkling and pest spraying functions separately [13–17]. Based on the research survey, it is understood that the agricultural sector is heavily involved in employing IoT sensors to collect data, which later helps in data mining and predicting using various artificial intelligence (AI) techniques. However, in recent times, low-cost data gathering electronic devices are becoming popular which include Arduino [18–22], Raspberry Pi [23–25], Zigbee [19,25–27], LoRa [24], WiFi [28], Bluetooth [22], Global System for Mobile (GSM) Communications [20], and General Packet Radio Service (GPRS) [29]. As well, there has been a rapid influx of commercial water sprinkling and pest spraying farming solutions embedded with above mentioned electronic devices into the market. These commercial systems demand an efficient energy storage system for their operation; with this, the use of rechargeable and portable batteries like nickel-cadmium (NiCd), and lithium-ion (Li-ion), etc., have become popular. More recently, the on-board solar powered-battery systems also emerged, making these commercial farming solutions energy efficient [19,20]. Besides energy management, water management is crucial; hence the on-board water and chemical storage tanks should be optimally designed considering the energy requirements. This suggests the need for effective control algorithms [12]. In ref [30], a decision tree approach was used to decide the seasonal irrigation frequency and water requirements for the farm based on forecasted soil moisture. Like this, there exist numerous prediction methods for soil moisture, see Figure 1.

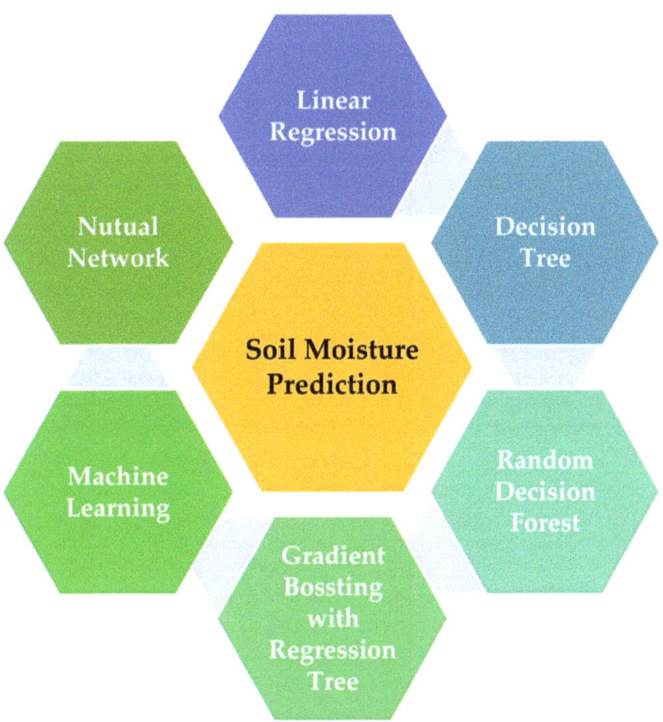

Figure 1. Commonly used methods for predicting soil moisture.

A fuzzy inference that uses IF-THEN (knows as Crisp rules) was also applied to equivalate the soil moisture compared with threshold values collected by sensors to decide whether to switch on an irrigation sprinkler system [31,32]. Like this, there are numerous approaches for controlling the sprinkler system see Table 1.

Table 1. Summary of the Literature with different control method in Smart Irrigation.

The Method Used to Control the Sprinklers of an Irrigation System	Reference
Supervisory Control and Data Acquisition (SCDA) system-uses the soil sensor to control and monitor the irrigation. The advantage of such an algorithm is it automatically carries out the data collection, preparation, and execution of the water balance management.	[33]
Wireless sensor nodes gather fully web-based system (online)—visualization of weather data and soil moisture, and spraying is done where is necessary. The system can send and receive messages and activates the irrigation system automatically.	[34]
Internet of Things (IoT) devices–farmers, can receive the data such as soil moisture, humidity, and whether forcing from different IoT sensors and manually or automatically spraying value is opened.	[35]

Studies also revealed that automation and path navigation are also the other two essential features for designing mobile robotic systems. In ref. [36], a farm machinery control system was developed to operate multiple tractors on a single large-scale farm by a remote Wireless Local Area Network (WLAN) monitoring system that automates navigation path control simultaneously. The control architecture for path tracking tasks for a mobile robot is usually based on rolling without slipping. However, this is not realistic for an off-road vehicle, such as a tractor or any other agricultural vehicle, as the study in [37] points out since farming vehicles face problems with sliding wheels and actuator delays. On the same note, this was also demonstrated many strategies that focus on predictive observer-based and adaptive control that can automatically direct farm machinery without an external sensor while dramatically reducing the precision loss for route monitoring.

From the above literature survey, it is clear that smart irrigation provides reliable operations for agricultural activities and efficiently delivers water. At the same time, systems related to pesticide spraying also offer greater efficiency. However, the studies on machinery that offer both functions (water sprinkling and pesticide spraying) are limited as per our knowledge. Hence, this study aims to build a multi-purpose smart farming robot (MpSFR) based on vehicular technology in irrigating the farm. The MpSFR functions with farm monitoring, preprocessing the data for decisions, control, and other irrigation management activities enhanced by soil moisture and plant health prediction. Besides, an experimental investigation is carried out in the test farm to verify the functions of the proposed MpSFR.

The novel contributions of this study are as follows:

- Photovoltaic powered battery-operated vehicular technology-based systems for smart irrigation;
- Sensor-based autonomous control is used to avoid collisions and obstacles during water sprinkling and pest spraying;
- Soil moisture-driven decisions for sprinkling water in the farm area and image-processing for pest detection to decide pest spraying.

The paper is organized as follows. In Section 2, the multi-purpose smart farming robot's system design is presented. Section 3 provides results and discussion based on the test farm investigation. Section 4 provides the challenges, future research suggestions, and conclusions.

2. Proposed Multi-Purpose Robot for Smart Farming

The proposed smart farming robot is inspired based on the vehicular design concepts, and with this, watering and pesticide spraying can be done remotely. The design consists of different components briefly discussed in Section 2.1. The layout of the proposed MpSFR is shown in Figure 2.

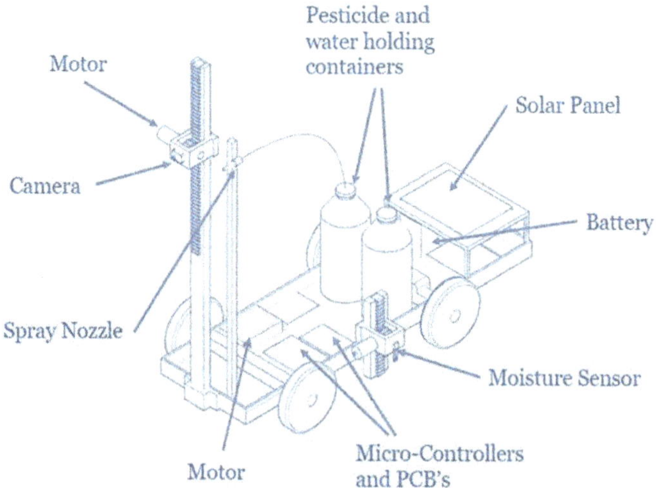

Figure 2. The layout of the Multi-Purpose Robot for Smart Farming (Note: PCB-Printed Circuit Boards).

2.1. Major Components

2.1.1. Soil Moisture Detection

Soil moisture is an integral part of the irrigation system, so data collection requires a sensor inserted near the plant. An on-board HL-69 soil hygrometer moisture sensor manufactured in China is integrated with the MpSFR, and this allows us to measure the soil moisture up to a depth of 0.2 m. The measured Soil Moisture Differences (SMD) are then used to predict whether irrigation is needed or not. The few prominent features of the HL-69 Soil Hygrometer are mentioned in Table 2.

Table 2. Features of the HL-69 Soil Hygrometer (Note: PCB: printed circuit board).

Parameter	Value/Description	HL-69 Soil Hygrometer
Operating voltage	3.3 to 5 Volts	
Output mode and interface	- Dual output mode - Digital output interface (0 and 1) - Analog output interface	
Installation	- A fixed bolt hole provision provides ease of the sensor installation	
Color indicators represent	- Red indicator represents the power - Green indicator represents the digital switching output	
Chip	LM393 comparator	
Dimension of the panel PCB	3 cm × 1.5 cm	
The dimension of the soil probe	6 cm × 3 cm	
Length of the cable	21 cm	

2.1.2. Spraying Mechanism

A spraying mechanism is being designed and developed using two servo motors. It can operate both in a vertical and horizontal motion. At the tip of the spraying nozzle, this spraying system is still kept intact. For this, we selected a full-cone even distribution spray type nozzle, and it can be easily mounted on a spray planter. The pounds per square inch (PSI) range is 20 to 60. The servo mechanism allowed the sprayer to have 180° horizontal freedom and 90° vertical freedom. Arduino Uno R3 microcontroller is used to control the two servo motors. The system is fabricated in a 3D printer using the polylactic acid (PLA) material can be seen in Figure 3.

In the system, we have three spraying modes, and accordingly, the servo motors have been initialized to operate. An infrared (IR) remote control is provided for the input control. The sprayer used can cover 156 m^2 with a given mode of operation, as in Table 3.

2.1.3. Mobile Control and Power Circuit Design

As mentioned earlier, the proposed MpSFR is based on vehicular design, and the chassis selection is crucial. For this, we used off-the-shelf "Toyabi" mobile chassis from China. The design consists of a pair of motors to control direction and motions. The user's movement input with the aid of remote control and movement is taken by manual mode. It is designed to test for an auto mode movement in the test farm, which follows a predefined pattern and is discussed further in Section 3.

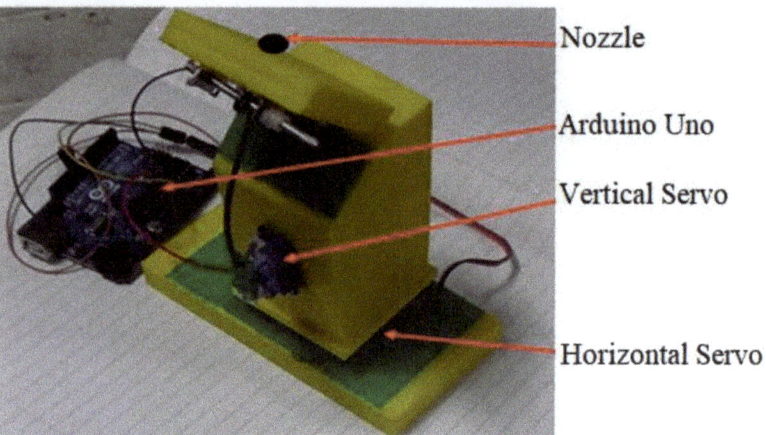

Figure 3. Spray mechanism decisions, which are based on the control of servo motors.

Table 3. Sprayer Mode and Operation.

Key Entry	Description	Mode
1, Select direction	To target a particular spot, drive the sprayer manually	Manual
2, Start	With slow horizontal movement, fast vertical spraying	Vertical
3, Start	Horizontal quick spraying with moderate vertical spraying	Horizontal

A consistent power supply is essential for any system, especially for battery-operated systems. A power circuit is designed using the H-bridge circuit and shown in Figure 4. Note that this may not provide the mechanism with adequate current whenever a full load requires from the motors. The direct power switch using bipolar junction transistor (BJT) is operated eventually to supply sufficient current. The power parameters and operating details are given in Table 4.

To operate the MpSER, the power transistors require a direct current (DC), which is about 200 mA DC at their base. This could not be possible with the microcontroller as it triggers and allows 20 mA output of DC at each output pin. Therefore, a BJT was placed between the pin of the microcontroller and the power transistor to draw current from the 5 V voltage common collector (VCC) supply of the microcontroller, which permitted the source to be up to 200 mA. When the pin is not high in the microcontroller, a 1 k Ohm resistor is placed to control the base's power transistor.

To make it easier for the user to access the receiver, the IR sensor is conveniently located at the rear side of the MpSFR. The two common heavy objects in the proposed design are the spraying tank and the battery, which are mounted in the center of the vehicle to increase stability by having the center of gravity in the center of the vehicle.

The servo motor which drives the vehicles has been mounted at the height at the very front so that water jets can cover vast distances of about 5 m. Under the top panel, the microcontroller and control circuitry are shielded to prevent any water spillage.

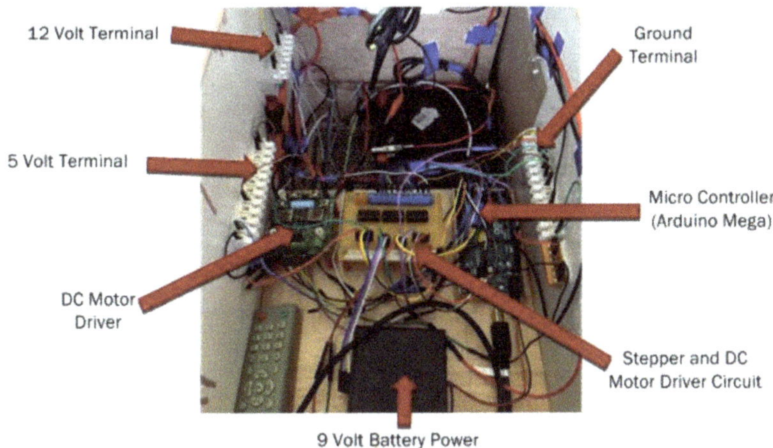

Figure 4. Controls and Power Circuit for multi-purpose smart farming robot.

Table 4. Components and their power parameters.

Components	Value/Description/Make
Photovoltaic panel	P_{max}: 20 W, V_{mpp}: 16.8 V, V_{oc}: 21.0 V, and I_{sc}: 1.29 A
Battery	Motor and spraying pump uses the rechargeable battery: 12 Vdc, 7 Ah
	MCU used the non-rechargeable battery, 9 Vdc
Arduino Uno R3 Board	Atmega328p, 9 Vdc
Infrared receiver	1838 T
Power Transistor	TIPL760B NPN
Micro Servo Motor	SG-90, 5 Vdc, Weight 9 g, torque 1.8 kgf.cm

Note: P_{max} = maximum power; V_{mpp} = voltage at maximum power point; V_{oc} = open circuit voltage; I_{sc} = short circuit current; Vdc = direct current voltage; MCU = microcontrollers; kgf.cm = kilogram-force centimeter.

Another reason was that the top panel has a coating of water protecting material that avoided any spillage onto the electrical and electronics components. For MpSFR, the wheel diameter is around 20 cm and is wide enough to navigate over rugged terrain. The process flow of the system is given in Figure 5.

2.1.4. Working of Water Pump

As the MpSFR is designed with the dual feature: (a) Water sprinkling for irrigation and (b) pesticide spraying. The MpSFR contains a water storage tank driven by a pump connected to the control board and battery. The pump is operated through a driver circuit and signals from the microcontroller. In this operation, the pump receives the command from the remote control.

Here, we used a 12 V DC motor pump and is sufficient to drive the spraying mechanism. The selected motor up can also lift the water to a maximum height of 5 m with a flow rate of 600 L/h. The MpSFR is designed to carry an on-board storage tank of 2.5 L capacity, and this is sufficient to cover at least 50 m^2 farm area.

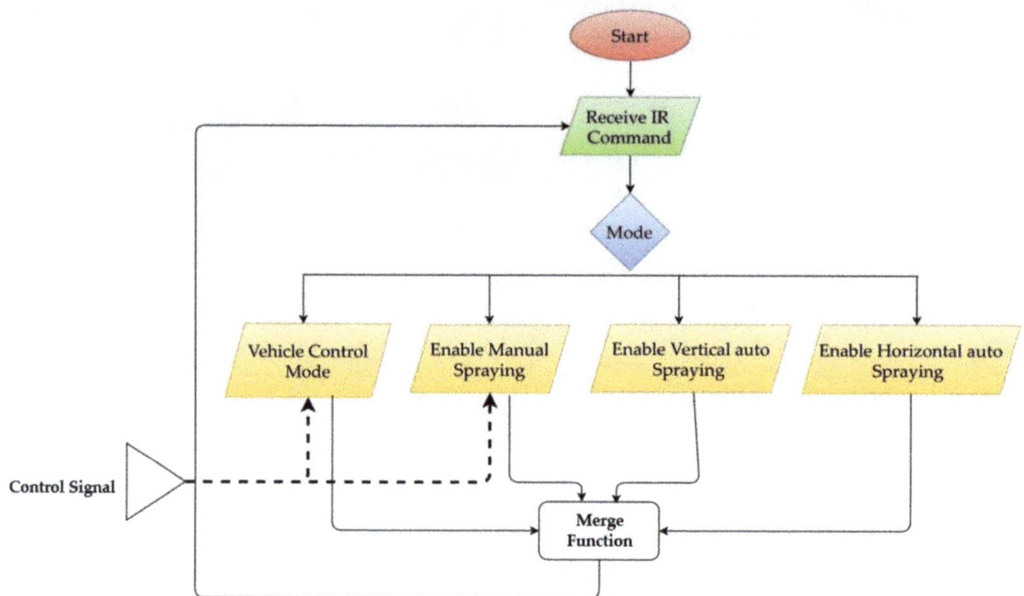

Figure 5. Program flow of the system (Note: IR: infrared; dashed lines represents the control signals to vehicle as well as manual spraying; continuous line represent the IR commands received).

2.1.5. User Interface through IR Remote Controller

The user interface application is developed, taking the compatibility measures suitable for a smartphone. An 1838T IR sensor can provide inputs to the controller on the board. For MpSFR operation, nozzle placement, pump control, and spraying mode option, a series of commands are given.

The Samsung WatchON program has been used for the propagation of IR signals as a remote control. The user interface consists of an option to drive the vehicle forward and backward. As well, pause and enter options for farming functions. The directional buttons are used to mount the servo motor system manually, and the middle button is used to turn the pump on the spray. To enter the mode that the user needs, the numbers on the bottom right are used.

2.1.6. Image Processing

Image processing is an essential feature that is used in pest detection in smart farming applications. With an MpSFR based irrigation system controlling and protecting the plants from various types of insects is the added advantage. Therefore, an IR FLIR A65 Thermal Camera is used, mounted in the designed vehicle as it is a light-weight design, so there was no major issue in implementing it. The FLIR A65 camera gives a good quality image of 320×256 pixels. An added benefit is that it can also produce a thermal image of 640×512 pixels.

2.2. Design of the Multi-Purpose Smart Farming Robot

Based on the schematic representation and the components discussed in Section 2.1. the MpSFR system is designed. Initially, a conceptual model was developed for testing purposes, and later it was developed with the correct sizes considering the weights and functions of the onboard components. In the conceptual model, the solar photovoltaic module is not used, but the final design of the MpSFR has an on-board solar photovoltaic array connected to the battery through a power converter, which helps in charging the battery and reduces dependency on grid power. Figure 6 shows the final prototype of the designed MpSFR. The designed MpSFR has some safety features such as obstacle detection and safe operation following the farm line path. Not only the safe operation in the MpSFR movement but also the controlled operation of sensor and camera. Here, the motors implant the soil moisture sensor into the ground and move the camera up and down to understand the soil moisture and pests better.

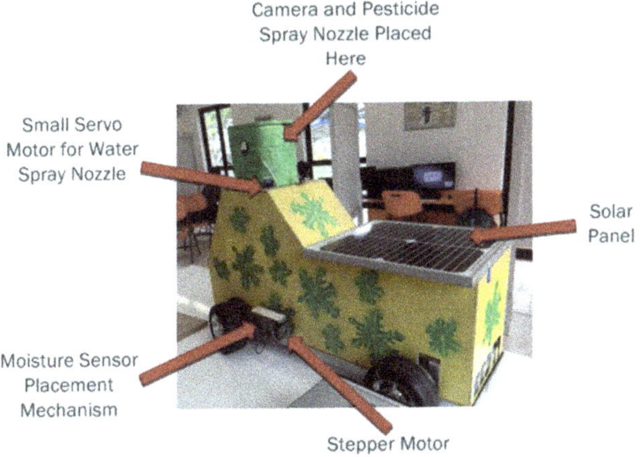

Figure 6. The prototype of the developed multi-purpose smart farming robot.

2.3. Methodology

This section provides the methodology that we used for verifying the operations of the designed MpSFR. In Figure 7, the methodological operational flow and system components integration is shown. We use the microcontroller during the operation backflow and at this point the electromotive force (EMF) need to be monitored. So, three diodes were inserted along with the motor's terminals in the control circuit configuration, which stops the back EMF from flowing. The BJT transistors in between also provided a type of separation between the low and high-power circuits.

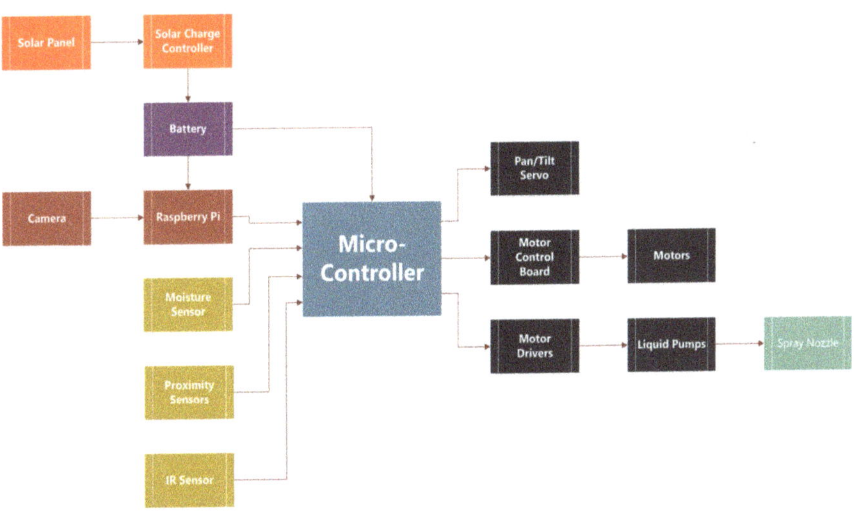

Figure 7. Methodological operational flow and system components integration in the proposed multi-purpose smart farming robot.

From Figure 7, it can be seen that the power circuit, sensing elements, motoring operations, and function delivery controls are mainly based on the microcontroller decisions. In Figure 8, we have presented the detailed methodology about the water sprinkling and pesticide spraying decisions initiated by the MpSFR based on its intelligent sensing capabilities.

Step by step operation is given below:

Step-1: The MpSFR starts heading towards the farm area, and at this step, it is made sure that the batteries are fully charged, also the storage tanks with water and required pesticides;

Step-2: Once the MpSFR reaches the first plant, the probes of the sensor placed in a water-sealed PCB package will be sticking out to measure the soil moisture at a depth of 0.2 m from the surface;

Step-3: The measured soil moisture data is compared with the standard values, and here if the observed soil moistures are lesser, then the watering decision is made where the pump motor starts operating to deliver the water to the plant through the sprinkler nozzle;

Step-4: Once the plant's watering is done or higher soil moistures are measured, the MpSFR starts operating its camera to check for the pests possible by applying image processing for the captured plant image. The pest detection, segmentation for the pest shape, and pest classification features allow the MpSFR to choose the required pesticide, and accordingly, the MpSFR sprays the pesticide;

Step-5: If no pest is identified, then the MpSFR will check for its available energy in the battery and the weather conditions (for example, favorable for solar charging or not). Here, if it is found that the battery energy level is lower and the solar charging conditions are not favorable, the MpSFR takes the help of an external charging plug;

Step-6: If the battery energy is available, then the MpSFR moves to the next plant. As well, in the favorable solar energy condition, the MpSFR continuous to move forward by checking the plants one by one.

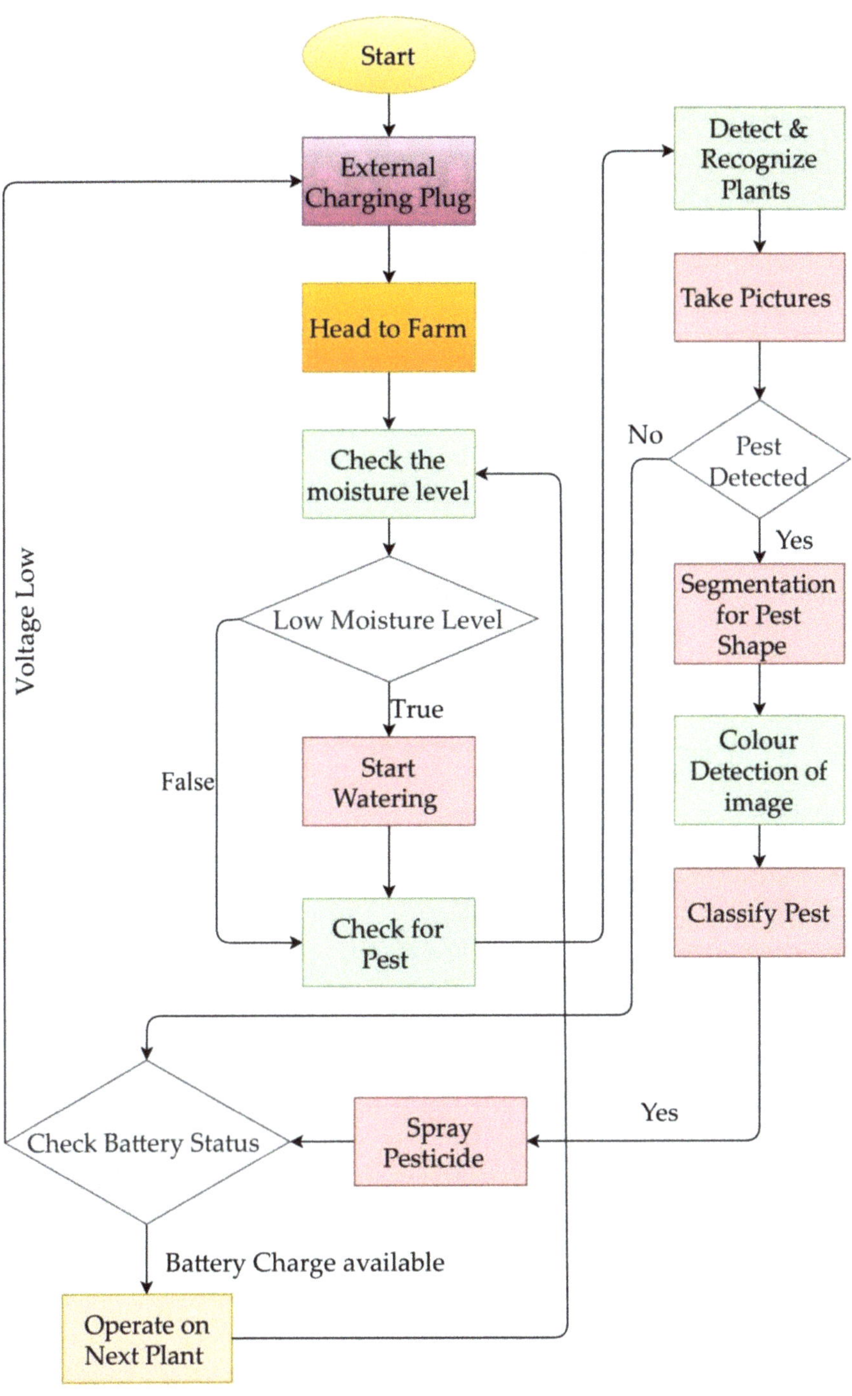

Figure 8. Flow chart showing the water sprinkling and pesticide spraying mechanism based on the soil moisture measurement and image analysis.

3. Results and Discussion

The designed MpSFR is experimentally verified by operating it in the test farm in the Papaya farm in Fiji, a Pacific Island nation. The main reason for selecting this location is that water availability in the dry season for Papaya production is one of the significant issues. As well, it is regarded as the agro-ecological constraints faced by the farmers around the Pacific Islands. However, many techniques and tools are developed and proposed to plan irrigation in the Pacific Islands. Still, due to various other reasons, their implementation has not seen a full potential. Here are some of the drawbacks and challenges: conditions of limited water resources in the wells, boreholes, etc., variability of rainfall, water stress indicators, irrigation efficiency, costs, incentives, and knowledge transfer issues. We believe the proposed concept of MpSFR could solve a few of the challenges and allow the farmers in Pacific Islands to schedule irrigation remotely. As mentioned earlier, in the Papaya farm, a scale-down test field (10 m × 8 m) whose area is approximately 80 m^2 is selected. The selected test field of the Papaya farm has 99 plants where each plant is separated by 1 m apart. In Figure 9, the pattern of mobile irrigation as per the considered Papaya test field is shown, which allowed us to run the experiment smoothly. The designed MpSFR followed the test field pattern considering the methodological process discussed in Section 2.3. The results of this experimental investigation were analyzed and discussed below.

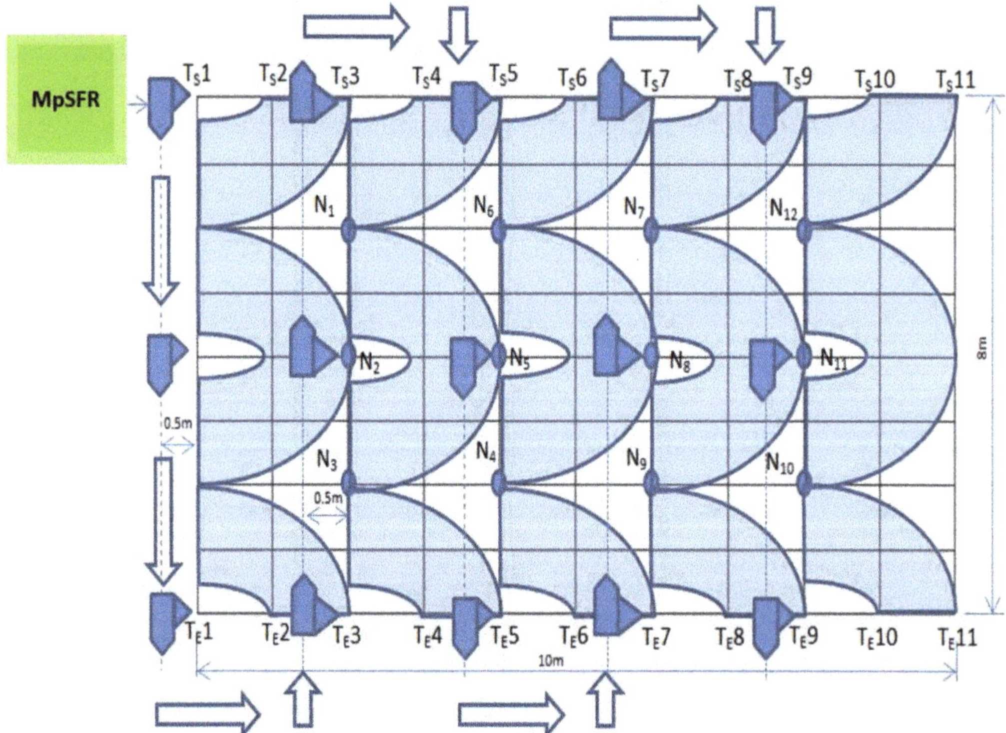

Figure 9. The pattern of smart irrigation in the considered Papaya test field (Note: T_S1, T_S2, etc. represents the starting position; T_S1, T_S2, etc. represents he end positions; N_1, N_2, etc. represents the node).

As shown in Figure 9; 11 × 2 = 22, radio-frequency identification (RFID) tags are placed on the test field along the two far end horizontal rows of the Papaya plant's location and marked as (T_S1–T_S11) & (T_E1–T_E11) for the irrigation starting/ending point for each column of the plants. Due to the no-contact and non-line-of-sight nature of this technology, RFID is very promising and economical. An MpSFR fitted with a single working RFID reader [8] can easily detect an RF tag. RFID automatically recognizes and monitors an entity using radio waves and consists of two main elements: a data-carrying unit (tag/transponder) and a reader (transceiver).

Usually, the RFID tag holds a serial number on an electronic microchip that recognizes an attached entity and attaches this microchip to an antenna that allows it to relay the identifying information to the reader that decodes the signals and transmits the information to a middleware (computer, robot control system, etc.). Middleware may be connected to other systems or networks that need more analytical information.

The MpSFR is placed close to the test field using a manual mode of operation near the RFID tag T_S1, as shown in Figure 9. Once the automatic irrigation mode is active, then the RFID reader will scan for the tag T_S1 if the scan confirms the availability of the tag T_S1, which confirms the starting point of the irrigation. After that, the controller scans the soil parameters (i.e., moisture and temperature) for node N_1 (see Figure 10 for the MpSFR measured soil moisture in the test farm). Obtained values are crossed-checked with the required crop parameters from the predefined database and check the weather forecast for the day. If the MpSFR decision is to water the plants for that particular zone (surrounding area; approximately $2 \times 2 = 4$ m^2), then the spray nozzle is oriented toward the plants and starts irrigation, keeping the flow distance at 2 m. Once the watering is finished for that particular zone (a shaded quarter circular areas as per Figure 9), the MpSFR starts following the imaginary line further (approximately 0.5 m keeping a distance from the 1st column of the plants, shown in Figure 9 as a dotted line on the left). Then the MpSFR reaches close to the N_2. Following the same procedure (having the information from N_2), MpSFR irrigates the next zone (i.e., $4 \times 2 = 8$ m^2; with a half-circular shaded area); if required, else skip the zone and reaches close to the N_3 and repeats the process. By this time, the MpSFR already arrived at the last row (end position for that particular column), confirmed by the scan data from tag T_E1. Then the MpSFR turns left and crosses a distance of 2 m; again, turn left to follow the next imaginary line. Tags T_E2 and T_E3 confirm the right path for the MpSFR. Based on the data provided by node N_4 the MpSFR either starts or skips irrigation for that particular zone and reached the end position for that respective column followed by the next column repeating the same procedure until it reaches the finish position, which confirms by the RFID tag. During this operation, the MpSFR uses an ultrasonic sensor to avoid collisions and obstacles and abort the mission once any obstacle is detected, followed by an alert message to the farm owner.

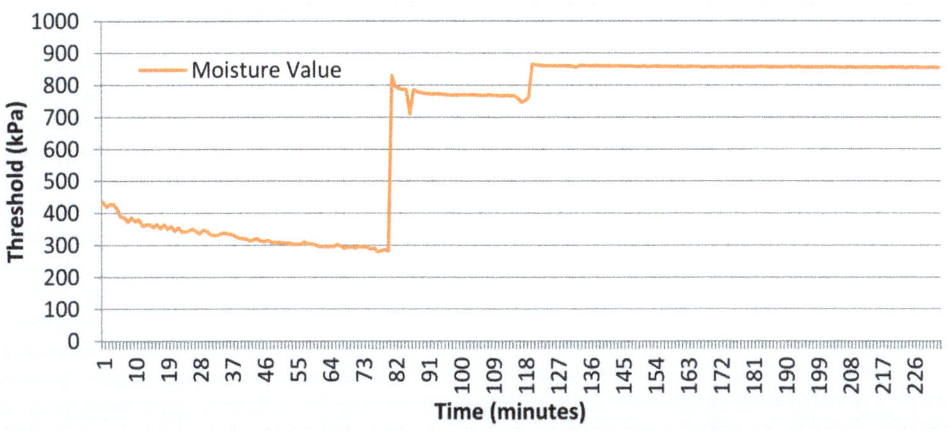

Figure 10. Measured soil moisture in the irrigation activity in Papaya test field by the sensor embedded in the MpSFR.

The water sprinkling distance range and the area are presented in Figure 11a,b. The range (Dx = 10.2 m) is the maximum horizontal distance reached by the MpSFR, given in Figure 11a. We used Equation (1) to estimate the MpSFR's water sprinkling distance. We obtained a maximum vertical angle of 45° with an initial velocity Vi of:

$$Vi = Vix \cos(45) \tag{1}$$

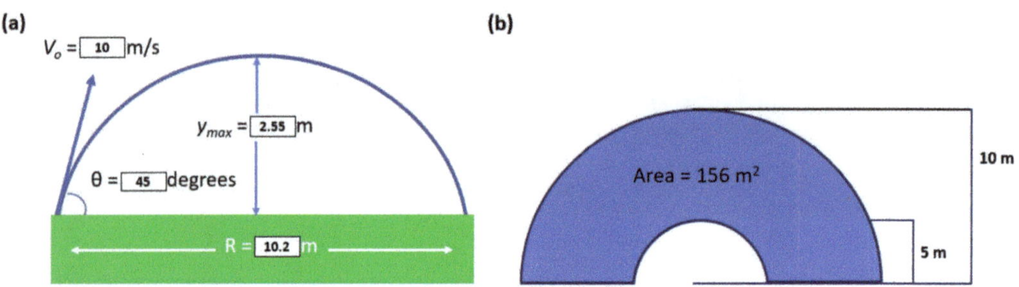

Figure 11. (**a**) Projectile motion illustration [38]; (**b**) area covered by the proposed MpSFR [38]. (Note: adopted from Author(s) own work).

The maximum area that MpSFR covers is given in Figure 11b. The field is accessible by a sprinkler with a horizontal reach of 180°, a vertical height of ±45°, a minimum horizontal reach of 0.5 m, and a maximum reach of 10 m. From Figures 10 and 11, it was observed that the proposed MpSFR has a favorable sprinkling range as well as the area covered. Based on the field testing, the soil moisture measurement results shown in Figure 10 reveal that the sensor trigger after 81 min requires water, and once the MpSFR does the watering, the soil gets wet, and irrigation stops after 800 thresholds.

Apart from the watering, the pesticide spraying results are also presented. As mentioned earlier, the MpSER was designed with a dual feature. Detecting insects by applying image processing is the smart feature that enabled this MpSFR suitable for modern farming. The pest detection experiments were conducted twice a month using the designed MpSFR, mainly due to Fiji's climate conditions. The analysis that is obtained from the image processing technique is shown in Figure 12.

From the experimental investigation in the Papaya field, it is understood that the MpSFR operates effectively by sprinkling water and spraying pesticide wherever it is needed. As well, as mentioned earlier, the MpSFR starts it working with a fully charged battery. The test run process in the Papaya fields was conducted, and the MpSFR was entirely operated on battery mode where solar panels charge the battery. Hence, this suggests looking into the battery characteristics, as shown in Figure 14.

Based on the observations from Figure 12, the leaf and the detected insects are labeled. Figure 13 provided the information on detecting 1 insect, which is done to have a detailed understanding of how the insect is detected in the field testing of the MpSFR.

The analyzed results show that the proposed MpSFR based on IoT and computer vision technologies can be considered a smart farming tool. The controller mounted on the MpSFR helps change the irrigation schedule based on the plant's watering needs rather than on a defined, predetermined schedule. The MpSFR controller automatically decreases the watering time, usually during the cooler months (these months generally need less water). On the other side, when outdoor temperatures rise or precipitation declines, the SMIS controller of the MpSFR varies the operating times or irrigation systems schedule to compensate for the fluctuation. With all the features mentioned earlier, the MpSFR automatically alters the irrigation schedules depending on site-specific factors, such as adjustments in soil type and local conditions, and irrigates/skips the corresponding plant according to the received signal.

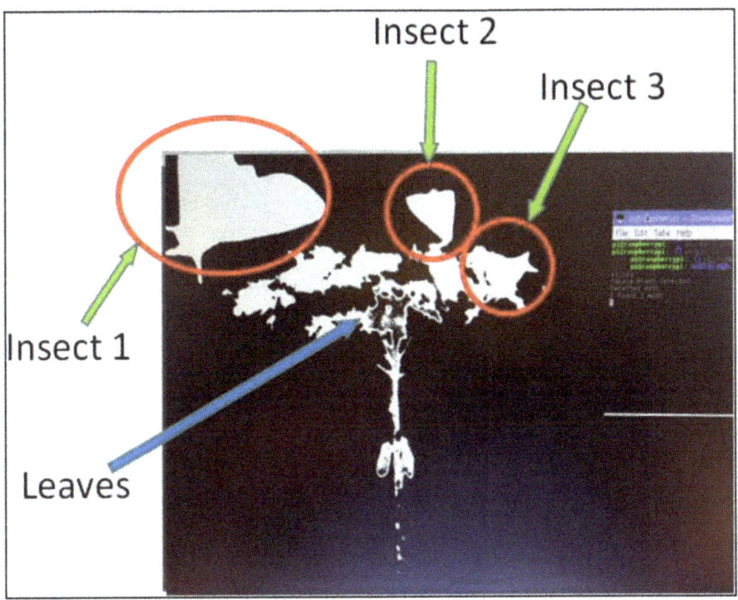

Figure 12. Pest detection while operating the MpSFR in the Papaya test field.

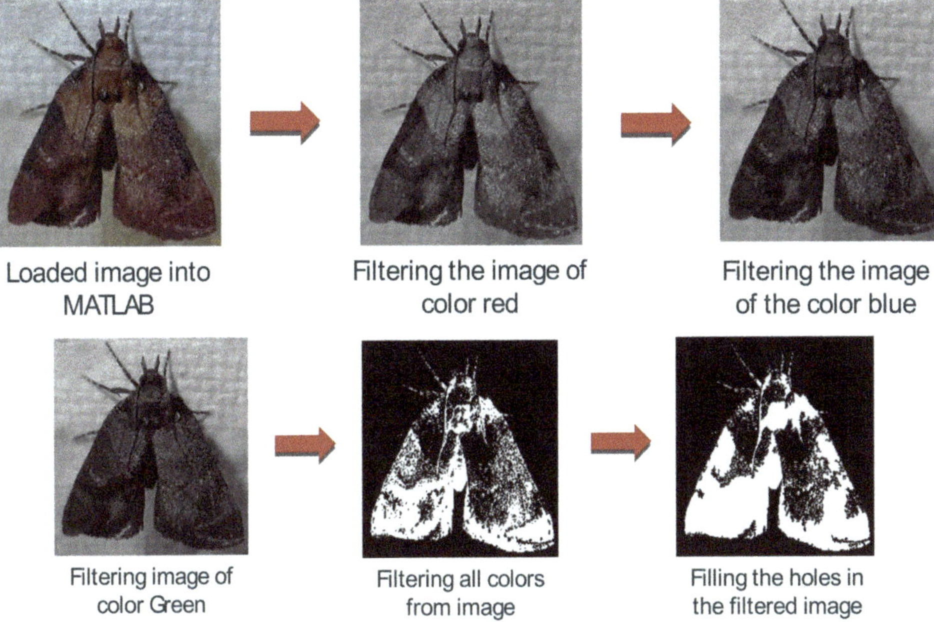

Figure 13. The process associated with detected pest with image processing.

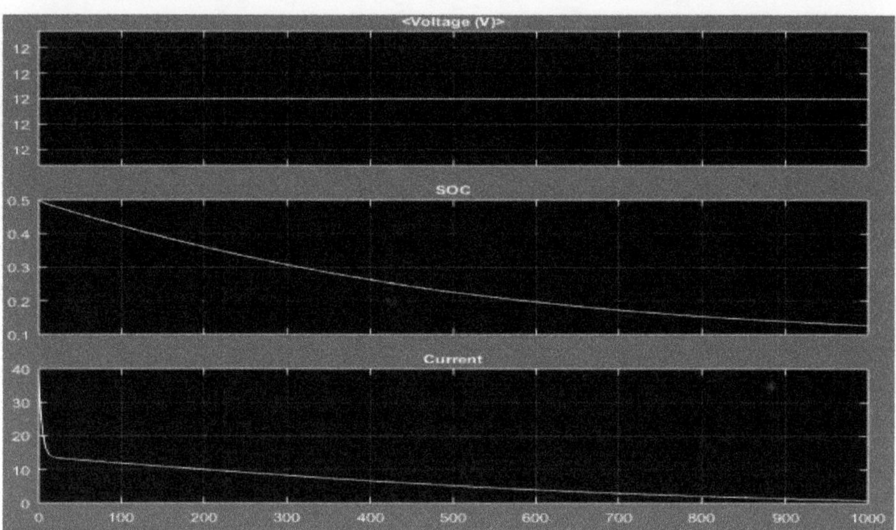

Figure 14. The discharging characteristics of the battery (Note: SOC-State of Charge).

4. Conclusions and Future Scope

A prototype of the automatic water sprinkling and pesticide spraying autonomous vehicle called multi-purpose smart farming robot (MpSFR) was designed and tested successfully for agricultural application. A robust power circuit is provided, which supplies voltages to all the components of the MpSFR without any interference, and this mainly due to the photovoltaic powered onboard battery system. The experimental investigation suggested that the MpSFR was able to sense the soil moisture and detect the pests from the Papaya test field. Based on the sensing and detection capabilities, the decisions on irrigation and pest spraying were made effectively. Overall, it is understood that the proposed MpSFR is smart, self-protective, and reliable in performing farming activities. Thus, developing farming robots like the MpSFR presented in this study would enable farmers to plan irrigation by reducing water and energy costs.

However, there exist a few research challenges that need further investigation. In the current stage, in Pacific Islands, farming is at an early stage to apply to automated decision-making and predictive solutions. The situation is similar in many developing nations. Hence there is a massive possibility for other digital technologies like information management systems and blockchain. We also expect that these digital technologies' applications need to be more robust when it comes to real-time operation. On the other side, the involvement of AI, IoT, Blockchain, etc., might not be cost-effective. Hence, understanding the techno-economic feasibility would be an important research area. Another research direction is on the system operation that should user-friendly; as well, as these systems operate in a remote environment, their operation and maintenance should be affordable.

Author Contributions: Conceptualization, A.A.C., U.M., N.M.K.; methodology, A.A.C., K.A.P., E.M., S.D., and N.M.K.; software, K.A.M., and U.M.; validation, A.A.C., and N.M.K.; formal analysis, A.A.C., K.A.P., E.M., and S.D.; investigation, A.A.C., K.A.P., E.M., and S.D.; resources, K.A.M., F.R.I., U.M., and N.M.K.; data curation, A.A.C., K.A.P., E.M., and S.D.; writing—original draft preparation, A.A.C., K.A.P., E.M., S.D., K.A.M., F.R.I., U.M., and N.M.K.; writing—review and editing, N.M.K.; visualization, A.A.C., and N.M.K.; supervision, K.A.M., U.M., and N.M.K.; project administration, K.A.M., U.M.; funding acquisition, A.A.C., K.A.M., and N.M.K. All authors have read and agreed to the published version of the manuscript.

Funding: This research project was funded by the School of Engineering and Physic (SEP) at The University of the South Pacific (USP).

Institutional Review Board Statement: Not applicable.

Informed Consent Statement: Not applicable.

Data Availability Statement: Data is contained within the article.

Acknowledgments: The research team would like to thank the School of Engineering and Physics, USP, Fiji, for offering excellent lab facilities and funding this project. We extend our gratitude to Kabir. A. Mamun and his family for allowing us to use the Papaya test field.

Conflicts of Interest: The authors declare no conflict of interest.

References

1. Eckstein, G.E. Water scarcity, conflict, and security in a climate change world: Challenges and opportunities for international law and policy. *Wis. Int. Law J.* **2009**, *27*, 409.
2. Mamun, K.A.; Islam, F.R.; Haque, R.; Khan, M.G.; Prasad, A.N.; Haqva, H.; Mudliar, R.R.; Mani, F.S. Smart Water Quality Monitoring System Design and KPIs Analysis: Case Sites of Fiji Surface Water. *Sustainability* **2019**, *11*, 7110. [CrossRef]
3. Islam, F.R.; Mamun, K.A. GIS based water quality monitoring system in pacific coastal area: A case study for Fiji. In Proceedings of the 2015 2nd Asia-Pacific World Congress on Computer Science and Engineering (APWC on CSE), Nadi, Fiji, 2–4 December 2015; pp. 1–7.
4. Prasad, A.N.; Mamun, K.A.; Islam, F.R.; Haqva, H. Smart water quality monitoring system. In Proceedings of the 2015 2nd Asia-Pacific World Congress on Computer Science and Engineering (APWC on CSE), Nadi, Fiji, 2–4 December 2015; pp. 1–6.
5. FAO. *World Agriculture: Towards 2015/2030—An FAO Perspective*; Earthscan Publications Ltd.: London, UK, 2014.
6. World Water Assessment Programme (Nations Unies). The United Nations World Water Development Report 2018 (United Nations Educational, Scientific and Cultural Organization, New York, United States). Available online: www.unwater.org/publications/world-water-development-report-2018/ (accessed on 2 November 2020).
7. Sutcliffe, C.; Knox, J.; Hess, T. Managing irrigation under pressure: How supply chain demands and environmental objectives drive imbalance in agricultural resilience to water shortages. *Agric. Water Manag.* **2020**, *243*, 106484. [CrossRef]
8. Pachapur, P.K.; Pachapur, V.L.; Brar, S.K.; Galvez, R.; Le Bihan, Y.; Surampalli, R.Y. Food Security and Sustainability. In *Sustainability: Fundamentals and Applications*; John Wiley & Sons Ltd.: Chichester, UK, 2020; pp. 357–374.
9. Namany, S.; Al-Ansari, T.; Govindan, R. Optimisation of the energy, water, and food nexus for food security scenarios. *Comput. Chem. Eng.* **2019**, *129*, 106513. [CrossRef]
10. Podder, A.K.; Al Bukhari, A.; Islam, S.; Mia, S.; Mohammed, M.A.; Kumar, N.M.; Cengiz, K.; Abdulkareem, K.H. IoT based smart agrotech system for verification of Urban farming parameters. *Microprocess. Microsyst.* **2021**, *82*, 104025. [CrossRef]
11. Ajaz, A.; Datta, S.; Stoodley, S. High Plains Aquifer–State of Affairs of Irrigated Agriculture and Role of Irrigation in the Sustainability Paradigm. *Sustainability* **2020**, *12*, 3714. [CrossRef]
12. Subathra, M.S.P.; Jegha, A.D.G.; Kumar, N.M.; Kumari, J.; Chakraborty, C.; Chand, A.A.; George, S.T. Evapotranspiration-based Autonomous Photovoltaic-battery Powered Drip Irrigation System with Fuzzy Logic Control Strategy for Energy and Water Management. *Microprocess. Microsyst.* **2021**, in press.
13. Qualls, R.J.; Scott, J.M.; DeOreo, W.B. Soil moisture sensors for urban landscape irrigation: Effectiveness and reliability. *J. Am. Water Res. Assoc.* **2001**, *37*, 547–559. [CrossRef]
14. Vasanth, A.; Grabow, G.L.; Bowman, D.; Huffman, R.L.; Miller, G.L. Evaluation of evapotranspiration-based and soilmoisture-based irrigation control in turf. In Proceedings of the World Environmental and Water Resources Congress 2008, Honolulu, HI, USA, 12–16 May 2008; pp. 1–9.
15. Parameswaran, G.; Sivaprasath, K. Arduino Based Smart Drip Irrigation System Using Internet of Things. *Int. J. Eng. Sci.* **2016**, *6*, 5518.
16. McCready, M.S.; Dukes, M.D.; Miller, G.L. Water conservation potential of smart irrigation controllers on St. Augustinegrass. *Agric. Water Manag.* **2009**, *96*, 1623–1632. [CrossRef]
17. Kumar, A.; Kamal, K.; Arshad, M.O.; Mathavan, S.; Vadamala, T. Smart irrigation using low-cost moisture sensors and XBee-based communication. In Proceedings of the Global Humanitarian Technology Conference (GHTC), San Jose, CA, USA, 10–13 October 2014.
18. Abaya, S.; De Vega, L.; Garcia, J.; Maniaul, M.; Redondo, C.A. A self-activating irrigation technology designed for a smart and futuristic farming. In Proceedings of the 2017 International Conference on Circuits, Devices and Systems (ICCDS), Chengdu, China, 5–8 September 2017; pp. 189–194.
19. Math, R.K.; Dharwadkar, N.V. A wireless sensor network based low cost and energy efficient framework for precision agriculture. In Proceedings of the 2017 International Conference on Nascent Technologies in Engineering (ICNTE), Mumbai, India, 27–28 January 2017; pp. 1–6.
20. Rajkumar, M.N.; Abinaya, S.; Kumar, V.V. Intelligent irrigation system—An IOT based approach. In Proceedings of the 2017 International Conference on Innovations in Green Energy and Healthcare Technologies (IGEHT), Coimbatore, India, 16–18 March 2017; pp. 1–5.

21. Santoshkumar; Udaykumar, R.Y. Development of WSN system for precision agriculture. In Proceedings of the 2015 International Conference on Innovations in Information, Embedded and Communication Systems (ICIIECS), Coimbatore, India, 19–20 March 2015; pp. 1–5.

22. Mesas-Carrascosa, F.; Santano, D.V.; Meroño, J.; de la Orden, M.S.; García-Ferrer, A. Open source hardware to monitor environmental parameters in precision agriculture. *Biosyst. Eng.* **2015**, *137*, 73–83. [CrossRef]

23. Balamurugan, C.; Satheesh, R. Development of Raspberry pi and IoT Based Monitoring and Controlling Devices for Agriculture. *J. Soc. Technol. Environ. Sci.* **2017**, *6*, 207–215.

24. Flores, K.O.; Butaslac, I.M.; Gonzales, J.E.M.; Dumlao, S.M.G.; Reyes, R.S.J. Precision agriculture monitoring system using wireless sensor network and Raspberry Pi local server. In Proceedings of the 2016 IEEE Region 10 Conference (TENCON), Singapore, 22–25 November 2016; pp. 3018–3021.

25. Maia, R.F.; Netto, I.; Tran, A.L.H. Precision agriculture using remote monitoring systems in Brazil. In Proceedings of the 2017 IEEE Global Humanitarian Technology Conference (GHTC), San Jose, CA, USA, 19–22 October 2017; pp. 1–6.

26. Heble, S.; Kumar, A.; Prasad, K.V.V.D.; Samirana, S.; Rajalakshmi, P.; Desai, U.B. A low power IoT network for smart agriculture. In Proceedings of the 2018 IEEE 4th World Forum on Internet of Things (WF-IoT), Singapore, 5–8 February 2018; pp. 609–614.

27. Sathish Kannan, K.; Thilagavathi, G. Online farming based on embedded systems and wireless sensor networks. In Proceedings of the 2013 International Conference on Computation of Power, Energy, Information and Communication (ICCPEIC), Chennai, India, 17–18 April 2013; pp. 71–74.

28. Kamelia, L.; Ramdhani, M.A.; Faroqi, A.; Rifadiapriyana, V. Implementation of Automation System for Humidity Monitoring and Irrigation System. *IOP Conf. Ser. Mater. Sci. Eng.* **2018**, *288*, 012092. [CrossRef]

29. Navarro-Hellín, H.; Torres-Sánchez, R.; Soto-Valles, F.; Albaladejo-Pérez, C.; López-Riquelme, J.; Domingo-Miguel, R. A wireless sensors architecture for efficient irrigation water management. *Agric. Water Manag.* **2015**, *151*, 64–74. [CrossRef]

30. Ferrandez, J.; Manuel García-Chamizo, J.; Nieto-Hidalgo, M.; Mora-Martínez, J. Precision Agriculture Design Method Using a Distributed Computing Architecture on Internet of Things Context. *Sensors* **2018**, *18*, 1731. [CrossRef] [PubMed]

31. Vaishali, S.; Suraj, S.; Vignesh, G.; Dhivya, S.; Udhayakumar, S. Mobile integrated smart irrigation management and monitoring system using IOT. In Proceedings of the 2017 International Conference on Communication and Signal Processing (ICCSP), Tamilnadu, India, 6–8 April 2017; pp. 2164–2167.

32. Pavón-Pulido, N.; López-Riquelme, J.A.; Torres, R.; Morais, R.; Pastor, J.A. New trends in precision agriculture: A novel cloud-based system for enabling data storage and agricultural task planning and automation. *Precis. Agric.* **2017**, *18*, 1038–1068. [CrossRef]

33. Capraro, F.; Tosetti, S.; Vita Serman, F. Supervisory control and data acquisition software for drip irrigation control in olive orchards: An experience in an arid region of Argentina. *Acta Hortic.* **2014**, *1057*, 423–429. [CrossRef]

34. Karimi, N.; Arabhosseini, A.; Karimi, M.; Kianmehr, M. Web-based monitoring system using Wireless Sensor Networks for traditional vineyards and grape drying buildings. *Comput. Electron. Agric.* **2018**, *144*, 269–283. [CrossRef]

35. Mat, I.; Kassim, M.R.M.; Harun, A.N. Precision agriculture applications using wireless moisture sensor network. In Proceedings of the 2015 IEEE 12th Malaysia International Conference on Communications (MICC), Kuching, Malaysia, 23–25 November 2015; pp. 18–23.

36. Jianjun, Z.; Min'gang, C.; Man, Z.; Su, L. Remote Monitoring and Automatic Navigation System for Agricultural Vehicles Based on WLAN. In Proceedings of the 4th International Conference on Wireless Communications, Networking and Mobile Computing, Dalian, China, 12–14 October 2008.

37. Lenain, R.; Thuilot, B.; Cariou, C.; Martinet, P. Mobile robot control in presence of sliding: Application to Agriculture Vehicle Path Tracking. In Proceedings of the 45th IEEE Conference in Decision and Control, San Diego, CA, USA, 13–15 December 2006; pp. 6004–6009.

38. Mehta, U.; Chand, P.; Mamun, K.A.; Kumar, S.; Chand, N.; Chand, V.; Sen, N.; Kumar, K.; Komaitai, H. Designing of a mobile irrigation system. In Proceedings of the 2015 2nd Asia-Pacific World Congress on Computer Science and Engineering (APWC on CSE), Nadi, Fiji, 2–4 December 2015; pp. 1–6.

 *agronomy*

Article

Economic Feasibility of Agrivoltaic Systems in Food-Energy Nexus Context: Modelling and a Case Study in Niger

Srijana Neupane Bhandari [1,*], Sabine Schlüter [1], Wilhelm Kuckshinrichs [2], Holger Schlör [2], Rabani Adamou [3] and Ramchandra Bhandari [1]

[1] Institute for Technology and Resources Management in the Tropics and Subtropics, TH Köln—University of Applied Sciences, Betzdorfer Strasse 2, 50679 Cologne, Germany; sabine.schlueter@th-koeln.de (S.S.); ramchandra.bhandari@th-koeln.de (R.B.)

[2] Research Centre Jülich, Institute of Energy and Climate Research—Systems Analysis and Technology Evaluation (IEK-STE), 52425 Jülich, Germany; w.kuckshinrichs@fz-juelich.de (W.K.); h.schloer@fz-juelich.de (H.S.)

[3] Faculty of Science and Techniques, Abdou Moumouni University of Niamey, Niamey BP 10662, Niger; rabadamou@wascal-ne.org

* Correspondence: srijana.neupane_bhandari@th-koeln.de; Tel.: +49-2218-2754-130

Abstract: In the literature, many studies outline the advantages of agrivoltaic (APV) systems from different viewpoints: optimized land use, productivity gain in both the energy and water sector, economic benefits, etc. A holistic analysis of an APV system is needed to understand its full advantages. For this purpose, a case study farm size of 0.15 ha has been chosen as a reference farm at a village in Niger, West Africa. Altogether four farming cases are considered. They are traditional rain-fed, irrigated with diesel-powered pumps, irrigated with solar pumps, and the APV system. The APV system is further analyzed under two scenarios: benefits to investors and combined benefits to investors and farmers. An economic feasibility analysis model is developed. Different economic indicators are used to present the results: gross margin, farm profit, benefit-cost ratio, and net present value (NPV). All the economic indicators obtained for the solar-powered irrigation system were positive, whereas all those for the diesel-powered system were negative. Additionally, the diesel system will emit annually about 4005 kg CO_2 to irrigate the chosen reference farm. The land equivalent ratio (LER) was obtained at 1.33 and 1.13 for two cases of shading-induced yield loss excluded and included, respectively.

Keywords: agrivoltaic; food-energy nexus; solar-powered irrigation; benefit-cost ratio; land equivalent ratio

Citation: Neupane Bhandari, S.; Schlüter, S.; Kuckshinrichs, W.; Schlör, H.; Adamou, R.; Bhandari, R. Economic Feasibility of Agrivoltaic Systems in Food-Energy Nexus Context: Modelling and a Case Study in Niger. *Agronomy* **2021**, *11*, 1906 . https://doi.org/10.3390/agronomy11101906

Academic Editor: Miguel-Ángel Muñoz-García

Received: 26 June 2021
Accepted: 20 September 2021
Published: 23 September 2021

Publisher's Note: MDPI stays neutral with regard to jurisdictional claims in published maps and institutional affiliations.

1. Introduction

Food and energy are two of the main needs for human wellbeing. Demand for both these resources is increasing globally, while it is expected to increase significantly in developing countries in the future. The situation is severe in countries in the Sahel region of Africa. Food security is an important issue there. Under extreme weather conditions, i.e., drought, food production can be secured and increased with energy-intensive methods, such as pressurized irrigation. Niger is one of such countries where energy and food security issues are predominant. The country has a high solar energy potential, which so far is little exploited. There are many challenges to provide enough food and energy for a growing population in the country. This will be further complicated due to predicted climate change impacts in the agricultural sector.

The use of an agrivoltaic (APV) approach could be an important contribution to overcome some of these challenges. Solar energy can produce electricity, which is needed for various process steps of farming (e.g., water pumping for irrigation, postharvest cold storage, etc.). According to the Fraunhofer Institute of Solar Energy Systems (ISE) [1], "the APV is a system technology that enables the simultaneous main agricultural production and secondary solar power generation on the same area and which seeks to optimally use the synergy effects and potentials of both production systems". This concept dates back to 1981 at Fraunhofer ISE in Freiburg [2]. There are concerns that the shadow of the solar system may hinder plant growth. To avoid these problems, the right choice of crops is necessary. For this purpose, the country's relevant food crops must be categorized with respect to shade tolerance. After this, an appropriate system configuration needs to be chosen, e.g., one of the following three [3]: (i) on the ground: crops (cereals, vegetables, fruits, etc.); (ii) above the crops: solar panels mounted on pillars; and (iii) process: Solar panels will get a part of the sunlight and leave the remaining light for the crop's growth.

The main objective of this study is to carry out an economic feasibility study of an APV system at a selected site in Niger (Dar Es Salam village in the Dosso region) and to present the quantitative results for different economic indicators. To compare the results, a reference farm size is used and different farming practices are analyzed: with rain-fed irrigation and diesel/solar only/APV powered irrigation. At the same time, socio-technical aspects (social-behavioral, rural development) are considered to interpret the economic results so that the stakeholders can make an investment or land use decision in a holistic way.

The study has some limitations too. It has been based on theoretical assumption, without having any experimental APV or irrigation system on the chosen site. For economic and farming calculations, multiple assumptions are made for the input parameters, both at cost and revenue sides. Many such data are based on farmer's experiences at the site, collected during the field survey. The results need to be verified with field experiments first before multiple such projects are implemented in similar villages.

Although the comparison of different farming systems might be seen by some experts as incompatible, the ultimate benefit of such APV projects should be seen as multi-sectoral. Such benefits (profits) should not be viewed only in the energy sector (investors or operators). The results of this study could be useful to support the decision-making of the farming communities on whether or not to take part in such APV projects.

This paper has been organized as follows. The next section describes the case study frame, with a brief literature review on APV systems and the country profile of Niger. The following sections present the methods used for feasibility analysis and the model. Afterward, results, discussions, and conclusions are presented.

2. Study Frame

2.1. Agrivoltaic (APV) Systems

As mentioned earlier, APV allows the simultaneous harvest of agriculture products and solar electricity on the same farm area by optimizing both production systems [1]. Brohm and Khanh [3] defined APV as the dual use of the same land area for solar electricity and agricultural production, including aquaculture. Others define APV as an emerging approach of harvesting energy and food together in a given land area to maximize land productivity with additional benefits including improved crop yield and socio-economic welfare of farmers [4–6]. Hernandez and colleagues confirmed that APV allows both agricultural production (food or energy crops) and solar energy generation within the same land area [7]. There are other similar definitions of APV systems in the publications from many other authors [8–13], etc. In a broader sense, all these definitions are similar, as they refer to the dual use of the same land area at the same time. Also, there are different terminologies evolved to describe this approach over the period. These include solar dual-use, agro-photovoltaics, agri-photovoltaics, agrivoltaic, etc. This term agrivoltaic (and in short APV) is used in this paper to represent this concept.

Several studies have already analyzed the performance of APV systems. Different aspects such as crop types, APV system configuration, irrigation water, etc. are common discussion points in most of these studies [14]. Vyas [15] discussed several advantages of the APV systems. Among others, the author highlighted the positive effects of shade beneath the solar modules on fresh salads and vegetables in desert climates. Sekiyama and Nagashima [13] aimed to identify the right solar photovoltaic (PV) system that could ease the stress between food and energy production in the same land area. They presented results by taking a case of an APV system in the cornfield. Santra et al. [16] integrated the rainwater harvesting channels with solar modules in their APV system. The collected water was used for module cleaning as well as for supplemental irrigation. Dupraz et al. [10] published a study in France comparing the results of separated energy and agri production vs. combined one. A comparison of land use for solar electricity (for electric mobility) vs. biofuel (as vehicle fuel) was also made.

Weselek et al. [6] carried out a systematic review to analyze the applications, challenges, and opportunities of APV systems. This was one of the most extensive reviews done in the field including results from a wide range of projects in many countries. The authors also summarized the effects of shading on different crops reported by several pieces of literature.

Some other findings on productivity and benefits were: the crops' growth rates underneath the PV modules were less, land productivity can increase by up to 70% when energy and crops are planted in the same field and APV systems enhance the farmer's economic benefits [9]. Dual benefits to the farmers are also reported at grape farms in India by Malu et al. [17]. Valle et al. [18] analyzed the effect of fixed type vs. tracking type PV systems on energy and lettuce production. An interesting finding was that tracked systems had not only increased electricity production, but also helped crop biomass growth due to slightly higher transmitted radiation to the crops (compared to the fixed types).

Further studies reported the land productivity increase with the dual use of crop and PV on the same land. These effects are expected further positive in semi-arid or arid areas. One of the reasons is the decreased water loss from the topsoil [19]. Decreased soil evaporation reduces the loss in agri-production [20]. Furthermore, the shade of PV modules during high solar irradiance hours could be beneficial for selected crops planted in these regions [21]. When it comes to the height of APV mounting structures, different heights are used depending upon the need for machinery used for farming underneath the PV modules. Varying examples of height for specific real or hypothetical projects are reported in the literature, e.g., 3 m in Japan [22] or 5 m in Spain [23].

APV's joint benefits for electrification and irrigation are presented by Ibrik [24]. By taking the arid region case at two rural communities in Palestine, the author analyzed how the diesel (as a fuel) dependency could be minimized by using such APV systems as extended solar microgrids. Pascaris et al. [25] analyzed the benefits and challenges that farmers perceive with the use of APV systems. Unlike the mainstream studies on APV, which are mostly project or pilot plant-based analyses, this study highlighted the generic barriers in the agriculture sector. The authors suggested the relevant stakeholders address these barriers for the successful and widespread implementation of APV systems. Moreda et al. [23] analyzed the profitability of a hypothetical APV system deployed on irrigated arable lands of southwestern Spain. The authors used the internal rate of return (IRR) as an economic indicator to check the profitability for different scenarios of crop types and solar modules orientation (south vs. southeast). Calculated IRR values were between 3.8% and 5.6%.

In terms of productivity, APV's benefits include crops protection from radiation stress, lower water demand due to reduced evaporation, higher crop yields in case of shade-tolerant crops, etc. [26]. This can contribute to enhancing economic development and food security specifically in rural areas, offering promising options for food production, water savings, and renewable energy production at the same time and location [27]. They can provide access to electricity, where this is not the case previously [7]. It helps to create new jobs in the farming and energy sector and to improve the socio-economic standard of living for people, especially farmers [28]. Generating solar electricity supports CO_2 emission reduction, thus it helps to meet the national climate change targets [3]. Water can be harvested and controlled for crop irrigation and other purposes by systematically using the APV infrastructure: PV arrays can act as irrigation run-off channels, which can drain the water directly on the crops [5]. The water used for cleaning the dust on solar panels can be used for irrigation of vegetation cultivated under the panels [29]. Having vegetation underneath and around solar panels can reduce the levels of dust and soiling on the panels [30].

Although the APV has significant advantages, there are some challenges too. There is a need for a more comprehensive understanding of technical and economic aspects together with the agricultural issues. Furthermore, factors affecting societal acceptance of these new applications need to be studied. From the project implementation perspective, APV systems are challenging because they strongly interact with agriculture, the local economy, and the stakeholders on-site [31]. The APV systems are costlier than ground-mounted PV systems. At the policy level, the issue of PV electricity sales to local consumers, integration to the grid, etc. are not addressed in many countries yet, unlike for PV-only solar power systems. Further, there could be practical obstacles (and additional costs) on the farming side. As these systems are relatively new, studies on more such projects at different scales (smallholder to utility-scale) are needed before answers can be found for these open questions.

The choice of suitable crops for APV applications is crucial, but it is also a challenge to assess in the African context without any best practice pilot projects available. As reported in different literature, most assessments on shading-tolerance and dual-use suitability have been made in Germany, France, Italy, China, and the USA. These projects give only an indication for their application in the context of Niger, because the climate is different except in part of the USA (Arizona). A fundamental effect of the shading caused by the PV modules is reduced water evaporation from the soil. Obviously, this has a positive effect, particularly in water-scarce arid regions, but it hinders the photosynthesis process. The following categorization (Figure 1) of suitable crops for APV has been derived by compiling the literature information.

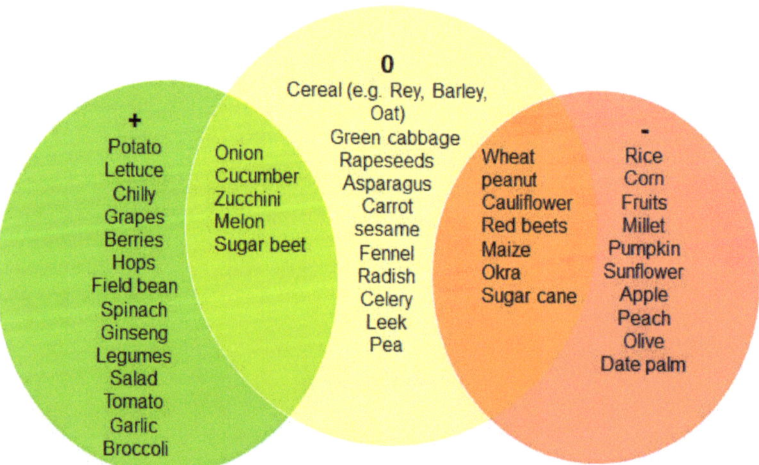

Figure 1. Suitable crops for solar dual-use (compiled from [1,3,10]).

Crops that perform better with shading are put under the "+" category (green), whereas the shade-intolerant crops are under "−" (red). The crops under the "0" category are indifferent to shade when the overall crop yield is considered. Although this categorization gives a first overview of crops selection, it might not apply to all the locations around the globe. Field experiments are necessary to verify this.

2.2. Niger—Country Profile

Niger is a landlocked country located in West Africa in the Sahel Zone. The country's terrain has predominately desert plains and sand dunes. The country is exposed to a mostly hot and dry desert climate with frequent sand storms. Niger is prone to natural hazards, mainly recurring droughts [32]. According to 2011 data, out of the country's surface area of about 1.27 million km², about 35.1% of the land was used for agriculture (12.3% arable land, 22.7% pasture, and 0.1% permanent crops). Forest covers only 1% of the total land. The remaining 63.9% is desert and built environment. In 2012, irrigation was available to only about 1000 km² farmland [32].

According to 2019 estimates, the country has about 23 million population [33]. An estimated 17% of people live in urban areas, while the remaining majority live in rural areas. The country's average annual population growth rate at 3.66% is one of the highest in the world [32].

In 2017, about 65.2% of the country's population had access to safe drinking water. This share was high in urban areas at 95.7% and low in rural areas at 59.2%. The remaining 34.8% of people were exposed to unimproved drinking water with about 4.3% in urban areas and 40.8% in rural areas [32]. The available total annual water resource (surface water and groundwater) of Niger was reported at about 34.05 billion m³ in 2017, out of which, only 1.751 billion m³ was total annual consumption in the same year [34]. Therefore, it is possible that groundwater could be extracted for irrigation in the country by implementing APVs.

Due to large family sizes, the farming land is divided into smaller plots. Children inherit it from their parents. Over the generations, plots sizes are reduced. These small landholdings limit mechanization and thereby hinder food production growth. Common agricultural products grown in the country are cowpea, cotton, peanut, millet, sorghum, cassava, and rice. The subsistence-based agricultural economy is frequently disrupted by droughts, which are common to the Sahel region of Africa. The monthly average precipitation for the capital city of Niamey is given in Figure 2. The rainy season is also the main agricultural season in the country. These climate extremes have a direct impact on the livelihood of local people, including the food and water supply.

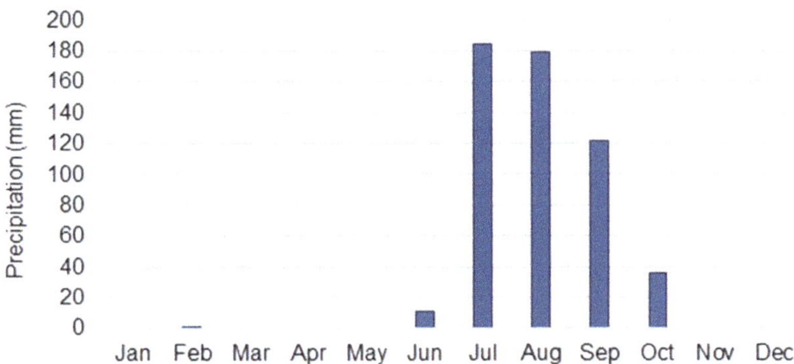

Figure 2. Monthly average precipitation in Niamey (data from [35]).

Due to its geographical location and hot-arid climate, there are not enough rivers or rivulets to divert water for agriculture. River Niger is the only available river in the country, even this river can be dry in some sections in the dry season (e.g., May). Solar electricity-driven groundwater pumping systems could be more reliable irrigation options.

More than 80% of the country's population is dependent on agriculture. Agri sector contributes to over 40% of the gross domestic product (GDP). Productivity of the agricultural sector is relatively low. A country's agriculture relies heavily on climatic conditions. When rainfall is less, the country suffers from additional food shortages. Even in normal rain conditions, the country relies on imports to meet national food demand [35]. The Sahel Sahara zone (north of Niger) is popular with livestock farming due to available land for pasture. Rain-fed crops and crop-livestock farming systems dominate the southern belt. The main crops there are dryland cereals (millets and sorghum) and legumes like cowpea and peanut. Millet is the main staple crop in Niger covering 65% of the cultivated land. The most productive (and irrigated) land is located along the Niger River [35], where another important cereal, rice, is produced.

Availability of more irrigation water would increase agricultural production. Solar-powered water pumps could be one of the solutions to support Niger's agricultural sector. However, business models and financing options are not yet established to offer this solution on a commercial scale [36].

Besides the cereals mentioned above, people in the areas close to the Niger River also produce many vegetables as cash crops. The study [37], in the suburb of Niamey, identified the production of many vegetables and their frequency of farming shown in Figure 3. This information, together with the field survey at the case study site, has been used to select the top five cash crops for the economic analysis in this study.

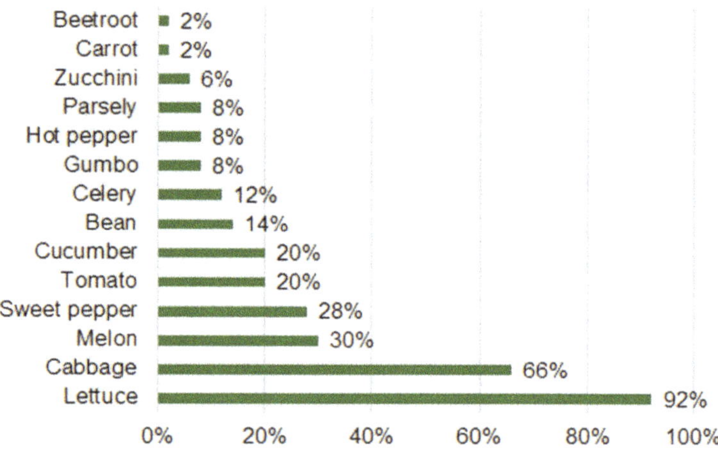

Figure 3. An example of different crops production in the suburb of Niamey (Data from [37]).

The national electrification rate is quite low, almost three out of four people do not have access to electricity [38]. The situation in rural areas is worse [39]. Currently, more than half of Niger's electricity supply is imported from Nigeria [40]. Even in areas with access to electricity, including the capital city Niamey, the supply is not reliable.

Indeed, off-grid solar systems could meet the energy demand in rural areas. Solar power could be combined with water pumping to meet the water demand for drinking and irrigation purposes. This approach of combining solar energy use with food production and water supply can be addressed by using the APV system, the focus of this study.

3. Materials and Methods

This section has been divided into two parts: first, the data collection steps are described followed by economic indicators and assumptions under considered scenarios.

3.1. Data Collection

Data collection was carried out by literature analysis combined with a field survey with a set of structured questionnaires. One part of the survey questionnaire was intended for local farmers, while the other part was for experts. Local university members in Niamey distributed the questionnaires among the target stakeholders in Niger (village farmers, academics, and students). A reference farming village was selected in Dar Es Salam in the Dosso Region of the country (about 75 km Southeast of Niamey). The survey was conducted in October 2020 and 124 households were visited.

At the farming side, the data include local common crop types, cropping seasons, land preparation practices, reference harvest quantity in reference farm size, agro-processing activities, postharvest management, local market for the products (or self-consumption), market price both for selling and purchasing of agri-products, costs for farming activities, etc. On the solar PV side, data include the cost of the solar plant (capital cost for installation, operation), electricity production, and selling prices (surplus selling and self-consumption). The climate data (solar radiation, precipitation, temperature, etc.) were taken from a meteorological/satellite database (Meteonorm V.8).

To analyze the data in line with the paper's objectives, a model was developed to evaluate the economic feasibility of a reference size project for both solar plant operators and farmers.

3.2. Crop Yield and Energy Production Combined

To increase land-use efficiency, energy and food production can be combined on the same land area. This can be measured by using the concept of land equivalent ratio (LER) [3]. The LER of an APV system can be defined as in Equation (1):

$$LER = \left(\frac{Crop\ yield\ in\ dual\ use}{Mono\ crop\ yield} \right) + \left(\frac{Electricity\ yield\ in\ dual\ use}{PV\ electricity\ yield} \right) \tag{1}$$

If LER is >1, APV system is more effective than producing only crop or only energy. Mixed cropping systems have usually LERs between 1.0 and 1.3 [10].

3.3. Indicators for Farming Side

The indicators used in economic modeling for the farming side are described below in Section 3.3, which are adapted from [41].

3.3.1. Gross Margin

The gross margin of farming is obtained by subtracting the variable costs from the value of production (i.e., revenue) by using Equation (2).

$$Gross\ margin = Value\ of\ production - Variable\ costs \tag{2}$$

Value of production is calculated by multiplying the produced quantity and price. Cost of production refers to the expenses involved in the production of crops. It is divided into two: variable and fixed costs. Fixed costs remain constant as production changes. These costs include the costs of farm equipment and water pumps/irrigation systems. Smallholder farmers have very small fixed costs, e.g., the cost for small agri-tools. Land purchase or leasing costs would also fall under this category, but such costs are not considered in the analysis, assuming farmers own the land. Variable costs vary when the production changes as this leads to a change in the amount of inputs used (and also the yield harvested).

3.3.2. Farm Profit

Farm profit is the money left after variable and fixed costs are paid back from the value of production as shown in Equation (3). This is an important indicator for farmers.

$$Farm\ profit = Value\ of\ production - Total\ costs\ (fixed + variable) \tag{3}$$

3.3.3. Benefit-Cost Ratio (B/C Ratio)

Benefit cost ratio is calculated by dividing the total income by total cost, as shown in Equation (4).

$$Benefit\ cost\ ratio = \frac{Value\ of\ production}{Total\ cost} \tag{4}$$

3.4. Economic Indicators for APV Side

For the APV side, further indicators are needed for the economic analysis because the cash flows occur over several years, and discounted cash flow analysis is necessary. The electricity generated from the PV system needs to be calculated by considering solar radiation data on-site and other parameters that affect the electricity yield (such as degradation rate).

Net present value is calculated by using Equation (5).

$$NPV = \sum_{n=0}^{N} \frac{R_n - C_n}{(1+d)^n} \tag{5}$$

where,

- d   Discount rate (%)
- NPV   Net present value (€)
- R   Revenue (€)
- C   Costs (€)
- n Numbers of years (0..to..N)

The present value of the total cost (C_{pv}) can be calculated by using Equation (6).

$$Cpv = C_{I,\,n=0} + \sum_{n=0}^{N} \frac{C_{OM}}{(1+d)^n} \tag{6}$$

where,

- C_I Total initial costs for PV and APV structure in year 0 (€)
- C_{OM} Annual operation and maintenance (OM) cost (€)
- N   Project lifetime, years

The annual energy yield from the PV system (E_n) can be calculated using Equation (7).

$$E_n = P_{peak}\, Q\, \frac{G}{I_{STC}} \left(1 - d_g\right)^n \tag{7}$$

where,

- d_g Energy yield degradation rate (0.5% per year)
- P_{peak} Installed solar capacity at APV plant (50 kW)
- Q   Performance ratio (value of only 60% is considered, decentralized grid)
- I_{STC} Constant value (solar irradiance at standard test condition, i.e., 1 kW/m^2)
- G   Solar radiation falling on solar panels at the project site (kWh/m^2.year)

Total electricity generated in all years (E_{total}) can be calculated by using Equation (8):

$$E_{total} = \sum_{n=1}^{N} E_n \tag{8}$$

The levelized cost of energy (LCOE) of a solar PV plant is the ratio between the total costs of the plant (€) and total electricity production (kWh), both over the economic lifetime of the project, as shown in Equation (9). It allows comparison of cost of electricity from different sources.

$$LCOE = \frac{C_{PV}}{\sum_{n=0}^{N} \frac{E_n}{(1+d)^n}} \tag{9}$$

The levelized cost of water (LCOW) can be also calculated in a similar manner, by dividing the present value of water pumping costs with total irrigation water demand throughout the project's economic lifetime.

3.5. Diesel Pumping System

With the help of literature, a pump size needed for the irrigation of reference farm size (0.15 ha land area) is calculated. The pumping system power requirement can be obtained by using a simplified Equation (10) [42]:

$$P = \rho g Q T_{dh} \eta \tag{10}$$

where,

- P   Hydraulic power required for pumping (W)
- ρ Density of water (1000 kg/m^3)
- g   Acceleration due to gravity (9.81 m/s^2)
- Q   Flow rate of the pump (0.001 m^3/s, assumed to be operated for about 3 h a day, available water for irrigation in a day will be about 7.2 L/m^2)
- T_{dh} Total dynamic head (m)
- η Efficiency of the pump

Considering the pumping height about 70 m and the pumping efficiency of 60%, the pumping power required (for 0.15 ha farming area) has been calculated at 1.15 kW.

The chosen pump size is 1.5 kW to match the next available market size. To operate this pump, about 2 L diesel is needed in one hour. Considering no or less irrigation water requirement during rainy days, between crop harvest and new seeding, etc., the pump will be operated only about 250 days in a year. Considering that the water will be pumped out from the existing well, no drilling costs are considered even if sometimes the drilling costs can be very high in a new location where the groundwater level is very low. The same assumption will be considered for the case of the solar power irrigation system, so that the results are comparable.

3.6. Field Survey and Cases under Analysis

The field survey confirmed that the case study village of Dar Es Salam has significant problems with water and energy supply. This has a direct impact also on agriculture and livestock production. Thus, one of the important issues, the irrigation water supply, have been analyzed to compare the farm production without and with irrigation systems. Two irrigation techniques-diesel and solar energy driven pumps have been considered for comparison.

Altogether four cases are considered. The first three are to analyze the benefits to farmers. The last one is to analyze benefits to PV investors as well as combined benefits to farmers and investors.

3.6.1. Case 1: Traditional Farming, Rain Fed Irrigation

In this case, the business as usual farming practice has been considered. Farmers grow their crops only during the rainy season. They continue farming the usual major cereals, i.e., millet and sorghum. They grow cowpea as intercrop and partly also peanut. Considering the total cost of farming and the total value of production of those produces, economic indicators are calculated and the results are presented. The assumptions considered for the economic analysis of this case are shown in Table 1.

Table 1. Basic assumptions for cases 1–3 (traditional farming, diesel, and solar PV).

Parameters	Case 1, Traditional	Case 2, Diesel Powered Irrigation	Case 3, Solar PV Irrigation
Land area	0.15 ha	0.15 ha	0.15 ha
Irrigation type	Rainfed	Diesel pumps for groundwater pumping from the well	Solar PV powered pumps for groundwater pumping from the well
Farming method	Traditional, human labor and animals use, 1 crop cycle per year	Cash crops (green vegetables), 4 crop cycles per year	Cash crops (green vegetables), 4 crop cycles per year
Main produces considered	Sorghum, millet, cowpea, and peanut	Salad, cabbage, tomato, mint, and okra	Salad, cabbage, tomato, mint, and okra
Other items not considered in revenue	-	Possible electricity generation and sale using the same diesel engine. Theoretically, it will be possible to schedule water pumping during the daytime and electricity generation in the evening.	-
Cost parameters	Farming land preparation, seeds, planting, fertilizers, harvest, etc.	Farming land preparation, seeds, planting, fertilizers, harvest, etc. Additionally, investment for diesel pump (depreciation of the first year), operating costs due to diesel consumption per year.	Farming land preparation, seeds, planting, fertilizers, harvest, etc. Additionally, investment for solar pump (depreciation of the first year), a small operating/maintenance (O/M) costs per year.
Cooperative or community loan share	50% of total cost	50% of the total cost	50% of the total cost
Loan interest rate	10% for a period of 8 months	10% for a period of 8 months	10% for a period of 8 months
Revenue	Self-consumption (market value set as pseudo selling price)	Selling the harvested cash crops to the market (e.g., in Kodo) or sales to local traders who re-sale the produce to the city of Niamey	Selling the harvested cash crops to the market (e.g., in Kodo) or sales to local traders who re-sale the produce to the city of Niamey
Target economic indicator	Gross margin, farm profit, and B/C ratio	Gross margin, farm profit, and B/C ratio	Gross margin, farm profit, and B/C ratio
Environmental indicator	-	CO_2 emission in a year from the diesel pump	-

3.6.2. Case 2: Irrigated Farming, Diesel Pump

In this scenario, irrigation is considered via a diesel pump by pumping water from a deep well. Due to water availability, in this case, farming of only high-value crops (cash crops such as salad, green vegetables) are considered. No cereals of case 1 are produced in this scenario. It is assumed that the farmers can buy these staple cereals with the money they would earn by selling the new cash crops. This case also considers the same land area of 0.15 ha for agricultural production. The following assumptions (Table 1) are considered for the calculations.

3.6.3. Case 3: Irrigated Farming, Solar Pump

In this case, crop types and cropping cycle assumptions are similar to those in case 2 above (irrigated farming, diesel pump). The only difference is that the diesel pump is replaced with a solar electric pump for irrigation water supply. To pump the same amount of water as in the diesel pumping case, a 2 kW solar PV system is considered (which is slightly higher than the diesel motor because of the intermittent nature of solar radiation). Solar PV electricity is used exclusively for water pumping for irrigation. The following assumptions (Table 1) are considered for the calculations.

3.6.4. Case 4: APV for Irrigation and Electricity Supply

In this case, an APV system is considered in the same reference farm size of 0.15 ha. In a land area of 0.15 ha, if all space would be covered with solar panels, a maximum of about 1500 m^2 solar panel surface area could be installed. However, literature suggests that only one third of the space could be covered with the APV's solar panels in order to optimize energy and food production together [6]. Also, it is necessary to have a gap between the rows of solar panels in order to allow rainwater to the field in rainy season. Therefore, in this case, only 500 m^2 area is assumed as potential solar panels area. Considering the commercially available solar panels in the market, about 10 m^2 area will be required to install 1 kW PV system. Assuming same values for Niger, about 50 kW solar PV system could be installed in 0.15 ha land. Such a system is much bigger than the one needed only for irrigation water pumping as mentioned earlier (case 3). This means, if the APV system is installed in the reference farm, only a small amount of energy is utilized in pumping, and a big share remains available for other use. This electricity could be supplied to the village by developing necessary distribution network. The alternative would also be to sell it directly to the national grid, where the required regulations allow this.

The crops and farming related assumptions made here are similar to the case 2 and 3 before. Additionally, a crop yield reduction due to solar shading is considered, assuming yield only at 80% compared to no shading case. On the revenue side, also the electricity sales have been considered. As a financial indicator for the investor, NPV results are presented.

4. Results and Discussions

4.1. Cost Calculation

The cost distribution of farming in all three cases is presented in Table 2. For easy comparison, the values in Table 2 refer to 1 ha farmland (and not for 0.15 ha case study plot).

In case 1, although the family members work in the field and there are mostly no cash payments involved, a small monetary value (of 91.50 €/ha) is assumed for part of the additional human labor that might be needed in different farming activities. For the manual agri-tools, a linearly depreciated annual value of 15.20 €/has been used. Not all farmers use inorganic fertilizer or pesticides. However, there is an increasing trend in their use. Only the main crops are considered as the farming produces and only one crop cycle of production is assumed in a year.

In case 2, for the diesel pump of 1.5 kW (required for 0.15 ha area) with a lifetime of 7 years, the initial cost of 1500 € has been considered. Annual depreciation for this machine and agri-tools are considered as fixed costs. When cash crops are planted, the overall need for labor and other farm inputs such as seeds, fertilizer, etc. is higher. Based on the farming costs in similar plots in the suburb of Niamey and considering further assumptions made above for case 1, the individual cost category is quantified. Unlike in case 1, the cost for irrigation needs to be considered here. This is calculated by using diesel consumption in a year and diesel price at a rate of 0.76 €/L.

In case 3, the initial cost of the PV pumping system is considered at 5000 € (for 2 kW system needed for 0.15 ha land). Here too, linear depreciation has been applied. Systems lifetime for PV has been considered at 25 years. The annuity value of the fixed costs for pumping was about 1428 €/ha and 1333 €/ha for diesel and PV irrigated systems respectively. Although the initial cost for a PV irrigation system is higher than a diesel system, the operating costs for the latter are far too high. Annual irrigation operating costs were about 7622 €/ha for diesel vs. 667 €/ha for PV. Diesel price is the main factor for high value.

Table 2. Farming cost distribution, cases 1–3 [14,43].

	Case 1, Traditional		Case 2, Diesel Powered Irrigation		Case 3, Solar PV Irrigation	
	CFA/ha-Year	(€/ha-Year)	CFA/ha-Year	(€/ha-Year)	CFA/ha-Year	(€/ha-Year)
Variable costs						
Human labor for different activities	60,000	91.50	1,000,000	1524.50	1,000,000	1524.50
Machinery use (hand tractor, plowing)	20,000	30.50	200,000	304.90	200,000	304.90
Seed	4000	6.10	600,000	914.70	600,000	914.70
Fertilizer	9000	13.70	700,000	1067.10	700,000	1067.10
Pesticides	1000	1.50	100,000	152.40	100,000	152.40
Miscellaneous (packaging, etc.)	3000	4.60	200,000	304.90	200,000	304.90
Irrigation use	0	0.00	5,000,000	7,622.40	437,307	666.67
Subtotal variable costs	**97,000**	**147.90**	**7,800,000**	**11,891.0**	**3,237,307**	**4935.2**
Fixed costs						
Depreciation of hand-made agro tools (for human labor)	10,000	15.20	10,000	15.20	10,000	15.20
Depreciation of diesel or solar-powered pump	0	0	937,086	1428.57	874,613.33	1333.33
Subtotal fixed costs	**10,000**	**15.20**	**947,086**	**1443.80**	**884,613**	**1348.60**
Financial costs	3567	5.00	291,570	444.00	137,397	209.46
Total costs	**110,567**	**169.00**	**9,038,655**	**13,779.00**	**4,259,317**	**6493.26**

4.2. Revenue Calculation

For case 1, the average production quantity and selling prices of these main crops are given in Table 3, as compiled by [43]. There might be few other intercrop vegetables planted parallel in few cases, this has been neglected in the revenue calculation. In all cases, land cost (e.g., rent or purchase) is not considered, because the assumption is that the farmers own small farmland of about 0.15 ha.

Table 3. Production quantity and selling price of main crops in the suburbs of Niamey ([43]).

Crop Types	Average Production (kg/ha)	Average Selling Price (€/kg)
Millet	449	0.38
Sorghum	305	0.35
Cowpea	186	0.40
Peanut	414	0.37

In the reference farm, it is assumed that sorghum, millet, and peanut can be planted each in one-third of the land. Cowpea is intercrop with millet and sorghum, this is why only two-third of the land is considered for its production. It should be mentioned that farmers do not earn the revenue in cash, rather they consume the production themselves, and thus there is no need of purchasing these cereals and peanuts from the markets. These crops are given the monetary value (€/kg) which is applicable in the village (few big farmers sell those produces and their selling price is taken as a reference).

In cases 2 and 3, with irrigation, different salads and green vegetables can be harvested multiple times a year (between 4 and 10 times). To stay on the safe side, only four harvests per year are assumed on average in the calculation. From 0.15 ha land area, about 60 plots (5 m × 5 m) are typically made. Here, about 20 plots are allocated for salad, and 10 plots are allocated for cabbage, tomato, mint, and okra each. The produce will be sold at the market in Kodo or even in Niamey through local traders. In the suburbs of Niamey, there is salad farming, where irrigation water is pumped from the Niger River. There is a common practice of dividing the land area into small plots with an area of about 25 m^2. Wholesalers or retailers buy the product there, mostly not based on harvest quantity, but based on the plot itself. Traders buy the plot for a crop cycle and instruct farmers when and how much of the crop is to be harvested. Selling price differs from season to season and from crop to crop. They vary quite a lot even from marketplace to marketplace. Some examples of the price variations around Niamey for 1 plot of land (about 25 m^2) are given below [43] (1 € = 655.96 CFA):

- Salad: 3000 to 15000 CFA/plot (3.81 to 22.87 €/plot)
- Cabbage: 2500 to 3500 CFA/plot (3.81 to 5.34 €/plot)
- Tomato: 3000 to 6000 CFA/plot (4.57 to 9.15 €/plot)
- Mint: 4000 to 12000 CFA/plot (6.10 to 18.29 €/plot)
- Okra: 3000 to 5000 CFA/plot (4.57 to 7.62 €/plot)

To make the calculations representative, market values from Niamey are adapted and a relatively low price is assumed for Dar Es Salam for the revenue calculation considering transport distance to the market, lack of cold storage at the village, etc. An average price assumed for the calculations is given in Table 4.

Table 4. Cash crop types, plots, and average selling price are considered [43].

Crop Types	No. of Plots (25 m^2)	CFA/Plot	€/Plot
Salad	20	5000	7.62
Cabbage	10	3000	4.57
Tomato	10	3500	5.34
Mint	10	6000	9.15
Okra	10	3500	5.34

4.3. Financial Indicators for Cases 1–3

For cases 1–3 mentioned above, the results for gross margin, farm profit, and B/C ratio are presented in Figures 4–6 respectively.

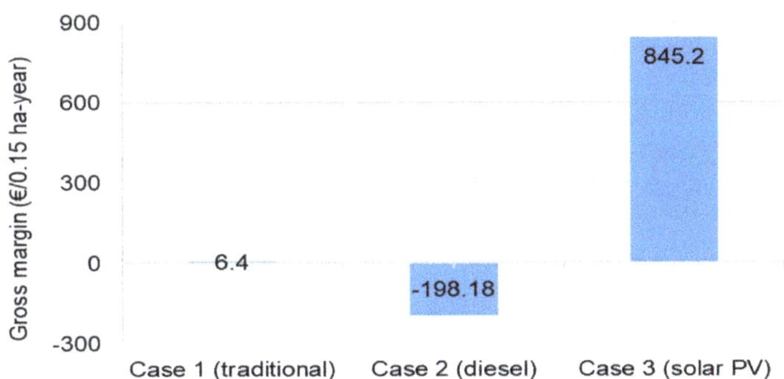

Figure 4. Gross margin for a reference farmer in a year (from 0.15 ha land).

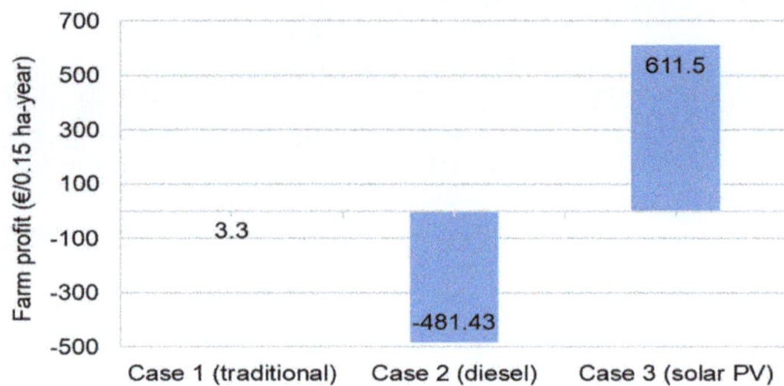

Figure 5. Farm profit for a reference farmer in a year (from 0.15 ha land).

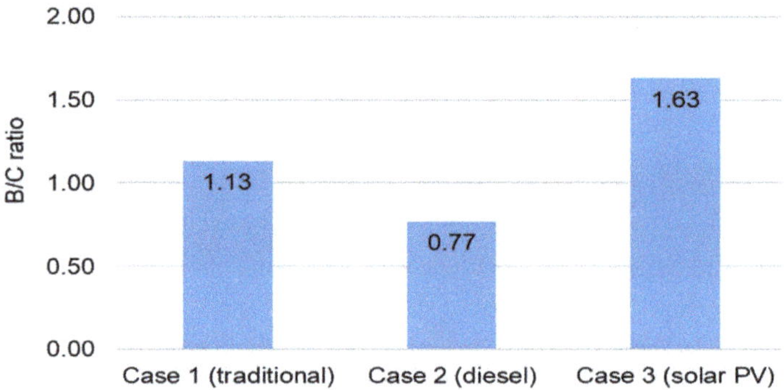

Figure 6. B/C ratio for a reference farmer in a year (from 0.15 ha land).

The economic indicator results are in line with the traditional farming practice in the village, i.e., farming is simply a subsistence agricultural economy. It is almost a net-zero profit activity but is a lifeline for survival.

In case 2, the results show that the gross margin and farm profit are both negative. The B/C ratio is less than one. All three indicators suggest that such a diesel pump-based irrigation system for the reference farm is not economically viable. One of the main reasons is the high cost of diesel. Besides the economic indicators, additional environmental impacts of diesel use for irrigation are calculated in terms of CO_2 emissions per year (for 0.15 ha). To irrigate 0.15 ha of land in a year, about 2700 m^3 of water is considered for irrigation and this leads to about 1500 L diesel consumption. Considering the CO_2 emission factor as 2.67 $kgCO_2$/L for diesel motors [44], the total annual CO_2 emission is calculated at about 4005 kg/year.

Among the discussed three cases, solar-powered irrigation systems perform the best. All indicators are positive. Due to relatively small PV system size and longer lifetime, the annual cost related to irrigation is much lower compared to case 2, with diesel.

The levelized cost of water (LCOW) has been also calculated by dividing the annuity value of total irrigation system cost divided by total annual irrigation water consumed. The obtained values are 0.11 €/m^3 vs. 0.50 €/m^3 for solar vs. diesel case respectively.

4.4. Case of Agrivoltaic (APV)

APV specific assumptions used for calculation in the model are given in Table 5.

Table 5. PV specific parameters.

Description	Value	Unit
Radiation at standard test condition, I_{stc}	1	kW/m^2
Solar yield degradation rate, d_g	0.5	%
Peak power of PV system, P_p	50	kW
Performance ratio, Q	60	%
PV size for irrigation pump, $P_{irrigation}$	2	kW
Discount rate in Niger, d	6 [45]	%
Grid tariff for electricity selling, P_{el}	0.10	€/kWh
Number of households	400	numbers
Annual electricity demand per household	300	kWh/year
Solar PV system cost	1200	€/kW
APV structure costs	50	€/kW
Repair, operation, maintenance costs	2	% of initial cost
Solar radiation data on an inclined surface (tilt at 15°)	2244 (Meteonorm V. 8)	kWh/m^2-year

The amount of electricity generation in the year will be about 67.32 MWh. Due to the considered degradation rate, the yield will be smaller in each following year, with a value for the final year at only 59.69 MWh. The irrigation water demand (for the same reference farm size) will be constant at 2693 kWh/year throughout these years. This leads to huge surplus electricity that needs to be used e.g., connection to central grid or its use in village electrification.

As the solar system size that can be installed in such a large land area is bigger compared to the one needed to pump irrigation water as in case 3 above, different scenarios are considered in this case. Based on the total cost and total revenue of the whole APV system, the economic indicators are calculated and the results are presented in the following two scenarios.

4.4.1. Scenario 1, Benefits to Investor

In this scenario, the APV system is installed by an investor. Farmer has to pay the electricity tariff for water pumping electricity needs. Practically, the farmer could switch from traditional cereals farming to cash crops farming and enjoy the higher revenue similar to cases 2 and 3 before.

For an investor, the total costs include the APV system investment cost as well as annual repair and maintenance costs. The revenue includes electricity sales to the grid and farmers. This village is not connected to the grid yet. Therefore, at the moment it is a hypothetical scenario that the mentioned revenue would be collected. However, if the villagers are supplied with the electricity after building the required local supply infrastructures, this revenue can be easily collected. The electricity selling rate considered (i.e., 0.10 €/kWh) is slightly below the current grid electricity tariff in Niger (about 0.12 €/kWh for consumers with demand in the range of 150–300 kWh in a year).

The project lifetime is relatively long, i.e., 25 years. Nominal annual revenue from the electricity sale is almost similar over the years (only affected by a slight decrease in yield, due to degradation rate of solar PV electricity yield caused by solar glass scratches, etc.). As the economic indicator for the investor, NPV is calculated by discounting the annual net cash flows. The calculated NPV is about 7779 € for the investor. With irrigation, the farmer will have higher farm profit (due to higher revenue as reported in case 3 before). This will be a win-win scenario for both investors and farmers.

4.4.2. Scenario 2, Overall Benefits (Combined of Both Farmer and Investor)

This scenario is the same as scenario 1, except for revenue calculations. In this case, the farmer's income from cash crops farming as well as investor's income from electricity sales are considered in the revenue calculation (discounted annual farm profit plus the investor's NPV, for the same accounting period of 25 years).

One further aspect of the negative shading effect on agricultural yield has also been considered. Crop production in this shading case is assumed at about 80% of the normal production. This reduction has an impact on the farmer's revenue. As expected, the combined NPV values are higher than in the individual case, even after considering the mentioned shading effect. The combined NPV values are compared in Figure 7.

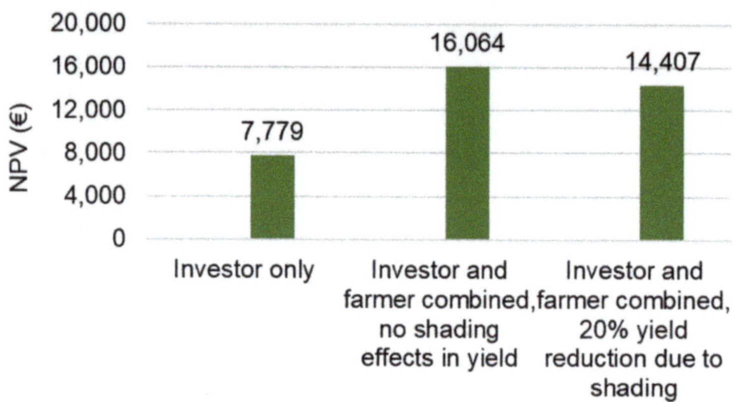

Figure 7. NPVs under different scenarios.

Overall, this can be interpreted that the APV system is economically profitable under the assumptions made in this study, both for farmers and PV investors.

For scenario 2 above, the LER can be calculated by using Equation (1). The obtained LER values are 1.33 and 1.13 for the case without and with yield reduction due to APV shading, respectively. The LER results show that the double use of land is more effective. Furthermore, when the APV system is installed, people will benefit from access to electricity, even if they will need to pay for the electricity. For the village's about 400 households, assuming annual electricity demand of up to 323 kWh/year (typical value in rural areas of developing countries lies about 300 kWh), only about 2 such APV systems would be able to supply the electricity needs of the village.

4.5. Sensitivity Analysis

As the study has many assumptions, it is necessary to analyze the sensitivity of some crucial parameters with regard to the financial indicators. For the following selected parameters, such analysis has been made.

4.5.1. Diesel Based Irrigation System, Case for Farmers

The price fluctuation of the cash crop can be much higher than currently expected. Therefore, a sensitivity analysis is carried out to observe the influence of selling price changes on the B/C ratio in (Figure 8). The results show that to reach the farming system B/C ratio to 1, the average selling price for all crops shall increase at least by 30%, which is currently not realistic.

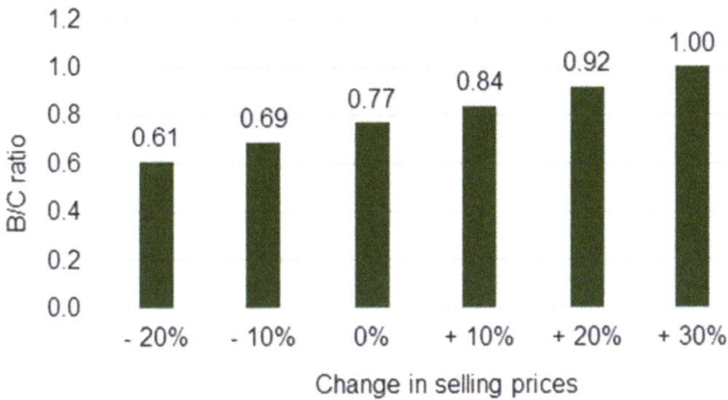

Figure 8. Effect of crop selling price changes to B/C ratio, case diesel.

The diesel price might vary in the market from the currently taken price of 0.76 €/L. Cost for diesel has a significant share in total cost, therefore the sensitivity analysis is carried out to observe the impact of diesel price variations on the B/C ratio. Results are presented in Figure 9. It can be interpreted that the diesel price shall drop below 40% from its current level to be the farming system at cost breakeven point. This scenario is also not realistic looking at the overall trends of rising petroleum prices in recent years.

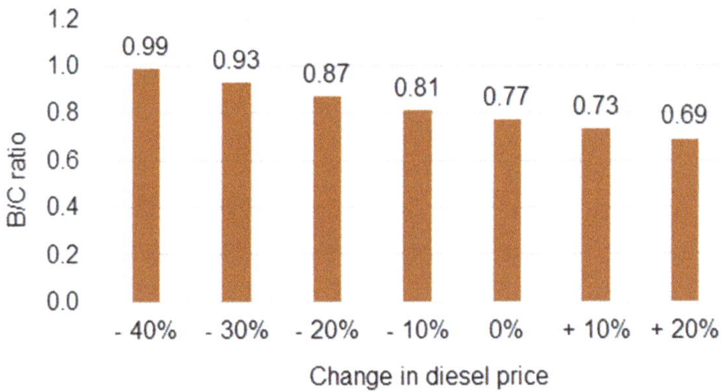

Figure 9. Effect of crop selling price changes to B/C ratio, case diesel.

4.5.2. APV System, Case for Investors

The calculated value of LCOE is at 9.11 €ct/kWh (with the energy yield discounted to its present value at a baseline discount rate of 6%). When the lesser discount rates e.g., 2% or 4%, and higher performance ratio (e.g., 70% or 80%), are considered, corresponding LCOE values will be lower, as shown in Figure 10.

On the revenue side, the most important parameter is the electricity selling tariff. Therefore, its influence on the NPV is calculated and the results are presented in Figure 11. If the selling tariff goes below 0.0911 €/kWh, the NPV will start to get negative. Therefore, this is the breakeven selling tariff.

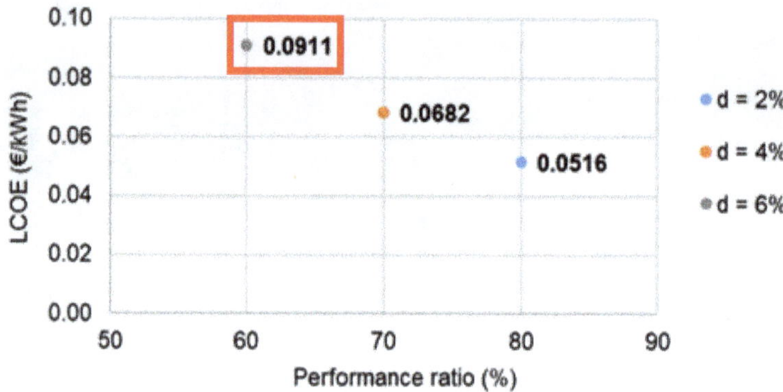

Figure 10. LCOE values under different discount rates and performance ratios.

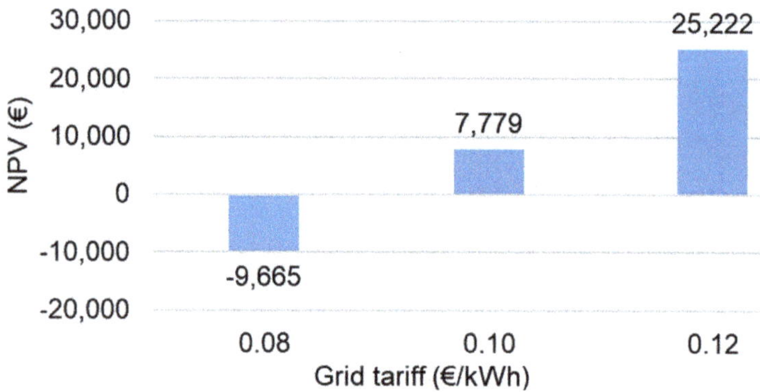

Figure 11. Effect of the electricity selling price to NPV, a case for investors.

On the cost side, the PV system cost is the most important parameter that determines the total system cost. The influence of change in PV system costs on the NPV is shown in Figure 12. It is expected that the solar systems will be cheaper in the coming years (if they follow the trend of the recent past). In that case, the APV system seems further promising.

In this case, the benefits to the farmer will not change much in comparison to the results presented above in Section 4.4 because this does not lead to significant extra cost for irrigation electricity (only a very small fraction of the APV electricity goes for irrigation of 0.15 ha land considered).

4.5.3. APV System, Both Cases (Combined Benefits and Shading Effect)

The results presented above in Figure 8 are based on the assumed value of a discount rate of 6% [45]. For APV projects with a long lifetime, the selection of the correct discount rate is important. A high discount rate leads to a smaller value of NPV as well as a higher value of LCOE. Figure 13 shows the calculated values of NPVs under the different scenarios for discount rates.

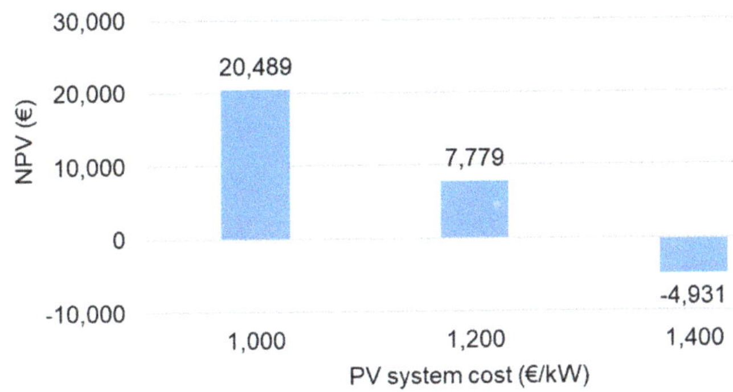

Figure 12. Effect of PV system cost to NPV, a case for investors.

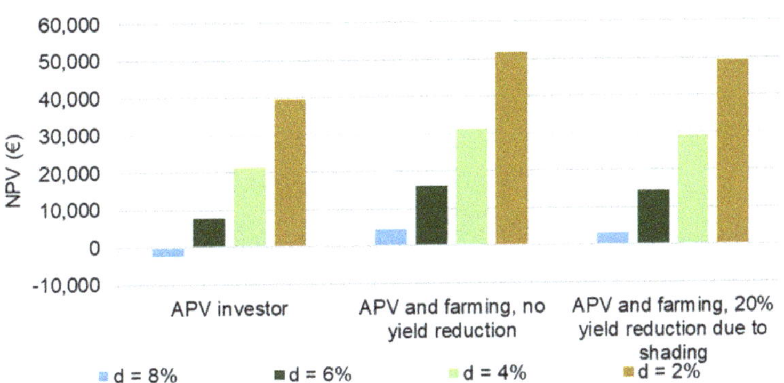

Figure 13. NPVs under different scenarios with varying discount rates.

Besides these direct monetary benefits to the investor, there are many indirect benefits of the APV. It could contribute to the GHG emission reduction year by year by avoiding the use of possible diesel-powered irrigation. In many regions of the world, diesel-powered irrigation is a common practice today. On top of that, when the electricity from APV is used to gradually replace the existing grid electricity in Niger, which is dominated by non-renewable energy, a significant contribution to the GHG emission reduction can be considered.

As expected, the LER results showed that the double use of land is more effective. Furthermore, when the APV system is installed, people will benefit from access to electricity, even if they will need to pay for the electricity (considered in the above analysis 0.10 €/kWh). Considering the village's household at about 400, and annual electricity demand of 323 kWh/year (typical value in rural areas of developing countries lies about 300 kWh, including Nepal), only about 2 such APV plants would be able to supply the lighting electricity needs of the village.

Another positive aspect of such APV would be the access to clean and sufficient drinking water to the villagers, which is currently a big problem as described in the previous chapter. Therefore, such an APV system's surplus electricity (after irrigation) could be used for additional water pumping for such water use purposes.

5. Conclusions

Based on the results described above, it can be concluded that the APV is a promising option in the village of Dar Es Salam. Implementation of APV could significantly increase the economic activities in the village, mainly in the field of small agricultural enterprises. In all four scenarios considered under APV, the results are positive and such a system seems to be an appropriate option to supply food and energy in the village. In a broad estimate, only two such APV systems would be able to supply the village's 400 households with electricity (about 323 kWh/year). APV systems are win-win options for both farmers and investors.

The analysis above is based on several stated assumptions. Therefore, the presented results are only valid, if these assumptions come true during the implementation of the real project. To validate the results presented in this study, it is necessary to install an APV system and perform the experimental analysis on-site. Based on these results, different business cases can be developed and practical business models can be developed for different interest groups: farmers, investors, and traders (in agri-value-chain). This experimental work shall be the next step in this field.

Author Contributions: Conceptualization, S.N.B., R.B., and R.A.; methodology, S.N.B., W.K., and H.S.; investigation, S.N.B.; writing—original draft preparation, S.N.B.; writing—review and editing, S.N.B., W.K., H.S., and R.B.; supervision, S.S. and R.A.; revision, S.N.B. and R.B.; project administration, R.B.; funding acquisition, R.B. All authors have read and agreed to the published version of the manuscript.

Funding: This research was funded by the Federal Ministry of Education and Research (BMBF) in Germany through its Project Management Agency Jülich (PtJ) under the framework of the RETO-DOSSO project, grant number 03SF0598A. The APC was funded by the same.

Institutional Review Board Statement: Not applicable.

Informed Consent Statement: Not applicable.

Data Availability Statement: Not applicable.

Acknowledgments: Thanks are also due to Adamou Hassane and his team at the University of Niamey who coordinated the field interview at the reference farm by using the questionnaire developed by the first author of this paper.

Conflicts of Interest: The authors declare no conflict of interest.

References

1. Fraunhofer ISE. *Harvesting the Sun for Power and Produce—Agrophotovoltaics Increases the Land Use Efficiency by Over 60 Percent*; Fraunhofer ISE: Freiburg, Germany, 2017.
2. Goetzberger, A.; Zastrow, A. On the Coexistence of Solar-Energy Conversion and Plant Cultivation. *Int. J. Sol. Energy* **1982**, *1*, 55–69. [CrossRef]
3. Brohm, R.; Khanh, N.Q. *Dual Use Approaches for Solar Energy and Food Production—International Experience and Potentials for Vietnam*; Green Innovation and Development Centre (GreenID): Hanoi, Vietnam, 2018.
4. Mead, R.; Willey, R.W. The Concept of a 'Land Equivalent Ratio' and Advantages in Yields from Intercropping. *Exp. Agric.* **1980**, *16*, 217–228. [CrossRef]
5. Dinesh, H.; Pearce, J.M. The potential of agrivoltaic systems. *Renew. Sustain. Energy Rev.* **2016**, *54*, 299–308. [CrossRef]
6. Weselek, A.; Ehmann, A.; Zikeli, S.; Lewandowski, I.; Schindele, S.; Högy, P. Agrophotovoltaic systems: Applications, challenges, and opportunities. A review. *Agron. Sustain. Dev.* **2019**, *39*, 35. [CrossRef]
7. Hernandez, R.R.; Armstrong, A.; Burney, J.; Ryan, G.; Moore-O'Leary, K.; Diédhiou, I.; Grodsky, S.M.; Saul-Gershenz, L.; Davis, R.; Macknick, J.; et al. Techno–ecological synergies of solar energy for global sustainability. *Nat. Sustain.* **2019**, *2*, 560–568. [CrossRef]
8. Barron-Gafford, G.A.; Pavao-Zuckerman, M.A.; Minor, R.L.; Sutter, L.F.; Barnett-Moreno, I.; Blackett, D.T.; Thompson, M.; Dimond, K.; Gerlak, A.K.; Nabhan, G.P.; et al. Agrivoltaics provide mutual benefits across the food–energy–water nexus in drylands. *Nat. Sustain.* **2019**, *2*, 848–855. [CrossRef]
9. Majumdar, D.; Pasqualetti, M.J. Dual use of agricultural land: Introducing 'agrivoltaics' in Phoenix Metropolitan Statistical Area, USA. *Landsc. Urban Plan* **2018**, *170*, 150–168. [CrossRef]
10. Dupraz, C.; Marrou, H.; Talbot, G.; Dufour, L.; Nogier, A.; Ferard, Y. Combining solar photovoltaic panels and food crops for optimising land use: Towards new agrivoltaic schemes. *Renew. Energy* **2011**, *36*, 2725–2732. [CrossRef]
11. Marrou, H.; Guilioni, L.; Dufour, L.; Dupraz, C.; Wery, J. Microclimate under agrivoltaic systems: Is crop growth rate affected in the partial shade of solar panels? *Agric. For. Meteorol.* **2013**, *177*, 117–132. [CrossRef]

12. Beck, M.; Bopp, G.; Goetzberger, A.; Obergfell, T.; Reise, C.; Schindele, S. Combining PV and food crops to agrophotovoltaic—Optimization of orientation and harvest. In Proceedings of the 27th European Photovoltaic Solar Energy Conference and Exhibition, EU PVSEC, Frankfurt, Germany, 24–28 September 2012.
13. Sekiyama, T.; Nagashima, A. Solar Sharing for Both Food and Clean Energy Production: Performance of Agrivoltaic Systems for Corn, A Typical Shade-Intolerant Crop. *Environments* **2019**, *6*, 65. [CrossRef]
14. Neupane Bhandari, S. *Feasibility Analysis of Agrivoltaic Systems: Case Study of Food Energy Nexus in Niger*; TH Köln—University of Applied Sciences: Cologne, Germany, 2020.
15. Vyas, K. Solar Farming with Agricultural Land. *Acta Sci. Agric.* **2019**, *3*, 23–25. [CrossRef]
16. Santra, P.; Singh, R.K.; Meena, H.M.; Kumawat, R.N.; Mishra, D.; Jain, D.; Yadav, O.P. Agrivoltaic system: Crop production and photovoltaic-based electricity generation from a single land unit. *Indian Farming* **2018**, *68*, 20–23.
17. Malu, P.R.; Sharma, U.S.; Pearce, J.M. Agrivoltaic potential on grape farms in India. *Sustain. Energy Technol. Assess.* **2017**, *23*, 104–110. [CrossRef]
18. Valle, B.; Simonneau, T.; Sourd, F.; Pechier, P.; Hamard, P.; Frisson, T.; Ryckewaert, M.; Christophe, A. Increasing the total productivity of a land by combining mobile photovoltaic panels and food crops. *Appl. Energy* **2017**, *206*, 1495–1507. [CrossRef]
19. Ravi, S.; Macknick, J.; Lobell, D.; Field, C.; Ganesan, K.; Jain, R.; Elchinger, M.; Stoltenberg, B. Colocation opportunities for large solar infrastructures and agriculture in drylands. *Appl. Energy* **2016**, *165*, 383–392. [CrossRef]
20. Amaducci, S.; Yin, X.; Colauzzi, M. Agrivoltaic systems to optimise land use for electric energy production. *Appl. Energy* **2018**, *220*, 545–561. [CrossRef]
21. Harinarayana, T.; Vasavi, K. Solar Energy Generation Using Agriculture Cultivated Lands. *Smart Grid Renew. Energy* **2014**, *5*, 31–42. [CrossRef]
22. Movellan, J. Japan Next-Generation Farmers Cultivate Crops and Solar Energy. 2013. Available online: https://www.renewableenergyworld.com/2013/10/10/japan-next-generation-farmers-cultivate-agriculture-and-solar-energy/ (accessed on 17 October 2020).
23. Moreda, G.P.; Muñoz-García, M.A.; Alonso-García, M.C.; Hernández-Callejo, L. Techno-Economic Viability of Agro-Photovoltaic Irrigated Arable Lands in the EU-Med Region: A Case-Study in Southwestern Spain. *Agronomy* **2021**, *11*, 593. [CrossRef]
24. Ibrik, I. Micro-Grid Solar Photovoltaic Systems for Rural Development and Sustainable Agriculture in Palestine. *Agronomy* **2020**, *10*, 1474. [CrossRef]
25. Pascaris, A.S.; Schelly, C.; Pearce, J.M. A First Investigation of Agriculture Sector Perspectives on the Opportunities and Barriers for Agrivoltaics. *Agronomy* **2020**, *10*, 1885. [CrossRef]
26. Hassanien, R.H.E.; Li, M.; Dong Lin, W. Advanced applications of solar energy in agricultural greenhouses. *Renew. Sustain. Energy Rev.* **2016**, *54*, 989–1001. [CrossRef]
27. NREL. Benefits of Agrivoltaics across the Food-Energy-Water Nexus. 2019. Available online: https://www.nrel.gov/news/program/2019/benefits-of-agrivoltaics-across-the-food-energy-water-nexus.html (accessed on 12 July 2020).
28. Njita, N.F. *Increasing Agricultural Land Use Efficiency and Generating Electricity Using Solar Modules*; Lappeenranta University of Technology: Lappeenranta, Finland, 2018.
29. Patel, B.; Gami, B.; Baria, V.; Patel, A.; Patel, P. Co-Generation of Solar Electricity and Agriculture Produce by Photovoltaic and Photosynthesis—Dual Model by Abellon, India. *J. Sol. Energy Eng.* **2019**, *141*, 031014. [CrossRef]
30. EERE. Farmer's Guide to Going Solar. 2020. Available online: https://www.energy.gov/eere/solar/farmers-guide-going-solar (accessed on 21 September 2020).
31. Ketzer, D.; Schlyter, P.; Weinberger, N.; Rösch, C. Driving and restraining forces for the implementation of the Agrophotovoltaics system technology—A system dynamics analysis. *J. Environ. Manag.* **2020**, *270*, 110864. [CrossRef] [PubMed]
32. CIA. The World Factbook—Niger. 2020. Available online: https://www.cia.gov/library/publications/the-world-factbook/geos/ng.html (accessed on 18 September 2020).
33. Knoema. Niger—Total Population. 2020. Available online: https://knoema.com/atlas/Niger/Population (accessed on 10 September 2020).
34. FAO. Aquastat: Country Statistics 2017—Niger. 2021. Available online: http://www.fao.org/aquastat/statistics/query/index.html?lang=en (accessed on 23 June 2021).
35. ICRISAT. Niger: Facts and Figures. 2020. Available online: http://exploreit.icrisat.org/profile/Niger/334 (accessed on 26 August 2020).
36. USAID. *Off-Grid Solar Market Assessment*; United States Agency for International Development (USAID): 2019. Available online: https://www.usaid.gov/sites/default/files/documents/1860/PAOP-Niger-MarketAssessment-Final_508.pdf (accessed on 14 August 2020).
37. Andres, L.; Lebailly, P. Peri-urban Agriculture: The Case of Market Gardening in Niamey, Niger. *Afr. Rev. Econ. Financ.* **2011**, *3*, 69–85.
38. Bhandari, R.; Sessa, V.; Adamou, R. Rural electrification in Africa—A willingness to pay assessment in Niger. *Renew. Energy* **2020**, *161*, 20–29. [CrossRef]
39. Tilahun, F.B.; Bhandari, R.; Mamo, M. Supply optimization based on society's cost of electricity and a calibrated demand model for future renewable energy transition in Niger. *Energy Sustain. Soc.* **2019**, *9*, 31. [CrossRef]

40. Bhandari, R.; Arce, B.E.; Sessa, V.; Adamou, R. Sustainability Assessment of Electricity Generation in Niger Using a Weighted Multi-criteria Decision Approach. *Sustainability* **2021**, *13*, 385. [CrossRef]
41. Kahan, D. *Economics for Market Oriented Farming*; Food and Agriculture Organization (FAO) of the United Nations: Rome, Italy, 2008.
42. World Bank. *Solar Pumping—The Basics*; The World Bank (WB): Washington, DC, USA, 2018.
43. ITT. *Final Report on the Project RARSUS*; TH Köln—University of Applied Sciences: Cologne, Germany, 2020.
44. ECG. *Feasibility Assessment for the Replacement of Diesel Water Pumps with Solar Water Pumps*; Emcon Consulting Group (ECG): Windhoek, Namibia, 2006.
45. World Bank. *International Development Association Project Paper on a Proposed Additional Credit and Grant to the Republic of Niger for the Electricity Access Expansion Project*. The World Bank: 2018. Available online: https://documents1.worldbank.org/curated/en/630161534524243997/pdf/NIGER-ELECTRICITY-PAD-08142018.pdf (accessed on 23 November 2020).

 *agronomy* 

Article

LoRa-LBO: An Experimental Analysis of LoRa Link Budget Optimization in Custom Build IoT Test Bed for Agriculture 4.0

Mahendra Swain [1], Dominik Zimon [2,*], Rajesh Singh [1], Mohammad Farukh Hashmi [3], Mamoon Rashid [4,*] and Saqib Hakak [5,*]

1   School of Electronics and Electrical Engineering, Lovely Professional University, Jalandhar 144001, India; er.mahendraswain@gmail.com (M.S.); srajssssece@gmail.com (R.S.)
2   Department of Management Systems and Logistics, Rzeszow University of Technology, 35-959 Rzeszow, Poland
3   National Institute of Technology, Warangal 506004, India; mdfarukh@nitw.ac.in
4   School of Computer Science and Engineering, Lovely Professional University, Jalandhar 144001, India
5   Canadian Institute for Cybersecurity, University of New Brunswick, Fredericton, NB E3B 5A3, Canada
*   Correspondence: zdomin@prz.edu.pl (D.Z.); mamoon873@gmail.com (M.R.); saqib.hakak@unb.ca (S.H.); Tel.: +91-7814346505 (M.R.)

Abstract: The Internet of Things (IoT) is transforming all applications into real-time monitoring systems. Due to the advancement in sensor technology and communication protocols, the implementation of the IoT is occurring rapidly. In agriculture, the IoT is encouraging implementation of real-time monitoring of crop fields from any remote location. However, there are several agricultural challenges regarding low power use and long-range transmission for effective implementation of the IoT. These challenges are overcome by integrating a long-range (LoRa) communication modem with customized, low-power hardware for transmitting agricultural field data to a cloud server. In this study, we implemented a custom-based sensor node, gateway, and handheld device for real-time transmission of agricultural data to a cloud server. Moreover, we calibrated certain LoRa field parameters, such as link budget, spreading factor, and receiver sensitivity, to extract the correlation of these parameters on a custom-built LoRa testbed in MATLAB. An energy harvesting mechanism is also presented in this article for analyzing the lifetime of the sensor node. Furthermore, this article addresses the significance and distinct kinds of localization algorithms. Based on the MATLAB simulation, we conclude that hybrid range-based localization algorithms are more reliable and scalable for deployment in the agricultural field. Finally, a real-time experiment was conducted to analyze the performance of custom sensor nodes, gateway, and handheld devices.

Keywords: localization; link budget; spreading factor; range; LoRa; node sensitivity; SNR

Citation: Swain, M.; Zimon, D.; Singh, R.; Hashmi, M.F.; Rashid, M.; Hakak, S. LoRa-LBO: An Experimental Analysis of LoRa Link Budget Optimization in Custom Build IoT Test Bed for Agriculture 4.0. *Agronomy* **2021**, *11*, 820. https://doi.org/10.3390/agronomy11050820

Academic Editor: Miguel-Ángel Muñoz-García

Received: 7 March 2021
Accepted: 19 April 2021
Published: 22 April 2021

1. Introduction

According to a United Nations report [1], the world population will reach 9.8 billion in 2050, representing a 25% increase over the current population. Additionally, the pattern of urbanization is anticipated to grow at a rapid pace, with approximately 70% of the world's population expected to be residing in urban areas by 2050 [2]. As a result, the amount of food production will be required to double by 2050 [3]. The size of the area of the Earth's surface remaining for agricultural usage is limited due to climate, temperature, soil quality, and topography [4]. Additionally, compared to past decades, the total utilization of agriculture for food production has declined. Thus, agriculture transformed into a means of living that was considerably more sustainable and able to generate food surpluses than hunting and gathering alone. Subsequently, agriculture has become an integral part of human life [5,6], and many advancements have since taken place in agriculture in terms of crops, fertilizers, pesticides, etc. Agriculture has become a means of survival and a key component of national economies. Hence, it can be said that the agricultural development of a nation speaks for the nation. Investments to improve agricultural activities also benefit the employees of the country.

In natural farming, farmers are required to visit fields to evaluate crop conditions. Moreover, 70% of farmers' time is spent in understanding the condition and monitoring of crops [7]. However, farming concerns can be overcome with smart agriculture, which aims to provide a sustainable solution with a low environmental effect. The recent advancements in communication and sensor technology allows the implementation of remote monitoring of crop fields from any location. At present, the Internet of Things (IoT) provides an opportunity to implement real-time monitoring of the agriculture field from any remote location [8]. The wide scope of the Internet provides an opportunity to effectively implement precision agriculture. For implementing the IoT in agriculture, the wireless communication protocol plays a crucial role in connecting the IoT server's agricultural fields. In addition, agricultural fields are generally located in areas with poor connectivity [9]. Thus, the IoT demands low power consumption and long-range transmission-based wireless communication protocols because the end devices are energy-constrained [10]. The evolution of the low power wide area network (LPWAN) is meeting the requirements of the IoT. Among LPWAN wireless technologies, long-range (LoRA) is delivering robust and advanced wireless connectivity for communicating data from sensor nodes to a cloud server with zero subscription charges [11,12]. Additionally, in the agriculture field, the deployment of nodes plays a significant role because it provides information regarding the number of nodes required in the field. The nodes should be embedded with a localization mechanism to send the data to the sink node and gateway using the shortest path. The localization of the sensor nodes in agriculture is also an important factor for identifying a node located within the large area of an agricultural field.

The contributions of the study are as follows:

a. An overview of LoRa and distinct localization algorithms, namely range-free and range-based, is provided in this article.

b. The customization of sensor nodes and the gateway was designed and implemented for monitoring agriculture.

c. LoRa and Wi-Fi communication for agriculture is also proposed.

d. Implementation of a localization algorithm for agriculture is presented, and we conclude that hybrid range-based localization algorithms are more reliable, scalable, and easy to deploy in the field.

e. The energy harvesting mechanism for the sensor nodes is presented, and was evaluated using the Cisco packet tracer.

f. To characterize the behavior of LoRa, we undertook a simulation using MATLAB.

g. A real-time experiment was performed using the customized sensor node and gateway. The sensor node was able to communicate with the cloud server through the LoRa-based gateway.

This paper is organized as follows: Section 2 provides the theoretical background with technical specifications of wireless communication protocols. Section 3 presents an overview of LoRa and localization algorithms. Section 4 discusses the customization of the sensor node and gateway, and the LoRa and Wi-Fi communication-enabled architecture for agriculture. Section 5 outlines the simulation of localization, energy harvesting, and MATLAB simulation. Section 6 describes the real-time experimental setup and discusses the results.

2. Theoretical Background

The Internet of Things (IoT) is a pioneering technology that provides an efficient and accurate automation solution for modernizing agriculture with minimal human interference [13]. The advancement of sensor technology, wireless communication technologies, and remote sensing technologies encourages precision agriculture (PA). In the attempt to enhance the yield and quality of crops, wireless transmission is necessary to transfer data to information processing centers. Wireless communication empowers the effective utilization of limited resources in agriculture, thereby allowing the development of agriculture in terms of reliable connectivity [14]. Generally, most agricultural fields are located in remote areas, where internet connectivity is unstable. To overcome these limitations, the implementation of a wireless sensor network (WSN) is required. In a WSN, open-licensed band communication protocols are embedded in the sensor node and the communication protocol for transmitting the agricultural field data to an area in which the internet connectivity is sufficiently strong to allow communication with the cloud server [15].

In [16], IoT- and WSN-based agriculture systems based on a CC3200 single chip for monitoring the humidity and temperature of the crop field were implemented. Additionally, a camera module was integrated with a CC3200 single chip to capture visuals and communicate a multimedia message (MMS) to a farmer. The WSN-based smart agriculture system was implemented by utilizing the Zigbee network and communicating the crops' status to farmers and a cloud server [17,18]. A Wi-Fi, GSM, and WEMOS D1 controller-based agriculture monitoring system has been implemented for monitoring the parameters of crops, such as pH, soil moisture, soil type, and weather conditions, and communicate these to a cloud server [19]. A low-cost information-based system was implemented for agriculture using 2G GSM and orang-pi [20]. In the agriculture scenario, wireless communication protocols such as ZigBee, Wi-Fi, and GPRS have been implemented for transmitting data to processing centers. Sensor-based data acquisition has been achieved using Bluetooth, ZigBee, and cloud servers. Because sensor nodes are energy-constrained devices, the implementation of GPRS and Wi-Fi communication protocols is challenging because they consume large quantities of power, as shown in Table 1. Because its communication is characterized by low power consumption, ZigBee is also implemented in agriculture. However, its transmission range is limited. Of the above-mentioned communication options, LoRa is a reliable and feasible communication technology used in agriculture to implement the effective WSN-based IoT. To implement the WSN, LoRa communication is chosen because it can transmit data to an Internet-connected area [21]. LoRa communication has been integrated into agriculture to capture and communicate real-time images of farms [22].

A wireless sensor network includes small, low-energy consumption sensor nodes for various applications. The task of localization is to determine the physical coordinates of the sensors. Because each application has specific requirements, several localization algorithms are used. WSN localization plays a significant role in the agriculture IoT. Tiny and cheap devices with low energy consumption and limited computing resources are being heavily used in agriculture applications. To deploy sensor nodes in fields, localization algorithms, such as time on arrival (TOA), time difference of arrival (TDOA), and received signal strength indication (RSSI), are required to estimate the number of sensor nodes and their position in a particular field [23]. A statistical method has been implemented to identify the non-line of sight (NLOS) nodes in the WSN network [24,25]. A combination of range-based, range-free-based, and hybrid-based localizations can be emphasized. A combination of range-based and range-free algorithms has been studied concerning the sensor node distance, density, and reliability. As discussed earlier, selecting accurate wireless communication is critical for overcoming the problems of power consumption and transmission range. A study of various communication techniques is tabulated below according to frequency band, network size, network topology, etc.

Table 1 discusses three emerging LPWAN technologies, namely LoRa, narrow band-IoT (NB-IoT), and SigFox, regarding the requirements of the IoT for a wide range of applications. Of these LPWAN technologies, LoRa is considered to be an independent network and can utilize the frequency bands without any cost [27]. NB-IoT is a licensed band, and is also a dependent network that charges for the use of the bands. The Sigfox network is deployed by network operators, and users are subscription-based. LoRa is one of the best candidates for long distance and low power transmissions [28].

Table 1. Technical specifications of communication protocols [26].

Parameters	Zig-Bee	Bluetooth	BLE	Wi-Fi	GPRS	LoRa	NB-IoT	SigFox
Frequency band	868/915 MHz and 2.4 GHz	2.40 GHz	2.40 GHz	2.40 GHz	900 to 1800 MHz	869 to 915 MHz	Licensed LTE frequency bands	868 to 915 MHz
Network size	Approx. 65,000	Approx. 8	Limited application	Approx 32	Approx 1000	10,000 no of (nodes per BS)	52,000 devices/channel/cel	1,000,000 no. of (nodes per BS
Network Topologies	P2P, tree, star, mesh	Scatter-net Topology	Star-bus topology	Point-to-hub topology	Cellular system Topology	Star-of-stars	Star topology	Star topology
Channel bandwidth	Equal to 2 MHz	1 MHz	1 MHz	22 MHz	200 kHz	<500 KHz	200 kHz	200 kHz
Power consumption in Txmode	Around 36.9 mW	Around 215 mW	Around 10 mW	Around 835 mW	560 mW	100 mW	NA	122 mW
Application	WPANs, WSNs, and Agriculture	WPANs	WPANs	WLANs	AMI, demand response, HAN	Agriculture, Smart grid, environment control, and lighting control	Smart metering, Tracking of persons, animals, or objects, etc.	Agriculture and environment, automotive, buildings, and consumer electronics
Limitations	Mandatory line-of-sight	Short communication range	Short communication range	High power consumption and high latency (13.74 s)	Power consumption problem	Network size(scalability), data rate, and message capacity	Incapable of a seamless handover between cells and does not provide low latency application	Low data rates

3. LoRa and Localization Algorithms

In this section, we provide an overview of LoRa and the localization algorithms of the WSN. In the Section 1, LoRa and its fundamental parameters are discussed. In the Section 2, the significance of localization and distinct types of localization algorithms are discussed.

3.1. Overview of LoRa

LoRa is a form of robust low-power wireless networking that is used for long-distance communication. LoRa utilizes chirp spread spectrum (CSS) modulation to modulate the Industrial, Scientific and Medical (ISM) bands [29]. Modulation of the chirp spread spectrum broadcasts a narrow band signal over broader channel bandwidth. LoRa operates on ISM bands such as 868 MHz in Europe, 995 MHz in North America, and 433 MHz in Asia [30]. LoRa can communicate over distances of between 10 and 40 km in rural areas, and urban area coverage is between 1 and 5 km. The LoRa protocol specification was developed by the LoRa Alliance, as shown in Figure 1. The LoRaWAN protocol comprises the MAC Layer and the Application Layer, and operates based on the LoRa physical layer. LoRaWAN is a network standard for telecom operators founded on the LoRa physical layer (PHY). It enables network services and encourages systems to transfer data to gateways wirelessly over a long range. LoRaWAN communicates between LoRa gateways and IoT devices via a star-network topology, and single hopping is allowed between a gateway and a LoRa device.

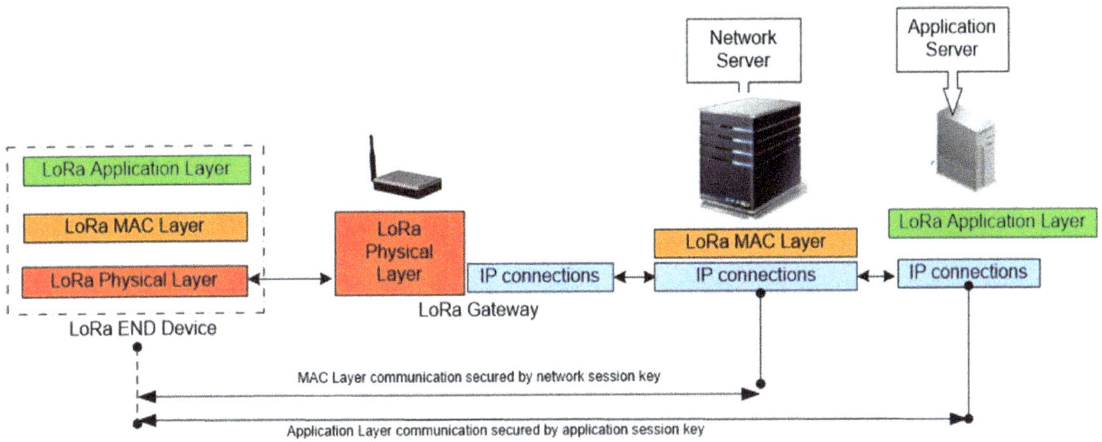

Figure 1. LoRa Physical, Application, and MAC layers of protocol architecture.

The analysis of LoRa communication in distinct applications can be undertaken based on the parameters of spreading factor (SF), link budget, signal-noise ratio (SNR), bandwidth (BW), link budget, receiver sensitivity, bit error rate (BER), and packet error rate (PER).

3.1.1. Spreading Factor

The initial frequency of the chirp is recognized as a symbol. The encoded bits in a symbol are configured by a unique parameter known as the spreading factor (*SF*). This indicates that a chirp with the spreading factor *SF* represents 2^{SF} bits per symbol, which implies that one symbol is described by multiple chips that are spread spectrum code pulses. *SF* is expressed as:

$$SF = \frac{Chirp\ rate}{symbol\ rate} \tag{1}$$

3.1.2. Signal-To-Noise Ratio (SNR)

SNR is the ratio of transmitted signal powers to the unwanted signal i.e., noise power. It is preferred that the SNR is minimized to ensure that demodulation at the receiver end is straightforward and the signal can be decoded correctly. To enhance LoRa performance, it uses forward error correction (FEC) techniques and the spreading factor, thus allowing significant SNR improvements. In particular, the SNR range is between -20 and $+10$ dB. The received signal is less distorted if the range is around $+10$ dB. LoRa has an SNR range of -7.5 to -20 dB.

3.1.3. Link Budget

The link budget cab be determined from transmitted power and node sensitivity, and is expressed as:

$$Link\ Budget_{(dBm)} = \left(Antenna\ transmitted\ power_{(dB)} \right) - \left(Node\ sensitvity_{(dB)} \right) \tag{2}$$

3.1.4. Sensitivity (S)

Sensitivity is defined as the ability of the receiver to amplify the weak signals that are obtained by the receiver. In LoRa, the spreading factor, noise figure, and bandwidth are considered to be inputs for providing the sensitivity as output. Sensitivity (S) is calculated as:

$$S = -174 + 10log_{10}\,BW + NF + SNR \tag{3}$$

where BW is the band width of the channel, NF is the noise figure gain in dB, and SNR is the signal to noise ratio power in dB.

Bit Rate/data rate is defined as the rate at which bits are transferred from one location to another. The bit rate (R_{bit}) of LoRa is expressed as:

$$Rbit = SF * \frac{BW}{2^{SF}} * CR \tag{4}$$

3.1.5. Bit Error Rate (BER)

BER is the percentage of bits that have errors compared to the cumulative number of bits received in a transmission. BER is represented as 10^{-4}. If BER is 10^{-4}, then it indicates that 10,000 bits have been transferred, and one bit has an error. A higher BER indicates that network performance is poor.

3.1.6. Packet Error Rate (PER)

PER is the total number of received packets divided by the number of error packets after forward error correction (FEC). A packet is a data unit used in a radio transmission that is subject to FEC.

3.2. Localization Algorithm

Localization is crucial for identifying the physical locations of sensors in the deployment area. Concerning the deployment land in agriculture, it is important to identify the required number of sensor nodes and the distances at which they need to be positioned so that sensor nodes can establish communication links among themselves. Here, we discuss localization algorithms to determine a suitable node architecture for farm land. Various parameters are evaluated using these algorithms, such as accuracy, reliability, and scalability. Localization algorithms are broadly classified into two types, namely, range-based and range-free-based, as shown in Figure 2 [23]. Moreover, these two types are classified into two kinds, namely, fully range-based and hybrid range-based.

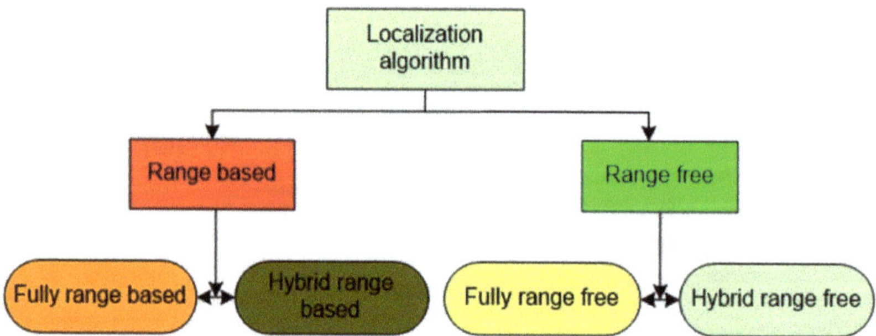

Figure 2. Classification of localization schemes for sensor networks.

(a)   Range-free localization algorithms

These algorithms determine the location of an unknown deployed sensor node [9]. Range-free methods utilize radio connectivity to communicate between nodes to identify their location. In range-free schemes, the angle of arrival (AOA), specific hardware, and distance measurement is not considered [31]. Range-free schemes comprise the centroid system, distance vector (DV) hop, approximate point in triangulation (APIT), and hop terrain, described as follows:

- DV—hop localization:
- In DV hop, the distance between the nodes is estimated using hop count, and the hop count of at least three anchor nodes is distributed across the network [32]. The hop count of a node is incremented by one when the neighbor node transmits the information to another neighbor node. The hop distance is evaluated as the distance between two nodes/number of hops.
- Centroid localization: This is the most basic scheme that uses anchor beacons, containing location information (X_i, Y_i) [31], where n is the number of the anchor nodes Ai.

$$(X, Y) = \left(\frac{\sum_{i=1}^{n} x_i}{n}, \frac{\sum_{i=1}^{n} Y_i}{n} \right) \tag{5}$$

- APIT: In APIT, the location information is obtained by anchor nodes through a global positioning system (GPS) and the unlocalized node receives the location information via overlapped triangles [33].
- Gradient: In the gradient algorithm, the unlocalized node utilizes the multilateration method to estimate the position of nodes. Moreover, it utilizes hop counting and the hop increment while being distributed to neighboring nodes.

(b)   Range-based localization algorithms

Range-based localization is based on angle estimation and distance estimation. The main techniques in this form of localization are time of arrival (ToA), angle of arrival (AoA), time difference of arrival (TDoA), and received signal strength indication (RSSI):

- ToA localization: This localization algorithm refers to the time of arrival, i.e., ToA, which refers to the time taken for the signal to travel from the sending node to the receiving node [34]. The distance is measured using roundtrip-time of flight (RTOF) to determine the distance between two nodes and is represented in Equation (6) as:

$$d = \frac{[(T3 - T0) + (T2 - T1)] * V}{2} \tag{6}$$

Sensor coordinates (x_1,y_1), (x_2,y_2), (x_3,y_3), and (x_4,y_4) determined using the response information and the TOA-based distance measurement method can be used to obtain the distances between them and the moving node S as d_1,d_2,d_3, and d_4, respectively. Given the coordinates of the moving node S, (x, y), the following equation can be used to calculate the location coordinates of the moving node:

$$d_i = \sqrt{(x - x_i)^2 + (y - y_i)^2} \tag{7}$$

where i = 1, 2, 3, 4.

- AODV localization: AODV is the routing protocol based on the distance-vector algorithm, which integrates the target serial number of DSDV and the on-demand routing discovery in DSR [35]. This protocol mainly includes routing discovery and routing maintenance, where the former is only requested to save the overdue routing.
- AOA: The location of an unlocalized node is estimated through the angles at the points at which the anchor signals are obtained [36,37]. Here the unlocalized nodes implement a triangulation procedure for estimating the location.
- RSSI: In this method, the estimation of the distance between receiver and transmitter is obtained by evaluating the signal strength at the receiver [38,39]. The power of the signal decreases when the distance between receiver and transmitter decreases.

A comparative analysis of fully range-based and hybrid range-based algorithms is shown in Table 2. Fully range-based algorithms are used to determine the distances or angles between nodes to allow an unknown node to be identified easily, whereas hybrid range-based algorithms use various distance and angle measuring methods. Different parameters, such as node density, range, scalability, and reliability of algorithms, are shown in

Table 2. These exploit geometry to improve hybrid AOA/TDOA-based localization (EATL). Fusion of RSSI and TDOA measurements from the wireless sensor network provides robust and accurate indoor localization (FRTL). Hybrid range-based algorithms perform better than fully range-based algorithms. Table 2 shows that hybrid range-based algorithms perform better in range combinations using TOA, TDOA, and RSSI. Scalability and accuracy of the fully range-based algorithms are comparatively lower than those of hybrid range-based algorithms.

Table 2. Localization algorithms (range free vs. hybrid range-based) [40].

Parameters	Fully Range Free Based Algorithms			Hybrid Range Based Algorithm		
	CA	*NCA*	*DV-HoP*	*ATPA*	*EATL*	*FRTL*
Node deployment	Both uniform and random	Random	Random	Random	Both uniform and random	Random
Node density	Low	Low	High	High	Low	Medium
Existence of obstacle	Yes	Yes	Yes	Yes	Yes	Yes
Anchor node presence	Yes	Yes	Yes	Yes	Yes	Yes
Range estimation	Computational	Computational	Computational	Computational	Computational	Computational
Range combination	Centroid	Centroid	TOA, TDOA	TOA	TDOA	RSSI
Localization co-ordinates	RD	3D	2D	2D	2D	2D
Scalability	Yes	Yes	No	Yes	Yes	Yes
Accuracy	Low	Low	Medium	Very High	Very High	High

To design a reliable and scalable sensor network architecture for smart monitoring, we proposed an architecture in which data nodes are deployed on agricultural land and agricultural field node localization is used to establish proper sensor coordination. For the deployment of data nodes to establish suitable communication among sensors, i.e., temperature, humidity, rainfall, altitude, pressure, and fire sensors in open-field lands, hybrid range-based localization techniques are preferred. Data nodes transfer information to the gateway and from the gateway to the cloud platform. A data logger was designed in which all of the sensor data can be stored. For the storage of data, both the local server and cloud server are preferred. The user on a remote device can visualize the data. From remote devices, via the Internet a farmer or any user can operate motors, sprinklers, and blowers connected to farming land.

4. Methods and Materials

4.1. Hardware

The sensors are interfaced and powered from the solar panel. All data is gathered at the sink node and is transmitted to the destination via a gateway. The hardware can be operated on a rechargeable Li-Ion 2200 mAH battery. Figure 3 consists of sensor modules, ADC, LoRa modem, and a LoRa helical antenna. The device can be deployed to collect sensor data from the field and transmit it via the Internet to the receiver device through the gateway, which is a customized ATMEGA 328 p board with an inbuilt Wi-Fi module and it is shown in the Figure 4.

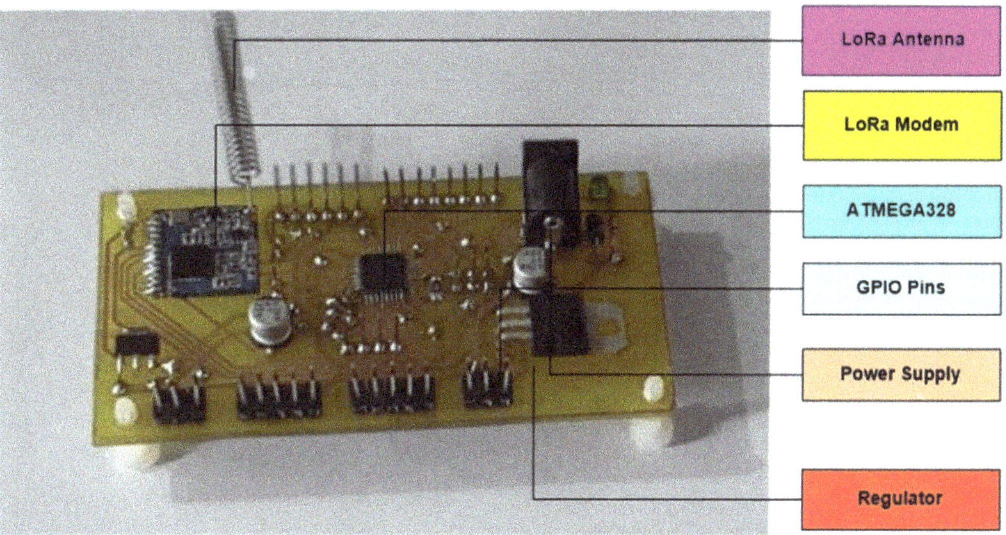

Figure 3. LoRa end node custom design for agriculture farm.

Figure 4. LoRa-based gateway.

The block diagram (Figure 2) is a depiction of the components used in the entire project. The connections were made so that the main component Raspberry Pi board is connected to the Arduino Uno board through the standard pyfirmata. The remainder of the sensors are connected with both the Arduino and Raspberry Pi models. The digital sensors are attached to the Raspberry Pi, and the analog sensors are connected to the Arduino Uno board. The sensors gather data in real-time. All of the outputs are stored on the Thingspeak server/Blynk on the cloud, and can be accessed at any point in time.

A circuit diagram of a reference model of the interface connections made in real time is shown in Figure 5. Firstly, the Arduino is connected to the Raspberry Pi using a USB cable. The gas sensor is connected to the Arduino at the analog A0 pin. The ultrasonic sensor is connected to pin no. 9 of the Arduino for trigger, and echo at pin no. 10. DHT11 is connected to pin 2 of the Arduino. Soil moisture is connected to analog pin A3 of the Arduino. The motor pump is connected to pin no. 8 of the Arduino.

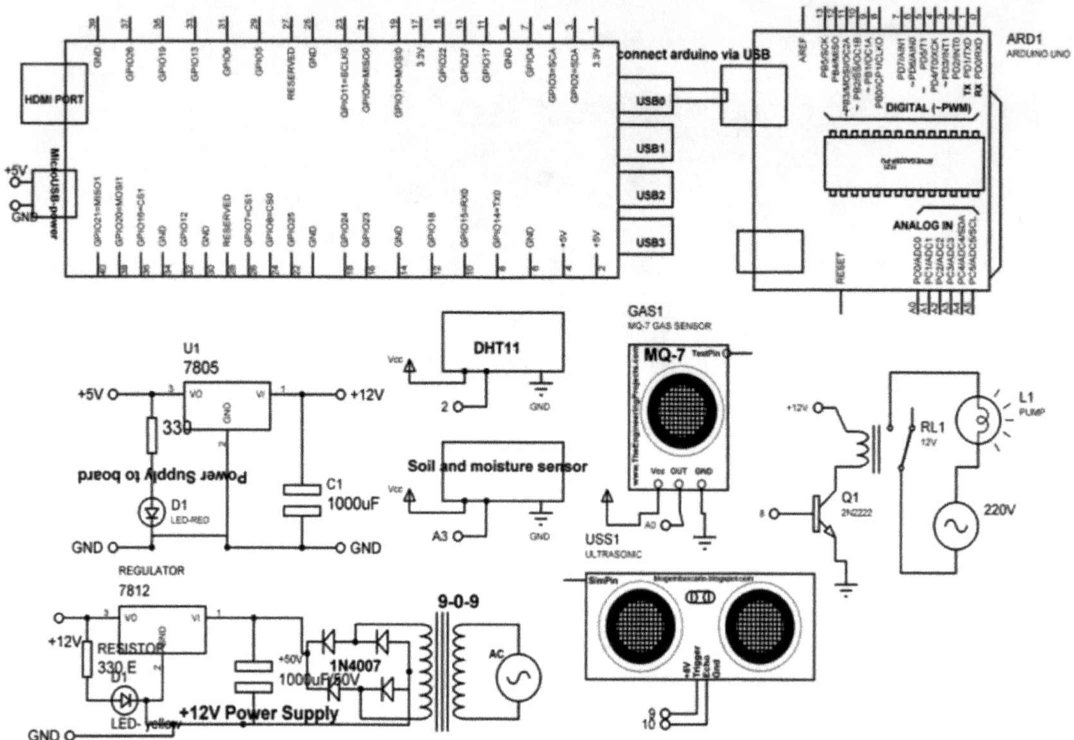

Figure 5. Circuit diagram of the sensor node.

4.2. Proposed LoRa Architecture

A cloud-based agriculture field-monitoring scheme was implemented to precisely monitor the temperature, humidity, and other required parameters to operate end devices in an agriculture field. This scheme facilitates the conservation of water and energy in a field. Here, an open agriculture field was studied. Figure 6 shows the architecture of the LoRa network. The proposed protocol can be adapted to a greenhouse, hydroponic agriculture, and vertical farming. Devices provide numerous data on agriculture parameters from the field, and analysis of the data allows the farmer to work more effectively. A LoRa-enabled sensor network was deployed in the agriculture field. A LoRa-enabled architecture for precise irrigation and monitoring has been demonstrated in which LoRa enabled each sensor node and could establish a communication link with another LoRa-enabled receiver. LoRa is a low-power operating technique that enables transmission and reception with a wide range of communication. Although many other protocols exist, they have significant limitations compared to LoRa.

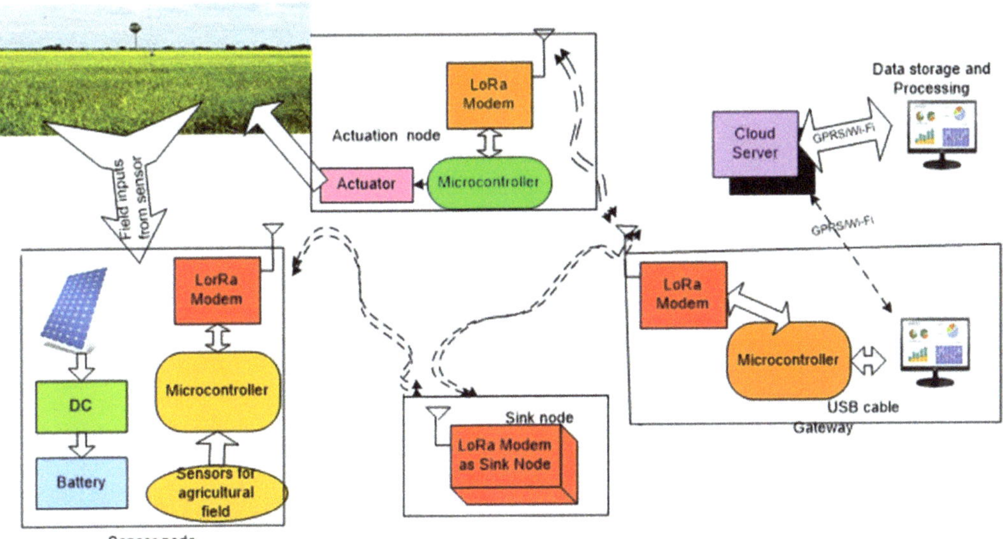

Figure 6. Block diagram of IoT-based farm automation using a solar panel.

5. Simulation

5.1. Localized Algorithm Simulation

A flow diagram of the sensor node localization is illustrated in Figure 7. Initially, the type of node distribution is chosen as per the application, which may be uniform or random. Furthermore, node density is determined to check the number of nodes present in the area (measured in square meters). The algorithm computes these input data and results in sensor node distribution patterns. After confirming the communication link among the sensor nodes, a hybrid range-based localization technique was chosen because it is more effective. Localization is crucial to determine the sensor node target tracking location. Here, we first grouped the algorithms into free range-based and hybrid range-based. Furthermore, we analyzed the suitable localization techniques. Range free-based localization is preferred because of its low power consumption, whereas hybrid range-based localization is preferred depending on the applications. Because the land is not uniform in nature, hybrid range-based localization is widely preferred in agriculture. Most IoT-based applications require sensor node localization because these applications are easy and convenient to monitor. Simulation studies were carried out to compute the localization of algorithms. The effect of the node density, data rate, and signal strength was analyzed to develop the optimal algorithm for our application. Furthermore, we undertook a comparative analysis of the range-based agriculture applications, and concluded that range free-based localization was more suitable.

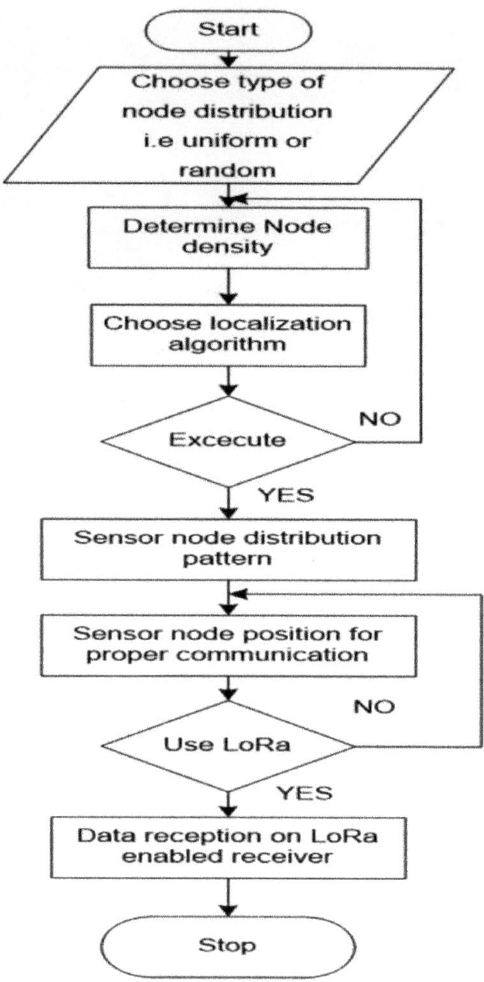

Figure 7. Flow chart for execution of localization of algorithms.

5.2. Simulation of Energy Harvesting

In agriculture, sensor nodes are deployed in an outdoor environment. Moreover, the sensor node's battery life represents a challenge because the life of the battery drains due to distinct environmental conditions. Thus, the optimal solution for implementing energy harvesting for sensor nodes involves renewable energy sources. Energy harvesting is the optimal mechanism used to power the activities of the sensor nodes, including sensing, preprocessing, and transferring data. The evaluation of sensor node battery life using solar panels and wind turbines was performed using a Cisco packet tracer. The Cisco packet tracer is a visualized-based simulation tool that encourages the user to implement distinct network-based simulations. The simulation model for evaluating the sensor node's energy through solar panels and wind turbine was implemented in the Cisco packet tracer environment and is shown in Figure 8.

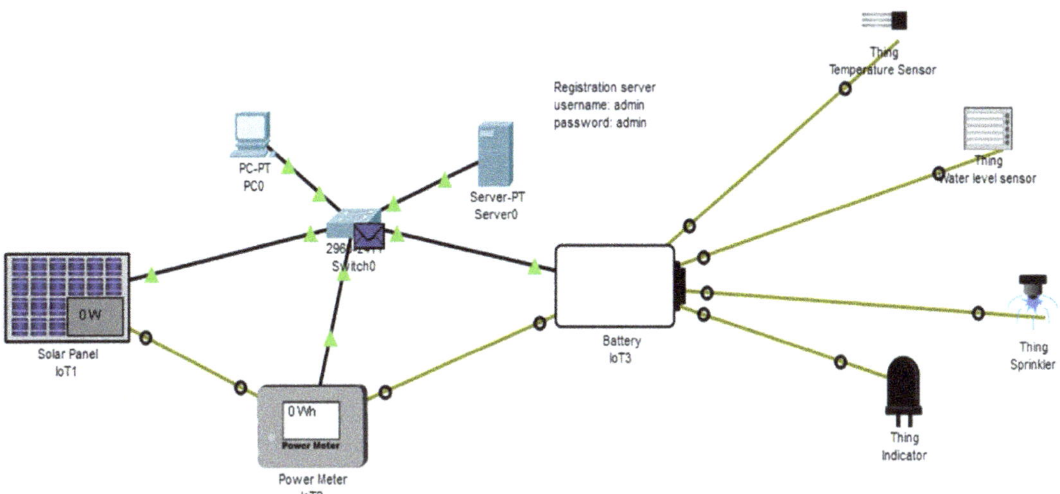

Figure 8. Sensor node communication establishment using LoRa link.

The lifetime of the nodes is presented in the simulation, as shown in Figure 10. Data transmission occurs from end nodes to the hub, the hub to a central server, and from mobile phones via cell towers. The simulation panel in Figure 9 presents the number of IoT devices connected to the network. Each renewable resource is connected using a power meter for calculating the power consumption. Timestamp details of the IoT devices are represented in the time column; the IoT 2 device is connected to the network with a time of 0.129 s. Devices IoT0 and IoT1 are currently connected within 1.011 s. Total device connection time is 77.660 s with a specific time gap or delay. Further power is transferred to the battery for storage purposes. A sample scenario of power of 11 kWh in the case of a turbine and 78 watts with 82 Wh was considered.

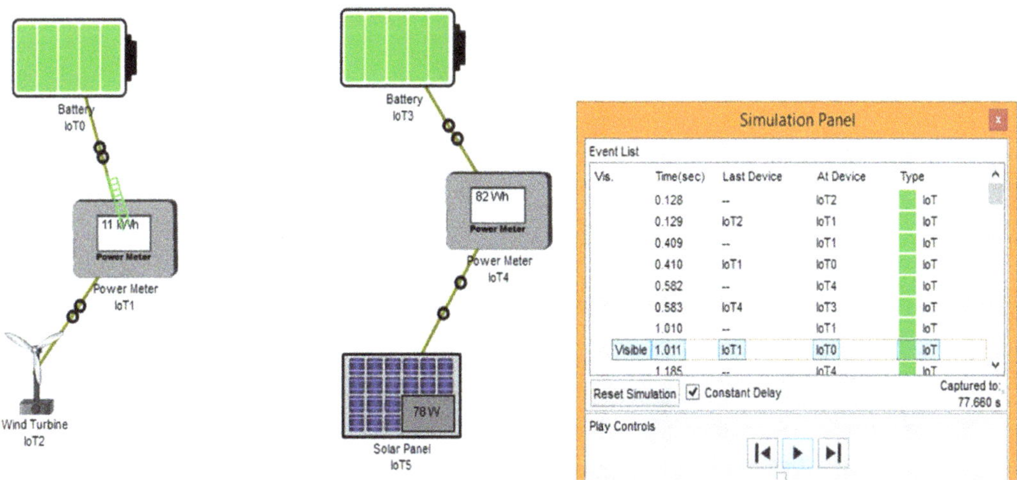

Figure 9. Sensor node energy conservation using renewable resources in an agriculture field.

The establishment of LoRa-based communication from the sensor node to the mobile phone is presented in Figure 10. The simulation model shown in Figure 8 signifies the communication link established among sensor end nodes and establishes a connection with the gateway to transmit the data to the central server. In the simulation panel, the data transmission time in seconds, from the router to the hub and the hub to smartphones, is displayed. The simulation signifies the on-air data transmission. Three events are noted as having a time of 218.952 s. Each of the three devices, i.e., wireless router, central office, and another wireless router, connected to the network at the same timestamp. This signifies, in LoRa communication, that the device synchronization time gap is extremely small, which is an additional advantage. As soon as the authentication key matches, the devices are synchronized and begin communicating.

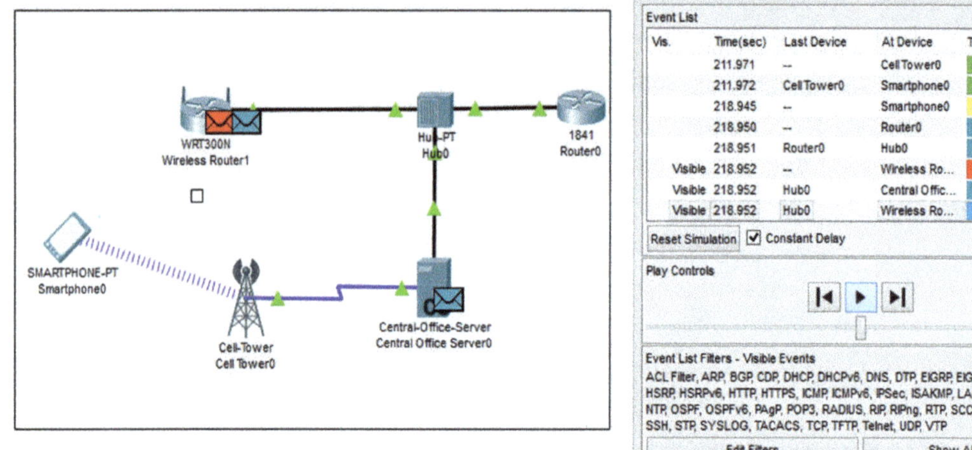

Figure 10. LoRa simulation platform with simulation parameters.

5.3. MATLAB Simulation

The MATLAB simulation platform was considered in this study for characterizing the behavior of the LoRaWAN network. A simulation model was developed in MATLAB for evaluating the effects of various parameters, such as SNR, bitrate, and SF. A sample LoRa testbed was developed on the MATLAB simulation platform. The simulation was implemented by considering the following features: a network with 10 to 100 nodes, one gateway, and one network server. The nodes were distributed both randomly and uniformly with a minimum period of 100 s. SF, transmission power, and gateway transmission power were defined as per the protocol. The SNR value was determined from the simulation result. We classified the simulation into cases as follows:

Case I:

LoRa trades the transmission and reception data rate for sensitivity within a given channel bandwidth. As shown in Table 3, LoRa implements an adaptive data rate by the utilization of orthogonal SFs. This allows the user to minimize the power consumption and optimize the network performance for a given bandwidth. The receiver maintains the mode of operation downlink, gate power at 27 dB, node sensitivity at -124 dB, operating frequency at 868 MHz, antenna gain at 10 dB, and node noise and node antenna gain at 30 dB. After performing the simulation, the obtained results are presented in Table 4. The results indicate that a change in antenna height from 1 to 7 m leads to significant changes in the coverage area.

Table 3. Adaptive data rate of LoRa.

SF	Chirps	SNR	ToA	Data Ate
7	128	−8.5	122 ms	6345 bps
8	256	−11	189 ms	4425 bps
9	512	−15.48	235 ms	2118 bps
10	1024	−18.5	381 ms	1233 bps
11	2048	−15.48	235 ms	2118 bps
12	4096	−18.5	381 ms	1233 bps

Table 4. Range and link budget at node sensitivity—124 dBm.

Mode (Down Link)	Gateway Height in Meter	End Node Height in Meter	Link Budget dBm	Range in Meter
Down Link	1	1	159	932
Down Link	2	1	159	1318
Down Link	3	1	159	1614
Down Link	4	1	159	1683
Down Link	5	1	159	2083
Down Link	6	1	159	2282
Down Link	7	1	159	2465

For the parameters, such as frequency of 868 MHz, node sensitivity of −124 dBm, gateway node power of 10 dB, and antenna gain of 27 dB, a relationship between the height of gateway and node with range is plotted in Figure 11.

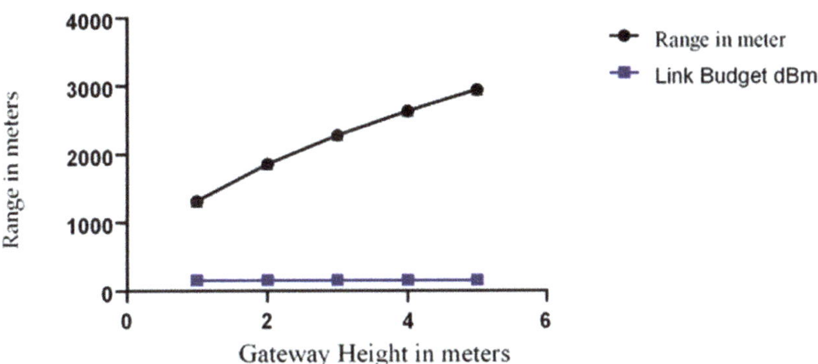

Figure 11. Gateway height versus range in meters.

The experimental data shown in Table 5 provides a brief overview of the correlation between the height of the gateway, the height of end nodes, and the range. During data reception at the receiver, the data were interpreted; the table's model signifies the downlink. The graph concludes that a constant link budget of 159 dBm range is directly proportional to the height of the end node and gateway. Thus, to achieve greater range coverage, it is considered that the end nodes should be mounted on cylindrical bars, as shown in Figure 11.

Table 5. Range and link budget at node sensitivity of −124 dB.

Mode (Down Link)	Gateway Height in Meter	End Node Height in Meter	Link Budget dBm	Range in Meter
Down Link	1	2	159	1319
Down Link	2	2	159	1863
Down Link	3	2	159	2282
Down Link	4	2	159	2635
Down Link	5	2	159	2946

Case II

The establishment of the LoRa communication link significantly depends on the gateway height and the end node heights. Experimental data are presented in Table 5. Here the node sensitivity is −124 dBm and the operating frequency is 868 MHz, and it can be observed that the signal coverage area changes in addition to the variation in heights of the gateway and end nodes. The end node height is 2 m from the ground. If the gateway height is adjusted from 1 to 5 m, the coverage area and signal strength change drastically change from 1319 to 2946 m. This signifies that maintaining all of the parameters at a certain value will enhance the link budget strength increases because there is less interference.

Case III:

In this case, the following parameters were considered: node sensitivity at −137 dBm and operating frequency at 433 MHz. Now, changing the value of the gateway height significantly impacts the range. Maintained a link budget of 151 dBm, it is observed that SNR decreases because there is a low possibility of interference. The mode of operation during the experiment is downlink, i.e., data received at the receiver side. During the experiment, it was observed that at a fixed node sensitivity of −137 dBm and antenna gain of 10 dB, the custom-built sensor node varies the value of the link budget (in dB) with a change in the wide-area range coverage at a frequency of 433 MHz. This is shown in Table 6 and Figure 10. In Figure 12, a relationship between antenna gain, link budget, and range is plotted. The blue line graph shows increasing antenna gain increases the range, whereas the red line shows increasing the range in response to the link budget in LoRa. By maintaining all of the parameter values, antenna height is directly proportional to the coverage range. This signifies that deploying the LoRa receiver at a certain height can enable a strong communication link.

Table 6. Range and link budget at node sensitivity of −137 dBm.

Mode (Down Link)	Gateway Height in Meter	End Node Height in Meter	Link Budget dBm	Range in Meter
Down Link	1	1	151	1681
Down Link	2	1	151	2377
Down Link	3	1	151	2911
Down Link	4	1	151	3362
Down Link	5	1	151	3758

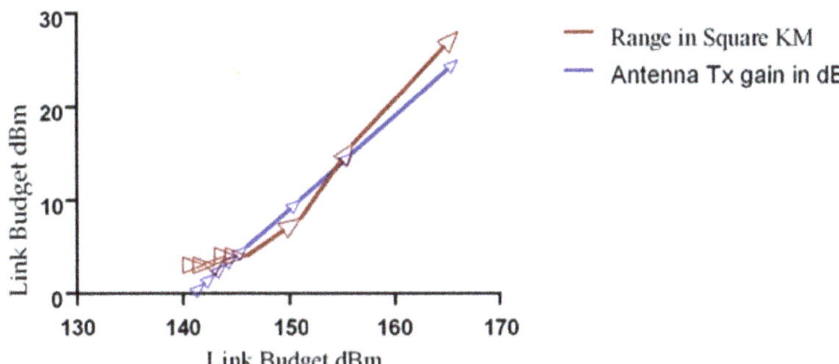

Figure 12. Link budget variation with range.

Case IV:

Transmission signal strength is optimized by setting the antenna gain from 1 to 25 dB. Here, the transmission range and link budget of the LoRa transmitter module were evaluated with a change in antenna gain. The mode of operation is uplink, as presented in Table 7. Antenna gain was changed by writing the MATLAB code for interpreting the signal strength at a frequency of 433 MHz. When gateway noise remains at 10 dB and node sensitivity at −137 dBm, the link budget increases from 141 to 166 dBm. Hence, it is concluded that by changing antenna gain, we can improve the coverage area. Thus, it is preferred to use an antenna with high gain.

Table 7. Link budget in dBm from 141 to 164 dBm for the transmitter LoRa module.

Mode (Up Link)	Antenna Tx Gain in dB	Link Budget dBm	Range in Meter	Range in Square KM
Up Link	0	141	945	3
Up Link	1	142	1001	3
Up Link	2	143	1061	3
Up Link	3	144	1123	4
Up Link	4	145	1190	4
Up Link	5	146	1260	4
Up Link	10	151	1681	8
Up Link	15	156	2241	16
Up Link	25	166	3986	28

Experimental parameters are presented in Table 8, and signify the importance of node sensitivity. We used a helical antenna in the design of the custom sensor node and gateway. Maintaining the uplink frequency at 433 MHz, by changing node sensitivity from −124 to −130 dBm, the sensor coverage distance increased from 945 to 3986 m. The data from Table 8 indicates that the change in node sensitivity is directly proportional to the range.

As the height of the gateway and sensor node changes, the range also varies. The variation in both gateway and end node height changes the range significantly during downlink mode at the receiver. In Figure 13, the red triangles denote the data rate, and black downward triangles signify the bit rate.

Figure 14a,b shows the relation of bit error rate, packet error rate, and symbol error rate with the spreading factor. As the spreading rate increases, there is an increase in the bit error rate. When the spreading factor is changed from 7 to 12, the bit error is minimized at SF = 6. Thus, there is a loss of data packets. It is preferred to send the data with the spreading factor so that bit error and packet error rate are minimized. To conclude, it is preferred to minimize the SF value to reduce the bit rate error. To maintain minimal collision among data packets, a lower value SF is preferred.

After computing all of the essential communication parameters of LoRa, it was found that increasing SNR affects the PER, SER, and BER. At −30 dB of SNR, BER is almost 1, which is not considered. It was also observed that with a decrease in the value of SNR to 0 dB, the error rate decreases.

Table 8. Range for the node sensitivity from −124 to −132 dBm.

Mode (Up Link)	Frequency	Node Sensitivity in dBm	Range in Meter	Range in Square KM
Up Link	433	−124	945	3
Up Link	433	−125	1001	3
Up Link	433	−126	1061	3
Up Link	433	−127	1123	4
Up Link	433	−128	1190	4
Up Link	433	−129	1260	4
Up Link	433	−130	1681	8
Up Link	433	−131	2241	16
Up Link	433	−132	3986	28

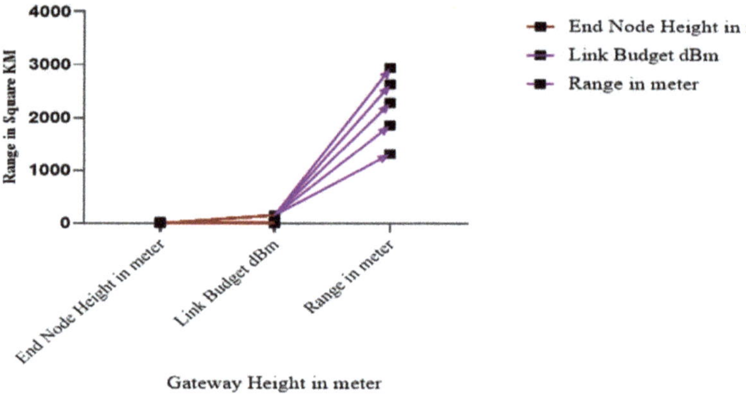

Figure 13. Link budget variation with range.

To calculate the data rate of LoRa, the input parameters such as CR, SF and BW are included in the equation. Figure 15 presents the data rate of LoRa from SF 7 to SF 12. The data rate is denoted in terms of bits per second (bps). In each SF, the data rate increases at BW 7, then increases exponentially after BW 8, and reaches a limit at BW 10. We can observe that SF 7 has an inverse effect on the data rate, because the data rate steadily decreases from SF 7 to SF 12. In SF 7, the data rate of the LoRa reached 22,000 bps, and in SF 12, the data rate was limited to 2000 bps. An increase in SF will lead to transmission of a low amount of data; thus, SF 7 is the optimal SF that needs to be considered for sending a large amount of data.

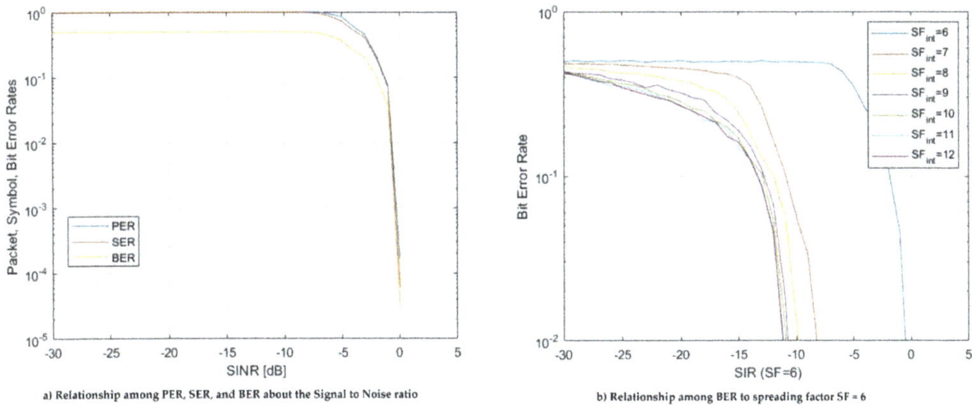

Figure 14. (**a**,**b**) PER, SER, and BER versus the signal to noise ratio and BER versus SF = 6.

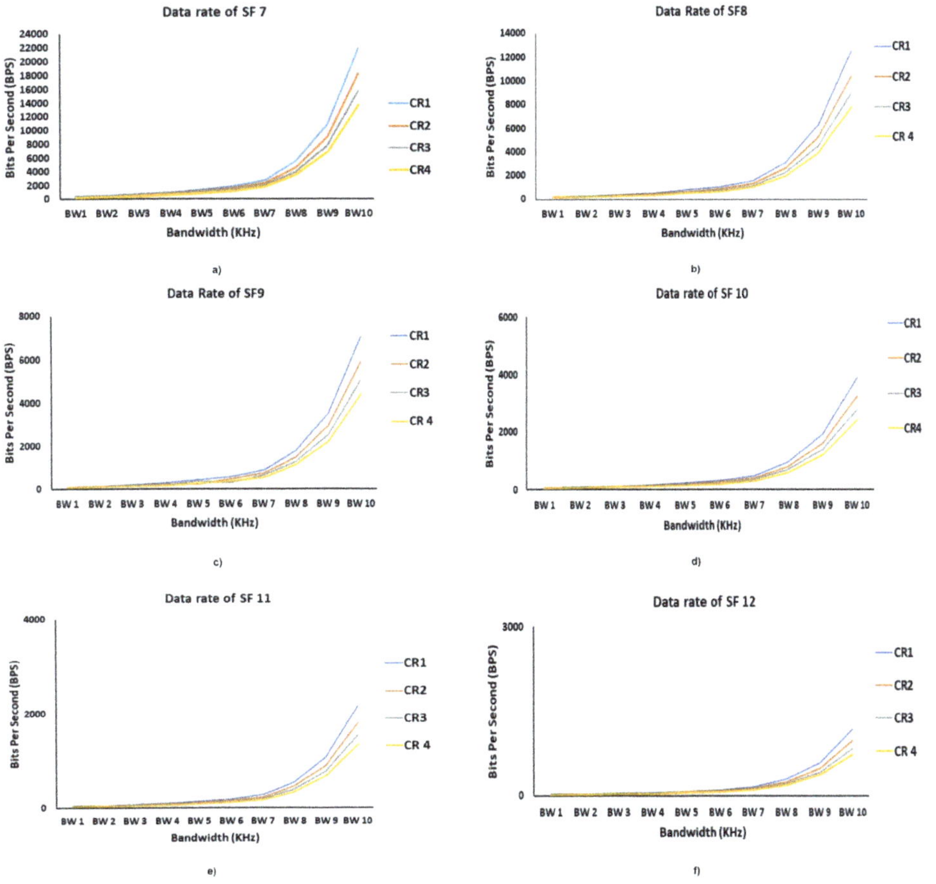

Figure 15. Data rate of LoRa from SF 7 to SF 12.

6. Results of the Experimental Setup

This section discusses the deployment of sensor nodes and the gateway in a real-time environment. We also present the sensor data recorded on the cloud server and compare previous studies with the proposed research in detail. To evaluate the coverage of LoRa, we deployed the sensor nodes in a testbed located in our university center. The gateway was placed at a distance of 1 km from the sensor nodes. To analyze the customized sensor node and gateway for crop field monitoring, the sensor node was deployed in the crop field, as shown in Figure 16a,b. The sensor node was embedded with sensors for temperature, humidity, soil moisture, and fire. A 433 MHz-based LoRa was embedded in the sensor node and gateway. LoRa is a form of transceiver communication. The LoRa-based gateway was positioned 1 km from the two sensor nodes. The gateway showed effective results in terms of accurately receiving the sensory data. The gateway was also embedded with an 8266 Wi-Fi modem to communicate the data over the cloud server through the Internet. The sensory data is visible in Figure 16c.

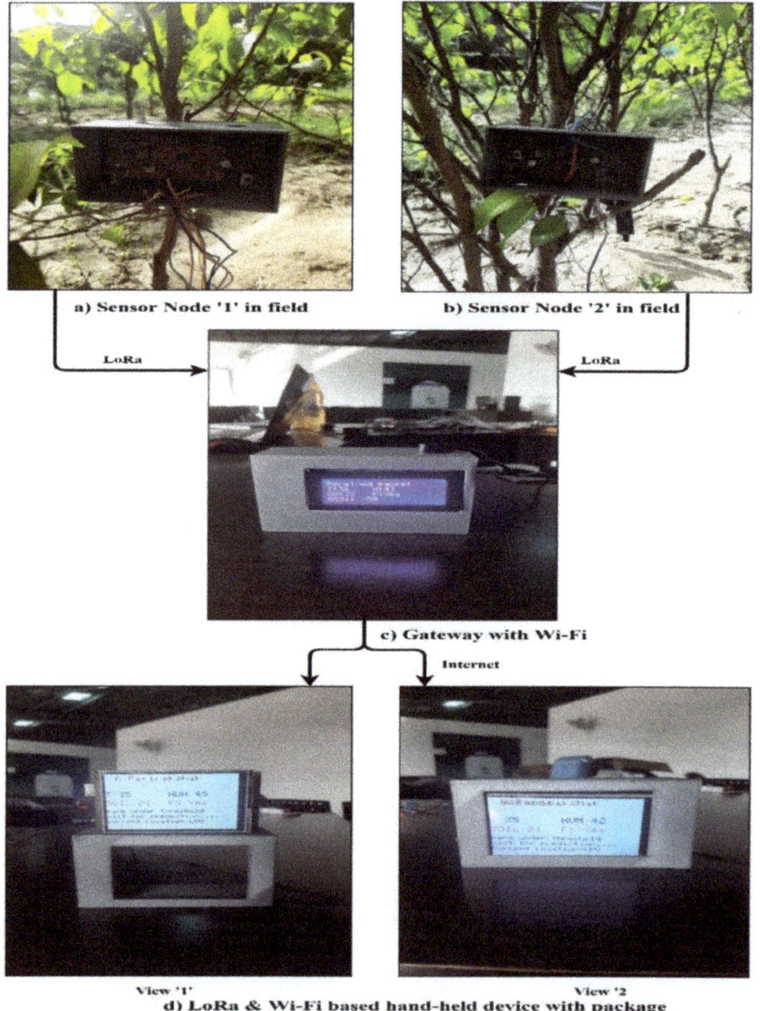

Figure 16. Real-time experimental setup of the sensor node and gateway.

Moreover, a hand-held device (or portable device) was also designed and implemented in real-time scenarios, shown in Figure 16d. This hand-held device will assist farmers in checking the field from a remote location and is easy to use. A hand-held device integrates LoRa communication and an ESP 8266 Wi-Fi module. These two communication modules assist in receiving the crop data regarding from the sensor node and gateway. The hand-held device is used to visualize the data of the crop field on a color LCD. The hand-held device shows the sensory data, including temperature, humidity, soil moisture, and fire detection. The gateway node logs the data on the cloud server through the Internet. Here, we used the Blynk cloud server to record the sensory data of the sensor node. The data regarding temperature and humidity can be seen in the Blynk dashboard, and is presented in Figure 17.

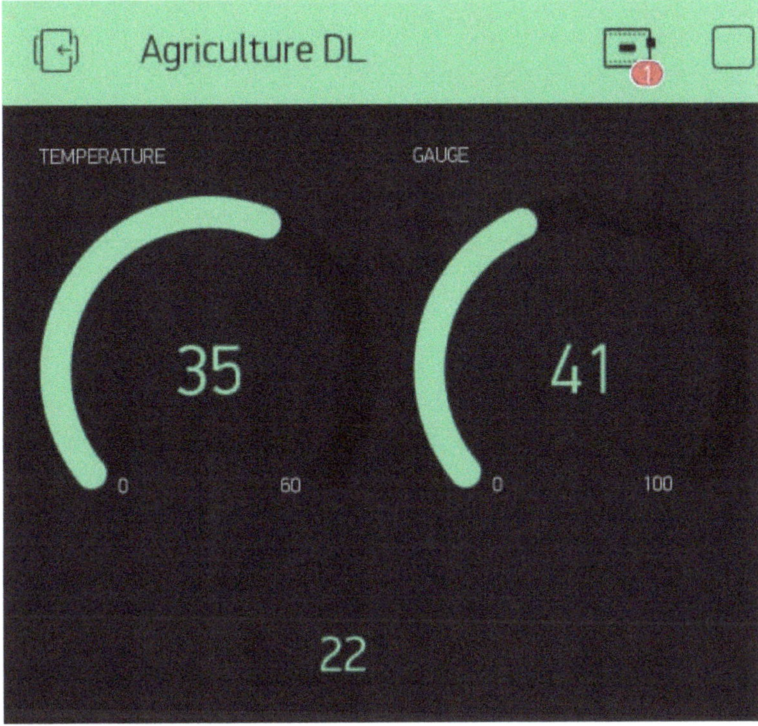

Figure 17. Blynk dashboard.

A comparative analysis of previous research on LoRa link budgets is depicted in Table 9. Evaluation parameters are the microcontroller unit (MCU), the communication protocol used, customized hardware designed for the sensor node, the gateway, proof of concepts, and simulation-based analysis used to validate the proposed study. The proposed study has the advantages of a communication protocol that uses both LoRa and Wi-Fi. This means the data can be collected on a local platform and the cloud platform. When designing sensor nodes, each sensor node and gateway were custom designed and built as per the requirements. Customization helped us to reduce the dimensions and architecture of the device. A simulation-based analysis was carried out to validate the proposed study, and further validated on the LoRa testbed.

Table 9. Comparison of the LoRa-based agriculture research with the present study.

Research	Communication Protocol	Custom End Node	Custom Gateway	Hand-Held for Farmer	Link Budget Validation	LoRa Simulation	Plot of Evaluation Metrics
[12]	LoRa	No	No	No	Yes	Yes	yes
[13]	LoRa	No	No	No	Yes	Yes	yes
[41]	WiFi	No	No	No	Yes	Yes	yes
[42]	WiFi	No	No	No	Yes	Yes	yes
Proposed study	LoRa + WiFi (with optimized embedded firmware)	Yes (customized)	Yes (custom design)	Yes (customized)	Yes	Simulation + validation on hardware	Yes

7. Conclusions

IoT is transforming all applications due to its unique feature of real-time monitoring. In agriculture, IoT can enhance crop yields by integrating distance sensors and wireless communication protocols. In agriculture, implementing internet-based sensor nodes is challenging due to the unavailability of nearby agriculture fields. To overcome this, we proposed an IoT-based WSN architecture for real-time monitoring of agricultural fields. A custom sensor node and gateway were developed and implemented for sensing and communicating real-time agricultural data. In agriculture, the sensor node's battery life plays a significant role. We implemented an energy harvesting mechanism using solar energy and a wind turbine. A MATLAB simulation was performed to evaluate the correlation of distinct parameters, such as link budget, spreading factor, and receiver sensitivity, in a custom-built LoRa testbed. Moreover, this article discusses the role of localization for deploying the sensor nodes in agricultural fields. The MATLAB simulation indicated that hybrid range-based localization algorithms are more reliable and are scalable for deployment in agricultural fields. Finally, a real-time experiment was performed to analyze the performance of the custom sensor node and gateway.

Author Contributions: M.S. and R.S. conceived the conception; M.S. and M.R. conducted literature collection and manuscript writing; D.Z., S.H., M.F.H. revised and polished the manuscript and provided valuable suggestions. All authors have read and agreed to the published version of the manuscript.

Funding: This research received no external funding.

Institutional Review Board Statement: Not Applicable.

Informed Consent Statement: Not Applicable.

Data Availability Statement: Not Applicable.

Conflicts of Interest: The authors declare no conflict of interest.

List of Abbreviations

AOA	Angle of Arrival
APIT	Approximate Point in Triangulation
BER	Bit Error Rate
BW	Bandwidth
CR	Code Rate
CSS	Chirp Spread Spectrum
DV hop	Distance Vector hop
FEC	forward Error Correction
GPS	Global Positioning System
GPRS	Global Packet for Radios Service
IoT	Internet of Things

ISM	Industrial, Scientific & Medical
LoRa	Long Range
LoRaWAN	LoRa Wide Area Network
LPWAN	Low Power Wide Area Network
MCU	Microcontroller Unit
MMS	Multimedia Message
NB-IoT	Narrow Band-IoT
NCA	Neighbor Constraint Assisted
NLOS	non-line of sight
PER	Packet Error rate
PA	Precision Agriculture
SF	Spreading Factor
SNR	Signal-to-Noise Ration
SINR	Signal into Noise Ratio
RSSI	Received Signal Strength Indicator
RTOF	Roundtrip-Time of Flight
TOA	Time on arrival
TDOA	Time difference of Arrival TDOA
WSN	Wireless Sensor Network
Wi-Fi	Wireless-Fidelity

References

1. World Population Projected to Reach 9.8 Billion in 2050 and 11.2 Billion in 2100. April 2021. Available online: https://www.un.org/development/desa/en/news/population/world-population-prospects-2017.html (accessed on 5 April 2021).
2. 68% of the World Population Projected to Live in Urban Areas by 2050 Says UN. April 2021. Available online: https://www.un.org/development/desa/en/news/population/2018-revision-of-world-urbanization-prospects.html (accessed on 5 April 2021).
3. Food Production Must Double by 2050 to Meet Demand From World's Growing Population. April 2021. Available online: https://www.un.org/press/en/2009/gaef3242.doc.html (accessed on 5 April 2021).
4. Ayaz, M.; Ammad-Uddin, M.; Sharif, Z.; Mansour, A.; Aggoune, E.H.M. Internet-of-Things (IoT)-based smart agriculture: Toward making the fields talk. *IEEE Access* **2019**, *7*, 129551–129583. [CrossRef]
5. Trivelli, L.; Apicella, A.; Chiarello, F.; Rana, R.; Fantoni, G.; Tarabella, A. From precision agriculture to Industry 4.0: Unveiling technological connections in the agrifood sector. *Br. Food J.* **2019**, *121*, 1730–1743. [CrossRef]
6. Horton, P.; Horton, P.; Horton, B.P.; Horton, B.P.; Horton, P.; Horton, P.; Horton, B.P.; Horton, B.P.; Horton, P.; Horton, P.; et al. Re-defining Sustainability: Living in Harmony with Life on Earth. *One Earth* **2019**, *1*, 86–94. [CrossRef]
7. Navulur, S.; As, C.S.S.; Prasad, M.N.G. Agricultural Management through Wireless Sensors and Internet of Things. *Int. J. Electr. Comput. Eng.* **2017**, *7*, 3492–3499. [CrossRef]
8. Goyat, R.; Rai, M.K.; Kumar, G.; Kim, H.J.; Lim, S.J. Improved DV-Hop Localization Scheme for Randomly Deployed WSNs. *Int. J. Sens. Wirel. Commun. Control* **2020**, *10*, 94–109. [CrossRef]
9. Jondhale, S.R.; Deshpande, R.S. Self-recurrent neural network based target tracking in wireless sensor network using state observer. *Int. J. Sens. Wirel. Commun. Control* **2019**, *9*, 165–178. [CrossRef]
10. Shafi, U.; Mumtaz, R.; García-Nieto, J.; Hassan, S.A.; Zaidi, S.A.R.; Iqbal, N. Precision agriculture techniques and practices: From considerations to applications. *Sensors* **2019**, *19*, 3796. [CrossRef]
11. Antony, A.P.; Leith, K.; Jolley, C.; Lu, J.; Sweeney, D.J. A Review of Practice and Implementation of the Internet of Things (IoT) for Smallholder Agriculture. *Sustainability* **2020**, *12*, 3750. [CrossRef]
12. Singh, R.K.; Aernouts, M.; De Meyer, M.; Weyn, M.; Berkvens, R. Leveraging LoRaWAN Technology for Precision Agriculture in Greenhouses. *Sensors* **2020**, *20*, 1827. [CrossRef]
13. Haseeb, K.; Ud Din, I.; Almogren, A.; Islam, N. An energy efficient and secure IoT-based WSN framework: An application to smart agriculture. *Sensors* **2020**, *20*, 2081. [CrossRef]
14. Farooq, M.S.; Riaz, S.; Abid, A.; Abid, K.; Naeem, M.A. A Survey on the Role of IoT in Agriculture for the Implementation of Smart Farming. *IEEE Access* **2019**, *7*, 156237–156271. [CrossRef]
15. Feng, X.; Yan, F.; Liu, X. Study of Wireless Communication Technologies on Internet of Things for Precision Agriculture. *Wirel. Pers. Commun.* **2019**, *108*, 1785–1802. [CrossRef]
16. Prathibha, S.R.; Hongal, A.; Jyothi, M.P. IoT based monitoring system in smart agriculture. In Proceedings of the 2017 International Conference on Recent Advances in Electronics and Communication Technology (ICRAECT), Bangalore, India, 16–17 March 2017; pp. 81–84.
17. Prakash, S. Zigbee based Wireless Sensor Network Architecture for Agriculture Applications. In Proceedings of the 2020 Third International Conference on Smart Systems and Inventive Technology (ICSSIT), Tirunelveli, India, 20–22 August 2020; pp. 709–712.

18. Saraf, S.B.; Gawali, D.H. IoT based smart irrigation monitoring and controlling system. In Proceedings of the 2017 2nd IEEE International Conference on Recent Trends in Electronics, Information & Communication Technology (RTEICT), Bangalore, India, 19–20 May 2017; pp. 815–819.
19. Ramachandran, V.; Ramalakshmi, R.; Srinivasan, S. An automated irrigation system for smart agriculture using the Internet of Things. In Proceedings of the 2018 15th International Conference on Control, Automation, Robotics and Vision (ICARCV), Singapore, 18–21 November 2018; pp. 210–215.
20. Saqib, M.; Almohamad, T.A.; Mehmood, R.M. A Low-Cost Information Monitoring System for Smart Farming Applications. *Sensors* **2020**, *20*, 2367. [CrossRef]
21. Siddique, A.; Prabhu, B.; Chaskar, A.; Pathak, R. A review on intelligent agriculture service platform with lora based wireless sensor network. *Life* **2019**, *100*, 7000.
22. Ji, M.; Yoon, J.; Choo, J.; Jang, M.; Smith, A. Lora-based visual monitoring scheme for agriculture iot. In Proceedings of the 2019 IEEE Sensors Applications Symposium (SAS), Sophia Antipolis, France, 11–13 March 2019; pp. 1–6.
23. Alrajeh, N.A.; Bashir, M.; Shams, B. Localization Techniques in Wireless Sensor Networks. *Int. J. Distrib. Sens. Netw.* **2013**, *9*, 304628. [CrossRef]
24. Aghaie, N.; Tinati, M.A. Localization of WSN nodes based on NLOS identification using AOAs statistical information. In Proceedings of the 2016 24th Iranian Conference on Electrical Engineering (ICEE), Shiraz, Iran, 10–12 May 2016; pp. 496–501.
25. Pak, J.M.; Ahn, C.K.; Shi, P.; Shmaliy, Y.S.; Lim, M.T. Distributed Hybrid Particle/FIR Filtering for Mitigating NLOS Effects in TOA-Based Localization Using Wireless Sensor Networks. *IEEE Trans. Ind. Electron.* **2016**, *64*, 5182–5191. [CrossRef]
26. Jawad, H.M.; Nordin, R.; Gharghan, S.K.; Jawad, A.M.; Ismail, M. Energy-Efficient Wireless Sensor Networks for Precision Agriculture: A Review. *Sensors* **2017**, *17*, 1781. [CrossRef] [PubMed]
27. Mekki, K.; Bajic, E.; Chaxel, F.; Meyer, F. A comparative study of LPWAN technologies for large-scale IoT deployment. *ICT Express* **2019**, *5*, 1–7. [CrossRef]
28. Chaudhari, B.S.; Zennaro, M.; Borkar, S. LPWAN Technologies: Emerging Application Characteristics, Requirements, and Design Considerations. *Future Internet* **2020**, *12*, 46. [CrossRef]
29. Lee, H.-C.; Ke, K.-H. Monitoring of Large-Area IoT Sensors Using a LoRa Wireless Mesh Network System: Design and Evaluation. *IEEE Trans. Instrum. Meas.* **2018**, *67*, 2177–2187. [CrossRef]
30. LoRa Modulation Basics. Available online: http://www.semtech.com/images/datasheet/an1200.22.pdf (accessed on 7 April 2021).
31. Huang, Q.; Selvakennedy, S. A range-free localization algorithm for wireless sensor networks. In Proceedings of the 2006 IEEE 63rd Vehicular Technology Conference, Melbourne, VIC, Australia, 7–10 May 2006; Volume 1, pp. 349–353.
32. He, T.; Huang, C.; Blum, B.M.; Stankovic, J.A.; Abdelzaher, T. Range-free localization schemes for large scale sensor networks. In Proceedings of the 9th Annual International Conference on Mobile Computing and Networking, San Diego, CA, USA, 14–19 September 2003; pp. 81–95.
33. Paul, A.K.; Sato, T. Localization in Wireless Sensor Networks: A Survey on Algorithms, Measurement Techniques, Applications and Challenges. *J. Sens. Actuator Netw.* **2017**, *6*, 24. [CrossRef]
34. Langendoen, K.; Reijers, N. Distributed localization in wireless sensor networks: A quantitative comparison. *Comput. Netw.* **2003**, *43*, 499–518. [CrossRef]
35. Mao, G.; Fidan, B.; Anderson, B.D. Wireless sensor network localization techniques. *Comput. Netw.* **2007**, *51*, 2529–2553. [CrossRef]
36. Srinivasan, A.; Wu, J. A survey on secure localization in wireless sensor networks. In *Encyclopedia of Wireless and Mobile Communications*; Auerbach Publications: Boston, MA, USA, 2007; p. 126.
37. Niculescu, D.; Nath, B. Ad hoc positioning system (APS) using AOA. In Proceedings of the IEEE INFOCOM 2003. Twenty-second Annual Joint Conference of the IEEE Computer and Communications Societies (IEEE Cat. No. 03CH37428), San Francisco, CA, USA, 30 March–3 April 2003; Volume 3, pp. 1734–1743.
38. Youssef, A.M.; Youssef, M. A taxonomy of localization schemes for wireless sensor networks. In Proceedings of the ICWN, Las Vegas, NV, USA, 25–28 June 2007; pp. 444–450.
39. Niu, R.; Vempaty, A.; Varshney, P.K. Received-Signal-Strength-Based Localization in Wireless Sensor Networks. *Proc. IEEE* **2018**, *106*, 1166–1182. [CrossRef]
40. Mesmoudi, A.; Feham, M.; Labraoui, N. Wireless sensor networks localization algorithms: A comprehensive survey. *arXiv* **2013**, arXiv:1312.4082. [CrossRef]
41. Roopaei, M.; Rad, P.; Choo, K.K.R. Cloud of things in smart agriculture: Intelligent irrigation monitoring by thermal imaging. *IEEE Cloud Comput.* **2017**, *4*, 10–15. [CrossRef]
42. Suma, D.N.; Samson, S.R.; Saranya, S.; Shanmugapriya, G.; Subhashri, R. IOT Based Smart Agriculture Monitoring System. *Int. J. Recent Innov. Trends Comput. Commun.* **2017**, *5*, 177–181.

Article

Effects on Crop Development, Yields and Chemical Composition of Celeriac (*Apium graveolens* L. *var. rapaceum*) Cultivated Underneath an Agrivoltaic System

Axel Weselek [1,*], Andrea Bauerle [2], Sabine Zikeli [3], Iris Lewandowski [2] and Petra Högy [1]

[1] Department of Plant Ecology and Ecotoxicology, Institute of Landscape and Plant Ecology, University of Hohenheim, 70599 Stuttgart, Germany; Petra.Hoegy@uni-hohenheim.de

[2] Department of Biobased Resources in the Bioeconomy, Institute of Crop Science, University of Hohenheim, 70599 Stuttgart, Germany; a.bauerle@uni-hohenheim.de (A.B.); iris_lewandowski@uni-hohenheim.de (I.L.)

[3] Center for Organic Farming, University of Hohenheim, 70599 Stuttgart, Germany; zikeli@uni-hohenheim.de

* Correspondence: a.weselek@uni-hohenheim.de

Abstract: Agrivoltaic (AV) systems increase land productivity through the combined production of renewable energy and food. Although several studies have addressed their impact on crop production, many aspects remain unexplored. The objective of this study was to determine the effects of AV on the cultivation of celeriac, a common root vegetable in Central Europe. Celeriac was cultivated in 2017 and 2018 as part of an organically managed on-farm experiment, both underneath an AV system and in full-sun conditions. Under AV, photosynthetic active radiation was reduced by about 30%. Monitoring of crop development showed that in both years, plant height increased significantly under AV. Fresh bulb yield decreased by about 19% in 2017 and increased by about 12% in 2018 in AV, but the changes were not significant. Aboveground biomass increased in both years under AV, but only increased significantly in 2018. As aboveground biomass is a determinant of root biomass at harvest in root vegetables, bulb yields may be further increased by a prolonged vegetation period under AV. Compound analysis of celeriac bulbs did not show any clear effects from treatment. As harvestable yields were not significantly reduced, we concluded that celeriac can be considered a suitable crop for cultivation under AV.

Keywords: agrivoltaic; agrophotovoltaic; Agri-PV; shading; crop performance; yields; product quality; organic agriculture; biodynamic agriculture; land productivity

Citation: Weselek, A.; Bauerle, A.; Zikeli, S.; Lewandowski, I.; Högy, P. Effects on Crop Development, Yields and Chemical Composition of Celeriac (*Apium graveolens* L. *var. rapaceum*) Cultivated Underneath an Agrivoltaic System. *Agronomy* **2021**, *11*, 733. https://doi.org/10.3390/agronomy11040733

Academic Editors: Miguel-Ángel Muñoz-García and Luis Hernández-Callejo

Received: 3 March 2021
Accepted: 7 April 2021
Published: 10 April 2021

1. Introduction

Agrivoltaic (AV) systems are currently being implemented in a number of countries as an approach for the dual use of arable land for renewable energy and agricultural production [1,2]. It has been shown that both land productivity and farm income can be increased by the additional energy generated through AV [1–5]. Recently, first concepts for the integration of AV into prospective farming systems—e.g., in combination with farming robots—have been proposed [6]. However, considering the land use conflict between food and energy production, a sustained adequate agricultural yield needs to be guaranteed if AV systems are to be used. This necessitates further field studies on the performance of agricultural crops under AV. The implementation of AV is currently being investigated in field trials by several researchers [2,5,7–9]. So far, a number of crops have been assessed for their suitability for cultivation underneath AV, including lettuce [8], corn [10], potatoes, winter wheat [9], and fruit vegetables (such as cherry tomatoes and chili peppers) [7]. Additionally, grass-clover has been investigated as a perennial forage crop [9]. These studies have shown that sufficient crop yields can be achieved in the partial shade of the photovoltaic (PV) modules of AV systems, but agricultural yield reductions of up to 20% can occur [8,9,11]. By contrast, in hot and dry weather conditions, reduced solar radiation and microclimatic alterations under AV (e.g., lower soil [8,9] and air temperatures [9] and potential advantages in water use efficiency [12]) can be beneficial for crop production and lead to increased yields [7,9].

The present study was conducted on-farm within a field trial with four different crops (celeriac, grass-clover, potato and winter wheat) cultivated underneath an AV system. The crops were cultivated as part of a crop rotation under organic management. This setup was chosen because, to date, no AV studies have been conducted under organic field management conditions. Furthermore, organic farming generally strives to reduce external inputs "by reuse, recycling and efficient management of materials and energy in order to maintain and improve environmental quality and conserve resources", as a matter of principle—as described by the International Federation of Organic Agriculture Movements (IFOAM) [13]. Thus, organic farming also addresses energetic self-sufficiency and the replacement of fossil energy resources. As such, AV would appear an appropriate approach in this context. Further details on the field trial were reported by Weselek et al. [9]. Harvestable crop yields decreased by 18.7% (wheat), 18.2% (potatoes) and 5.3% (grass-clover) in 2017, but increased by 2.7% (wheat) and 11% (potatoes) in 2018. Grass-clover yields in 2018 were reduced by 7.8% [9]. The results were linked to quite distinct climatic differences between the years; 2018 brought lower precipitation, higher temperatures and greater solar irradiance. In the same time frame, 246 MWh of energy were generated by the AV facility in the first cropping year, which corresponds to about 83% of the electrical yield a conventional ground-mounted PV installation covering the same area would have achieved [14]. Hence, even with a reduction of harvestable crop yield of 18.7% for winter wheat in 2017, overall land productivity was increased by about 56% in comparison to single crop and PV production [14]. The results further emphasized findings from previous studies [3] on the benefits of AV regarding land use and land productivity.

As a recent study showed, long term land productivity and market certainty are often seen as the main arguments favoring the implementation of AV from farmers' perspective [15]. This emphasizes the need for agricultural field trials. However, experimental data on the impact of AV on crop production are scarce; few data are available for field vegetables and, in particular, root vegetables. In 2017, vegetables were cultivated on a total area of 2.2 million hectares in Europe [16]. As comparatively high market revenues can be achieved with vegetables, the impacts of AV on cultivation and harvestable crop yields will be of major interest. Celeriac (*Apium graveolens* L. var. *rapaceum*), also known as turnip-rooted celery or knob celery, is a celery variety cultivated primarily in Central and Eastern Europe [17,18]. In contrast to common celery (*Apium graveolens* var. *graveolens*) and leaf celery (*Apium graveolens* var. *secalinum*), this biennial crop forms large bulbs in the first cropping year—which consist of hypocotyl, tap root and stem in equal proportions [17]. Celeriac bulbs have white flesh and can be used both cooked and raw. In 2018, organic celeriac was cultivated on a total of about 219 hectares in Germany, producing 6853 tons of harvested celeriac bulbs [19].

The aim of our study was to investigate how celeriac (a common field vegetable) would be affected if it were cultivated underneath the solar panels of an AV system (Figure 1). In addition to examining parameters such as crop development and yields, the study examined, for the first time, how altered microclimatic conditions in the partial shade of the AV facility affected the chemical composition—and consequently, the quality—of celeriac.

Figure 1. Celeriac plants growing underneath the agrivoltaic (AV) facility in 2017. The reference site is located behind the facility. (source: Bauerle/University of Hohenheim).

2. Material & Methods

2.1. Site Description & Field Experiment

Celeriac was cultivated as part of an on-farm field experiment using a four-year crop rotation (along with winter wheat (*Tricticum aestivum*), potato (*Solanum tuberosum*) and grass-clover) [9]. The field trial was performed on a commercial farm managed according to biodynamic principles (Demeter) as described in [9]. Details on the design of the AV facility were described by Trommsdorff et al. [14]. In both 2017 and 2018, celeriac was grown on a strip 24 m long and 19 m wide under an AV system, with a total size of 0.3 ha. Additional celeriac was grown on an adjacent reference area (REF) without solar panels (Figure 2). To avoid any shading of the REF site, it was located at a distance of 20 m from the AV facility. On both sites, four trial plots of 1 m^2 were defined. To reduce border effects—in particular under the AV facility—the plots were located at least 4 m from the sites' borders. Celeriac plantlets (*Apium graveolens* L. var. *rapaceum*, Goliath variety) were sown in seed trays and planted out around development stage 13 (according to BBCH (Biologische Bundesanstalt, Bundessortenamt und CHemische Industrie) scale for root and stem vegetables [20]) at a density of 45,000 plants per hectare. In both years, planting took place on 5 May. The celeriac cropping area was fertilized with 15 t composted cow manure per hectare between mid-February and mid-March. Biodynamic preparations (20 l per hectare each of horn manure and horn silica) were sprayed according to Demeter guidelines twice a year. Weed control was mainly conducted by currycombing before planting (twice) and hoeing after planting (up to four times). Additional hand weeding was performed if weed pressure became high within the rows. In 2017, the preceding crop was perennial grass-clover; in 2018, it was potato. For further information on field management, see [9].

2.2. Microclimate

Microclimate was monitored via eight microclimate stations (i.e., four per treatment) on the celeriac cropping area, each assigned to one of the trial plots. Each microclimate station was equipped with different sensors and recorded various microclimatic parameters. Air temperature and humidity were measured at a height of 2 m using a VP-4 sensor. Soil temperature and moisture were measured at a depth of approximately 25 cm using a 5TM sensor. Due to tillage operations, soil sensors were only installed during the celeriac cropping period from 8 June to 10 October in 2017, and from 9 May to 22 October 2018.

Photosynthetic active radiation (PAR) was estimated by photosynthetically active photon flux density (PPFD) using a QSO-S sensor. All parameters were recorded with data loggers (EM50G). Data loggers (and the sensors mentioned above) were obtained from METER Group AG (Munich, Germany). In addition to the data collected in the field trial, meteorological data for comparison were obtained from Agricultural Meteorology Baden-Wuerttemberg, published by the Agricultural Technology Centre Augustenberg (LTZ) [21]. The weather station nearest to the field trial was located at Billafingen (47.83° latitude 9.13° longitude), 2 kilometers away. Mean monthly temperature and accumulated precipitation are shown in Figure 3 (data taken from Billafingen weather station [21]). Note that values recorded in the field trial cannot be directly compared with those recorded at the weather station, as they are located at different spots and their instruments have not been calibrated. Furthermore, in 2018, no values were recorded at our field trial from 11 to 13 December due to a power outage.

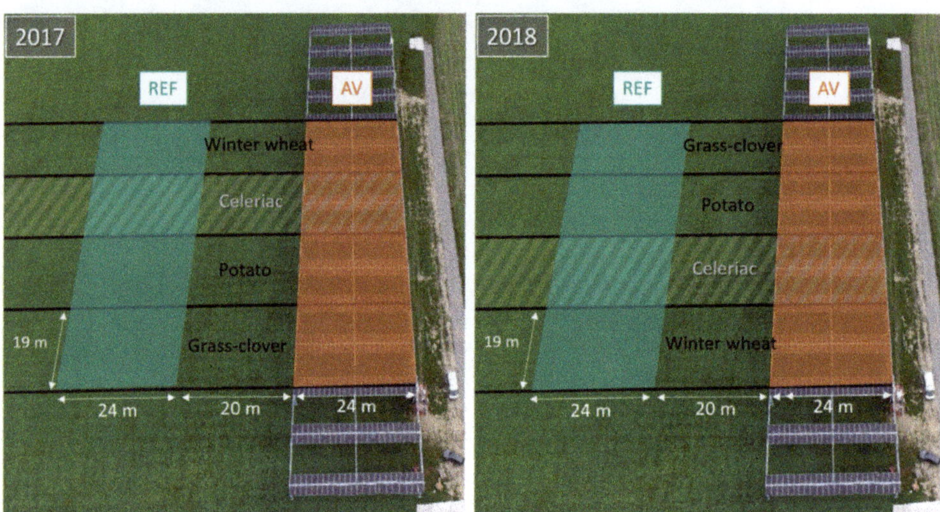

Figure 2. Setup of the field experiment in 2017 and 2018 with location of celeriac within the crop rotation. The experimental site was split into a reference (REF) and agrivoltaic (AV) site. The diagram is a schematic illustration and not to scale. (image source: BayWa r.e., modified).

Climatic conditions varied greatly between the two years. In 2017, annual accumulated precipitation was 1351 mm, annual solar was radiation 1180 kWh/m^2, and mean annual temperature was 8.6 °C. In 2018, accumulated precipitation was 916 mm, annual solar radiation was 1204 kWh/m^2, and mean annual temperature was 9.7 °C.

2.3. Crop Monitoring & Harvest

Crop development was monitored over two growing seasons, beginning in May (both years) immediately after the celeriac was planted and lasting until shortly before final harvest. The last monitoring dates were 26 September in 2017 and 18 October in 2018. In each of the defined plots, 12 individual plants were selected and tagged. Of these, 10 plants were monitored and two were kept as backup in case of plant losses. Crop development was monitored every two weeks. Crop height was measured using a folding rule. Leaf area index (LAI) was measured using a plant canopy analyzer (LAI-2200C, LI-COR Biosciences, Lincoln, Dearborn, MI, USA). On each monitoring date, twelve single measurements were taken per plot: six measurements between plants within the rows, and six measurements between rows. The final harvest was performed on the farm's actual harvest dates. The 12 selected plants in each plot were harvested manually. Each celeriac plant was separated into aboveground and belowground biomass. Remaining roots were roughly removed from the bulbs. The aboveground biomass from each plot was weighed and subsequently dried for 48 h at 60°C to determine dry matter yield. Diameter and weight of each celeriac bulb was measured. For the analysis of chemical composition, bulbs were peeled, washed with distilled water and ground (Thermomix, Vorwerk, Wuppertal, Germany). The resulting fibrous pulp was freeze-dried at 0.34 mbar and −32°C until completely dry and then stored at −20°C for further analysis.

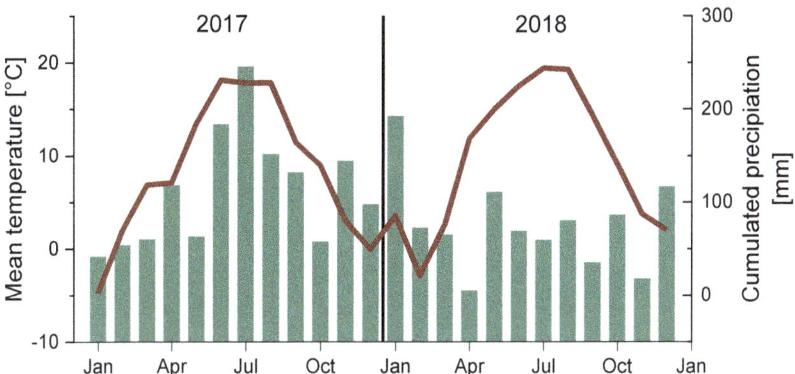

Figure 3. Monthly mean temperature (red curve) and monthly accumulated precipitation (cyan bars) in 2017 and 2018. Data from Agricultural Meteorology Baden-Wuerttemberg, Billafingen weather station.

2.4. Analysis of Chemical Composition

For chemical analysis, the freeze-dried samples were ground to a fine powder (MM400, Retsch, Haan, Germany) using ceramic grinding jars to avoid any heavy metal contamination. Before analysis by ICP-OES and ICP-MS, samples were digested by microwave pressure digestion (UltraCLAVE III, MLS, Leutkirch, Germany) according to method 10.8.1.2 of the Association of German Agricultural Analytic and Research Institutes (VDLUFA) [22]. For analysis of Al and Si, samples were additionally digested with 0.5 M hydrofluoric acid to avoid silicate formation. The minerals Al, B, Ba, Ca, Cu, Fe, K, Mg, Mn, Na, P, Zn, and Si were analyzed by ICP-OES (5100 ICP-OES, Agilent, Santa Clara, USA) according to EN standard 15621:2017-10 [23]. The trace elements and heavy metals Cd, Co, Cr, Mo, Ni, Pb, Se, Fe, Cl, and I were analyzed by Inductively Coupled Plasma Mass Spectrometry ICP-MS (NexION 300X ICP-MS, PerkinElmer, Waltham, MA, USA) according to the VDLUFA (Verband deutscher landwirtschaftlicher Untersuchungs- und Forschungsanstalten) method 17.9.1 [24]. For Cl analysis, samples were extracted in simmering water according to VDLUFA method 2.2.2.2 [25]. For iodine analysis, samples were extracted with 0.5% ammonium hydroxide according to VDLUFA method 2.2.2.3 [26]. Carbon and sulfur were analyzed based on the Dumas combustion method [27]. Crude protein, crude fat and crude fiber were determined (using a Fibertherm apparatus, C. Gerhardt, Königswinter, Germany) following the European Commission (EC) regulation No. 152/2009 III [28]. For the calculation of crude protein, the N concentration was multiplied by a conversion factor of 6.25. Neutral detergent fiber (amylase treated, after ashing; aNDFom), acid detergent fiber (after ashing; ADFom), and acid detergent lignin (ADL) were determined according to VDLUFA methods 6.5.1, 6.5.2 and 6.5.3 [29–31], respectively, using a Fibertherm apparatus (Fibertherm, C. Gerhardt, Königswinter, Germany). In 2017, aNDFom, ADFom and ADL were not analyzed due to insufficient sample material.

2.5. Statistical Analysis

Statistical analysis was conducted according to the method of Weselek et al. [9]. The experimental setup can be considered a single replicate of a strip-plot design with two treatments, AV and REF. Note that a true replicate for the treatment would require another

AV system. The data analysis was carried out with SAS software version 9.4 (SAS Institute Inc., Cary, NC, USA) using the following model for crop development:

$$y_{ijkl} = \mu + b_{kij} + \tau_i + \varphi_j + (\tau\varphi)_{ij} + e_{ijkl},$$ (1)

where b_{kij} is the fixed effect of lane k in treatment i at day j, τ_i is the i-th treatment effect, φ_j is the j-th day effect and $(\tau\varphi)_{ij}$ is the interaction effect of day j and treatment i. e_{ijkl} is the repeated measurement error of observation y_{ijkl} with a first-order autoregressive variance-covariance structure of error effects from the same measuring point. Note that the variance of repeated measurements on the same plot underestimates the true error variance, and thus all tests are too liberal.

As harvestable crop yield was measured in two successive years but only once per year, an analogous model to (1) can be fitted replacing day j by year n:

$$y_{inkl} = \mu + b_{kin} + \tau_i + \rho_n + (\tau\rho)_{in} + e_{inkl},$$ (2)

where ρ_n and $(\tau\rho)_{in}$ are the effects of the n-th year and its interaction effects with treatment. All other effects are defined analogously to model (1).

Analysis of chemical composition was conducted accordingly for each parameter:

$$y_{inkl} = \mu + b_{kin} + \tau_i + \rho_n + (\tau\rho)_{in} + e_{inkl}$$ (3)

where significant differences were found via an F test, a multiple t-test (Fisher's LSD test) was performed. Results of multiple t-tests are presented as a letter display.

3. Results & Discussion

3.1. Microclimate

An overview of the results of microclimate monitoring is presented in Table 1. Photosynthetic active radiation was, on average, reduced by about 29.5% under AV, which is within the range of the results from previous modeling and field studies, where reductions of irradiance ranged from 12% up to more than 60%, depending on the setup of the AV system [3,32,33]. Soil temperature was reduced by 1.2 °C in 2017 and 1.4 °C in 2018. This is in accordance with findings from Marrou et al. [32], who also found soil temperature to be reduced under AV. In 2017, yearly mean soil moisture was 1.9% higher under AV, while it decreased by about 3.1% in 2018. In both years, yearly mean air humidity was 2.8% higher in AV compared to REF. No differences between the treatments were found in yearly mean air temperature. In contrast, Marrou et al. [32] did not find any differences in aerial microclimate (temperature and humidity) between AV and unshaded control. The results also reflect the differences between the years—as also shown by the weather data recorded at the weather station in Billafingen (see Section 2.2.)—with comparably high temperatures and dry conditions in 2018. The yearly mean air temperature was 1.7 °C higher in 2018 compared to 2017. Air humidity and soil moisture were lower in both treatments compared to 2017. Additionally, photosynthetic photon flux density was slightly increased in 2018. Further details on microclimate monitoring have been reported [9].

Table 1. Yearly averages of air temperature and humidity, soil temperature and moisture as well as photosynthetic active radiation expressed by photosynthetic photon flux density (PFD) under the agrivoltaic system (AV) and on the reference site (REF) in 2017 and 2018.

		Air Temperature [°C]	Humidity [%]	Soil Temperature [°C]	Soil Moisture [%]	PPFD [μmol/m²s]
2017	REF	8.7	79.1	18.4	25.2	469.4
	AV	8.7	81.9	17.2	27.1	336.7
2018	REF	10.4	71.6	19.2	20.9	497.9
	AV	10.4	74.4	17.8	17.8	344.5

3.2. Crop Development

Celeriac growth and development was monitored on 10 days in 2017 and 11 days in 2018 (due to later harvest date).

After planting in May, the plantlets established quite slowly in 2017 in both treatments, which may be explained by the subsequent low precipitation of about 50 mm in May (Figure 4a). This also led to a certain amount of plant loss (not quantified). Consequently, crop development was delayed for several weeks until shoot growth started: mean plant height remained constant on the first four monitoring days and had even decreased slightly at the end of June. After the pronounced period of drought in May, monthly precipitation was between 150 and 250 mm from June to August. Nevertheless, it took until the middle of July before the celeriac plantlets had recovered, at which point shoot height gradually increased, reaching a maximum crop height of 35.7 cm under AV and 29.4 cm for REF 130 days after planting (DAP).

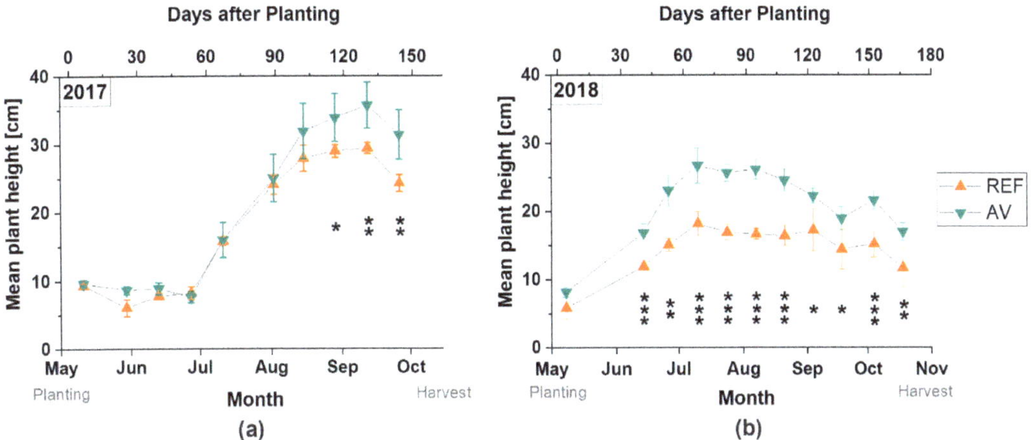

Figure 4. Mean plant height under AV (cyan triangles) and on REF (orange triangles) in 2017 (**a**) and 2018 (**b**). Significant differences are indicated by stars (* $p < 0.05$; ** $p < 0.005$; *** $p < 0.0005$). Standard deviation is depicted by error bars.

By contrast, in 2018, plants had already doubled their height by the second monitoring date in mid-June (Figure 4b). At this point, the celeriac cultivated under AV was 30% higher than on the REF site, while in 2017 growth had just begun in both treatments. The mean maximum crop height of 26.6 cm in AV and 16.7 cm in REF was recorded at 66 DAP. Plant height then decreased until final harvest. In May 2018, accumulated precipitation was 100 mm—more than twice as high as in May 2017. After that, however, monthly precipitation in 2018 remained below 100 mm until December (Figure 2) and consequently aboveground plant growth had stopped by mid-July in both treatments.

As a result, final plant development was better in 2017 than in 2018, although plantlet establishment was less problematic in 2018. The potatoes, which were planted shortly before the celeriac, were also found to have a lower initial plant height during the first weeks after emergence in 2017 than in 2018 [9].

In addition to year-related effects, crop height was also affected by treatment: celeriac plants were significantly higher under AV than in REF on three monitoring dates in 2017 and ten of the eleven dates in 2018 (Figure 4a,b). Differences in crop height between the treatments were more pronounced in 2018 than in 2017: averaged over all monitoring dates, crop height in AV was 30.6% higher than REF in 2018, but only 14% higher in 2017. In 2017, the mean difference in crop height between the treatments slowly increased from the 5th monitoring date (67 DAP) onwards, reaching a maximum (at final harvest, 144 DAP) of +7.2 cm in AV. In 2018, mean difference in crop height between the treatments was highest on the 6th monitoring date (94 DAP) at +9.5 cm in AV and then slowly decreased to +5.4 cm at final harvest (166 DAP). These treatment-related differences within the years corresponded to the general crop development, as described above. In 2017, crop height (and also difference between the treatments) increased from July onwards; meanwhile, in 2018, the crop reached maximum height by the middle of the growing season and then decreased until final harvest. However, the results show that crop height was increased by AV in both years. Similar results have been found for potatoes and winter wheat [9], where crop height was significantly increased by AV in both 2017 and 2018. As discussed by [9], increases in crop height are most probably due to shading; under AV, PAR was reduced by 30% on average in both years [9] (Table 1). These findings are in line with results from experiments with artificial shading, in which the canopy height of wheat [34,35] and potato [36] was increased by shading. Increased elongation growth under decreased light intensities can be interpreted as a shade-adaptive response by the plants in order to capture more light [37,38].

Leaf area index (LAI) was measured on seven monitoring days in 2017 and ten monitoring days in 2018. In 2017, no measurements were possible until the end of June as the plantlets were too small. LAI values differed only slightly between the years (Figure 5). As discussed above, the LAI values also indicated delayed development of the plantlets in 2017, which began to grow slowly from the end of June onwards (Figure 5a). On the other hand, in 2018, LAI values of approximately 2.5 had already been recorded in June (Figure 5b). Variations in LAI between the monitoring dates may be explained as an artifact caused by the occasional occurrence of weeds and the senescence of outer leaves—which may have led to lower LAI values being recorded from time to time. In 2017 in particular, leaves showed clear signs of Septoria leaf spot infection caused by the fungus *Septoria apiicola*, which led to early leaf senescence and consequently to a certain amount of loss of outer, older leaves. This explained the trend of declining LAI values from September onwards. As a similar effect of premature leaf senescence was observed in both treatments, the impact of uneven rain distribution under AV [9] can be excluded as the cause of the infestation by fungal leaf disease, based on the present data. However, infestation and pathogenesis were not monitored explicitly, and should be addressed in more detail in future—particularly as humidity was shown to be slightly higher under AV (Table 1). In 2018, celeriac leaves were still green at final harvest and did not show any signs of Septoria leaf spot infection. This can be seen from the LAI values, which were more or less constant until harvest. Similarly to crop height, LAI increased under AV, but the increase was only significant on one monitoring date in 2017 (166DAP) and four monitoring dates in 2018 (66, 94, 136 and 166 DAP). An increased leaf area under AV has also been found in lettuce [8], winter wheat, potatoes and grass-clover [9]. In lettuce, changes in total leaf area were linked to an increment in individual leaf area (width and length), as well as to altered leaf angles. However, the number of leaves was reduced by shading and depended on the level of shading applied [8]. In our experiment, the determinants of increased LAI could not be clearly specified, as leaf number and other leaf morphology characteristics were not monitored. In general, an increase in leaf area can be interpreted as a further physiological adaptation to diminished light availability under AV, in addition to increased crop height. Both strategies focus on intercepting more light to maintain sufficient photosynthetic performance [37].

As discussed above, both crop height and LAI of celeriac cultivated under AV were increased. Enhanced vegetative growth, as a consequence of decreased light intensities, can be interpreted as a shade-adaptive response aimed at enhancing light adsorption [37,38]. At the same time, increased elongation growth in response to shading is considered a shade-avoidance strategy, predominantly found in species less adapted to shaded environments [37,38]. Increased specific leaf area and leaf area ratio—both of which describe the relation of leaf area to plant biomass—can enhance the shade tolerance of plants [37,39]. Although the specific leaf area and leaf area ratio could not be deduced from the LAI measurements in our trial, the results indicated that the celeriac—and also crops like potatoes and wheat [9] —adapted to the reduced irradiation underneath the PV panels of the AV facility through a combination of shade-adaptive mechanisms.

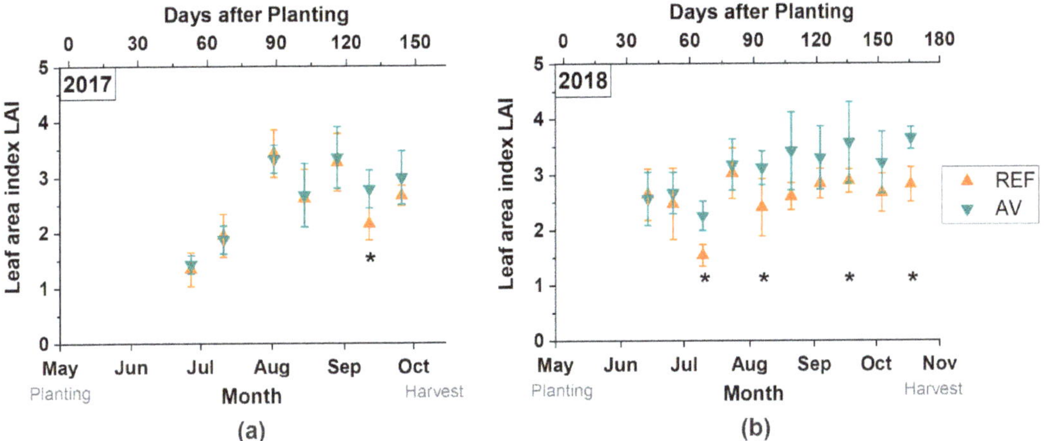

Figure 5. Leaf area index (LAI) of plants grown under AV (cyan triangles) and in REF (orange triangles) in 2017 (**a**) and 2018 (**b**). Significant differences (* $p < 0.05$) are indicated by stars. Standard deviation is depicted by error bars.

3.3. Bulb Yields and Yield Components

The celeriac was harvested on 10 October in 2017 and 22 October in 2018 in both treatments. The early harvest date in 2017 was due to the fact that no further yield increases were to be expected on account of early leaf senescence (see also Section 3.2.). However, the date was still within the common celeriac harvest period. Aboveground biomass was increased by AV in both years, but only significantly in 2018 (Figure 6a). Dry matter yield of aboveground biomass was 0.37 t ha^{-1} in REF and 0.55 t ha^{-1} in AV (+48%; $p = 0.082$) in 2017, and 1.1 t ha^{-1} in REF and 1.4 t ha^{-1} in AV in 2018 (+31.9%; $p = 0.0045$). Interestingly, aboveground biomass was higher in 2018, although crop height was higher in 2017. We postulate that this was caused by the very distinct weather conditions in the two years, which affected both aboveground biomass and crop height in different ways. First, initial shoot growth was virtually zero in the first few weeks after planting in 2017. We assume that this period conferred a crucial growth advantage in 2018, leading to higher final shoot biomass in that year. Second, the dry weather conditions in summer 2018 may have led to a decrease in turgor pressure as a response to drought stress, leading to more wilting of leaves. As crop height was measured without lifting up individual leaves, this will also have led to lower crop heights being recorded. This explanation is supported by the finding that, in 2018, crop heights had decreased by the middle of July with the onset of drought stress. Furthermore, hanging leaves will also have led to an enlarged leaf rosette, explaining why LAI was higher in 2018 despite lower crop heights. The third—and presumably most crucial—factor was disease; aboveground biomass was lower in 2017 due to infection with Septoria leaf spot, leading to early leaf senescence and consequently a certain loss of matured leaves.

Celeriac bulb yield was 11.9 t ha^{-1} in REF and 9.7 t ha^{-1} in AV (-18.9%; $p = 0.15$) (Figure 4) in 2017, and 9.6 t ha^{-1} in REF and 10.8 t ha^{-1} in AV (+11.8%; $p = 0.49$) in 2018 (Figure 6b). Neither the differences between the treatments nor those between the years were significant. The yields in both years and treatments were low in comparison with the national average for organically cultivated celeriac, which was 29.6 t ha^{-1} in 2017 and 31.1 t ha^{-1} in 2018 [19]. In general, celeriac is considered drought-sensitive, with drought stress leading to small, poorly developed bulbs [17,18]. Therefore, it can be assumed that the dry weather conditions in spring 2017 and especially summer 2018 probably led to comparatively low bulb yields. This is particularly probable given that the celeriac plants in our trial were not irrigated. The bulbs were poorly developed in both years and treatments (Figure 7). Average individual bulb weight was 196 g (REF) and 158 g (AV) in 2017, and 186 g (REF) and 197 g (AV) in 2018. Average bulb diameter was 7.3 cm (REF) and 6.6 cm (AV) in 2017 and 7.5 cm (REF and AV) in 2018. Both average weights and diameters can be considered undersized. To meet the criteria of the wholesaler the farm supplies, celeriac must fulfill the class 1 UNECE (United Nations Economic Commission for Europe) standard for root and tubercle vegetables [40]. In addition, the bulbs must have a minimum weight of 350 g if only the bulbs are sold, or a minimum size of 60 mm if whole plants (bulb including leaves) are sold. Taking these criteria into consideration and assuming only bulbs (without leaves) are sold: of the 48 bulbs harvested from each treatment in 2017, only one (AV) and four (REF) were actually marketable. In 2018, the respective numbers were zero (AV) and one (REF). Extrapolated to a hectare, marketable bulb yields would consequently have been only 0.5 t ha^{-1} (AV) and 2 t ha^{-1} (REF) in 2017, and 0 t ha^{-1} (AV) and 3.6 t ha^{-1} (REF) in 2018. If sold as whole plants, marketable bulb yield would have been 7.8 t ha^{-1} (AV) and 10.6 t ha^{-1} (REF) in 2017, and 10.3 t ha^{-1} (AV) and 8.8 t ha^{-1} (REF) in 2018.

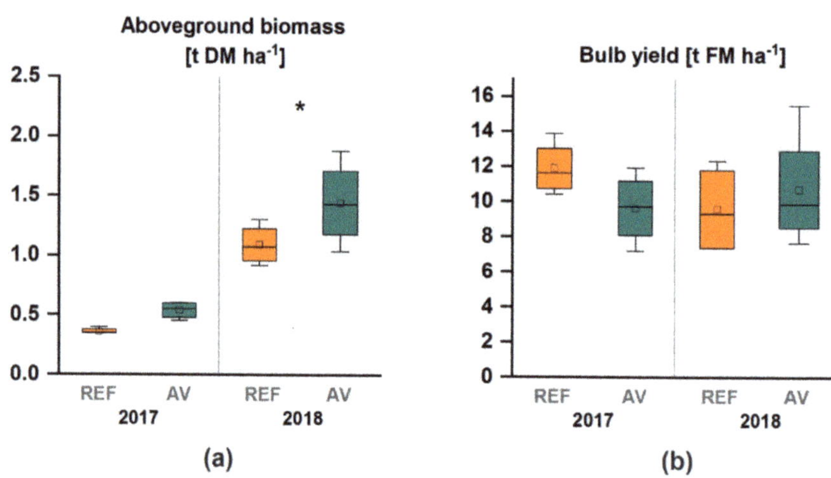

Figure 6. Celeriac aboveground biomass (t dry matter ha^{-1}) (**a**) and bulb yield (t fresh matter ha^{-1}) (**b**) in REF and AV in 2017 and 2018. Significant differences are indicated by stars (* $p < 0.005$).

As mentioned above, yield variations within the years differed between the treatments. Averaged over both treatments, bulb yield was higher in 2017 (10.8 t ha^{-1}) than 2018 (10.2 t ha^{-1}). While bulb yields from the REF site were 2.3 t ha^{-1} lower in 2018 than 2017, yields on the AV site actually increased by about 1.1 t ha^{-1}. Lower yields under AV in 2017 were most probably caused by the reduction in solar radiation (about 30%) (Table 1). In contrast, the yield increases under AV in 2018 indicate that the celeriac plants benefitted from shading that year. It can be assumed that, in 2018, drought, intensive solar radiation, and high temperatures counterbalanced the adverse effects of shading on celeriac productivity. However, it remains unclear whether this was caused directly (by attenuating irradiation) or indirectly (by altering microclimatic conditions to provide a more favorable microclimatic environment for celeriac growth). It was expected that soil moisture would increase under AV, as a reduction in evapotranspiration in the partial shade of AV panels was already reported [12]. However, soil moisture under AV only increased in 2017; it was actually reduced in 2018 (Table 1). Therefore, soil moisture can be excluded as a potential explanation for increased crop yields under AV in 2018.

Figure 7. Celeriac harvested in 2017. The plants all come from one plot. In each treatment, four plots (with 12 plants each) were harvested. Most bulbs are comparatively small and some leaves are already senescent. (Source: Bauerle/University of Hohenheim).

As shown, soil temperature was reduced by AV (Table 1). Although this was the case in both years, we assumed that reduced soil temperature and a direct reduction in solar radiation under AV were the determining factors that diminished the adverse effects of excessive irradiation, heat, and drought on crop yields in 2018. Furthermore, increases in aboveground biomass, as mentioned above, may also have led to higher bulb yields in 2018 through higher amounts of assimilates being translocated from the shoots to the bulbs.

While no comparable data is available for celeriac, studies with other root vegetables (e.g., carrot, parsnip, radish and beetroot) have shown that shoot and storage-root weights are linearly correlated, with slight differences between species [41–43]. Biologically, celeriac is comparable to beetroot and radish, species in which the storage organ also develops from the hypocotyl. We therefore hypothesized that the significantly higher aboveground biomass under AV in 2018 was a determining factor for the higher bulb yield compared to REF. As the vegetation period in 2018 was prolonged due to the later harvest date, the period for the translocation of assimilates from the shoot to the storage root was also extended. This raises the question of whether delaying the harvest could have facilitated mobilization of the full assimilate potential stored in the shoot, increasing bulb weights and yields under AV. A study with lettuce cultivated underneath an AV system found that a delayed harvest date led to yields comparable to the unshaded control [44]. However, in the case of celeriac, a further increase in bulb yields through a prolonged vegetation period would be limited by environmental conditions. Mild autumnal temperatures are required for translocation of assimilates to the storage organ to continue. In addition, the 2017 results showed that infestation with fungal diseases can also become a limiting factor, leading to premature leaf senescence and preventing further yield increases.

Relative changes in harvestable yields of winter wheat and potatoes cultivated under AV were comparable to those of celeriac in the present study. While in 2017, yields decreased by about 18–19%, they increased by about 3% (wheat) and 11% (potato) in 2018 [9]. Accordingly, all annual crop species investigated in the field trial showed comparable responses to cultivation in the altered environment underneath the AV facility. Moreover, yield fluctuations between the years were less pronounced under AV, as was the case with celeriac. This supports the hypothesis that cultivation under AV can be advantageous in dry weather conditions and may have yield-stabilizing effects in the long term [9,33], but further trial years are needed for validation. The results indicated that—outside of dry climates where a general reduction in sun exposure can be beneficial and certain crops adapted to shaded conditions—leaf vegetables may be particularly suitable for cultivation under AV [2]. The increases found in above ground biomass and in growth parameters like crop height and LAI will become directly relevant for harvestable yields. This is supported by findings in lettuce, where cultivation under AV led to increased yields of some cultivars, and was also linked to increased leaf area [8].

3.4. Chemical Composition

Chemical composition analysis of celeriac bulbs revealed that most of the parameters analyzed were affected more by year than by treatment (Tables 2 and 3). The test of fixed effects revealed that all determined parameters were significantly affected by year, except S, Mg, Mn, and Se. No significant differences were found for Si, Co, and I, as the concentrations were below the detectable thresholds of 150 mg kg^{-1} (Si), 0.02 mg kg^{-1} (Co), and 0.5 mg kg^{-1} (I) given in Table 3. Dry matter (DM) content was significantly lower ($p < 0.0005$) in 2017 than in 2018 (9.9% (both AV and REF) in 2017 compared to 14.4% (AV) and 13.6% (REF) in 2018). In 2018, DM content was significantly lower in REF ($p = 0.002$) than in AV. No significant differences between treatments were found for crude protein and crude fat (Table 2). Crude protein was slightly higher in AV than in REF in 2017. However, crude protein in AV was lower than in REF in 2018, which may be explained by a dilution effect, as yields in AV were lower in 2017 and higher in 2018. Crude fat was affected by year, but virtually unaffected by treatment. Both crude fat and protein were significantly higher in 2018 than in 2017. Fresh matter (FM) protein content (2017: 0.99% AV, 0.94% REF; 2018: 1.17% AV, 1.20% REF, data not shown) was lower than the reference values of 1.2–1.5% stated in the literature [17,18]. FM crude fat content (2017: 0.22%, AV and REF; 2018: 0.28% AV and REF) was also slightly lower than literature values (0.3–0.4%) [17,18]. Carbon content was significantly lower under AV in both years, indicating that less carbon was allocated from the shoots to the bulbs, despite higher shoot biomass. This may be due to generally lower photosynthetic assimilation of carbon dioxide as a consequence of lower irradiance and/or diminished translocation to the storage organs, which would support the hypothesis that maturation is delayed under AV (see Section 3.3). This would also explain the higher C content in 2018: prolongation of the vegetation period, increased irradiation (and consequently photosynthetic performance), and increased aboveground biomass (and consequently translocation potential) may have led to higher amounts of assimilates being translocated to the bulbs. The C/N ratio was higher in 2017 than in 2018 in both treatments (data not shown), which can be explained by the higher N content in 2018. The C/N ratio under AV was at 24.7, significantly lower ($p = 0.012$) than in REF (26.6), in 2017, and at 21.7, slightly higher ($p = 0.45$) than REF (21.3) in 2018.

Table 2. Concentration of crude protein, crude fat, neutral detergent fiber (aNDFom), acid detergent fiber (ADFom), acid detergent lignin (ADL) and macroelements C, S, Ca, K, Mg, Na, P (in % dry matter DM). Significant differences ($p < 0.05$) are indicated by different letters. *p*-values correspond to the test of fixed effects year, treatment (Trt) and their interaction. SEM = Standard error of means.

Treatment	[% DM]												
	Crude Protein	Crude Fat	aNDF$_{om}$	ADF$_{om}$	ADL	C	S	Ca	K	Mg	Na	P	Cl
						2017							
AV	10*a*	2.25*a*	-	-	-	39.5*a*	0.09*a*	0.31*a*	4.09*a*	0.2	0.31*a*	0.58*a*	0.08*a*
REF	9.4*a*	2.18*a*	-	-	-	40.1*b*	0.09*a*	0.34*b*	3.9*a*	0.21	0.31*a*	0.59*a*	0.08*a*
						2018							
AV	11.7*b*	2.78*b*	13.3	9.5*a*	2.06	40.7*c*	0.08*b*	0.28*c*	2.19*b*	0.19	0.16*b*	0.33*b*	0.05*b*
REF	12.1*b*	2.83*b*	16.1	10.5*b*	3.0	41.1*d*	0.08*ab*	0.3*ac*	2.25*b*	0.18	0.22*c*	0.3*b*	0.06*ab*
SEM	0.196	0.075	1.348	0.072	0.327	0.149	0.002	0.005	0.076	0.01	0.018	0.01	0.009
						p-value							
Year	<0.0001	<0.0001	-	-	-	<0.0001	0.0612	0.0001	<0.0001	0.082	<0.0001	<0.0001	0.0152
Trt	0.6643	0.8713	0.1918	0.0002	0.0985	0.0048	1.0	0.0021	0.4225	0.9	0.0844	0.3628	0.7760
Trt*Year	0.0397	0.4254	-	-	-	0.6843	0.3166	0.2009	0.1221	0.7071	0.1065	0.1832	0.2694

Table 3. Concentration of microelements (ppm dry matter (DM)). Significant differences ($p < 0.05$) are indicated by different letters. *p*-values correspond to the test of fixed effects year, treatment (Trt) and their interaction. SEM = Standard error of means.

Treatment	[ppm DM]															
	Al	B	Ba	Cu	Fe	Mn	Zn	Si	Cd	Co	Cr	Ni	Mo	Pb	Se	I
2018																
AV	3.16*a*	33*a*	8.04*a*	16*a*	21.3*a*	36.4*a*	29.3*a*	<150	0.67*a*	<0.02	0.23*a*	0.92*a*	0.05*a*	0.08*a*	0.03	<0.50
REF	3.02*a*	30.8*a*	11.1*b*	17.4*a*	21*a*	60.2*b*	31.2*a*	<150	1.33*b*	<0.02	0.08*ab*	1.49*b*	0.04*a*	0.11*b*	0.03	<0.50
2017																
AV	1.28*b*	27*b*	2.47*c*	14.1*b*	30*b*	42.1*ac*	25.9*b*	<150	0.5*a*	<0.02	0.02*b*	1.19*a*	<0.02*b*	0.06*c*	0.02	<0.50
REF	2.7*ab*	24.6*b*	3.8*c*	12.8*b*	28.4*b*	48.1*c*	25.7*b*	<150	0.96*c*	<0.02	0.03*b*	2.15*c*	<0.02*b*	0.1*ab*	0.02	<0.50
SEM	0.53	1.02	0.876	0.515	1.404	2.462	1.069	-	0.079	-	0.059	0.089	0.003	0.006	0.007	-
p-value																
Year	0.0626	0.0001	<0.0001	<0.0001	0.0001	0.2261	0.0019	-	0.0064	-	0.0484	0.0004	<0.0001	0.0029	0.163	-
Trt	0.2705	0.0467	0.0299	0.8685	0.6533	0.0001	0.4452	-	<0.0001	-	0.2875	<0.0001	0.2872	0.0002	0.9202	-
Trt*Year	0.1818	0.8578	0.3487	0.0225	0.6657	0.0046	0.3385	-	0.231	-	0.1996	0.0521	0.2872	0.4191	0.7898	-

Fiber content (aNDF$_{om}$, ADF$_{om}$, ADL) was lower under AV, but only significantly for ADF$_{om}$ (1.0% absolute decrease). For aNDF$_{om}$ and ADL, the standard error between the plots was comparatively high. As data on aNDF$_{om}$, ADF$_{om}$ and ADL only exist for one year, the results should be treated with care; year or yield effects cannot be excluded. However, this may provide further evidence that carbon metabolism is affected by AV through altered carbon assimilation as well as delayed translocation to the bulbs. Apart from Na and Ca, none of the macroelements were significantly affected by treatment. Concentration of Ca was increased by 0.03% (absolute) in 2017 and Na by 0.06% (absolute) in 2018 on the REF site. Concentration of all macroelements was higher in 2017 than 2018 in both treatments. This effect was significant for all elements, except S and Mg. As the trace elements Si, Co, I, and Mo (2018 only) were under the detectable thresholds (Table 3), no differences were detected. Al, B, Ba, Cu, Fe, Zn, Cr, Mo, and Cl were affected by year, but not by treatment. In 2018, lower concentrations were found throughout, except for Fe, which increased. No differences were found in Se concentrations. Mn content was lower in AV in both years but only significantly so in 2017. Ni decreased in AV in both years. Both Cd and Pb content were significantly lower under AV in both years. In general, celery is known to accumulate heavy metals such as Cd and Pb [45]. However, the detected concentrations were far below the acceptable maximum concentrations (0.20 mg kg^{-1} FM (Cd) and 0.1 mg kg^{-1} FM (Pb)) [46]. The treatment-related differences in Cd and Pb concentrations may be explained by differences in soil levels. Soil sample analyses showed that Cd and Pb soil levels were slightly lower on the AV site (data not shown). Overall, concentrations of most minerals and elements analyzed for both treatments were within the range stated in the literature [18].

However, this was only the case in 2017; in 2018 significant reductions were found, as described above. It is generally known that nutrient uptake (and, consequently, concentrations of various minerals and trace elements) is reduced under drought [47]. Hence, the significantly lower mineral content in 2018 was most probably caused by low soil water status as a consequence of dry weather conditions in summer, leading to impaired root uptake and translocation to the shoots.

The results show that cultivation underneath an AV system had only a slight effect on the chemical composition of celeriac. Concentrations of C, Ni and Mn were decreased by AV; all other parameters were mainly affected by year. The fiber fractions aNDF$_{om}$, ADF$_{om}$ and ADL were only measured in 2018. Apart from the results shown here, no comparable data on the effects of microclimatic alterations (particularly shading) on the quality characteristics of celeriac are available. Most studies featuring celeriac and celery focus on the accumulation of furanocoumarins [48–50] and quality parameters such as content of vitamins and secondary plant metabolites [51–54]. These were not the subject of our study. In general, celery is considered to be nitrate-accumulating [55]. Nitrate is thought to have a negative effect on human health [55,56]. Nitrate concentrations in crops are affected by a number of factors, including N fertilization and environmental factors (e.g., light intensities) [55,57,58]. In several crops, including blade celery (*Apium graveolens* L. var. *dulce*), nitrate concentrations have been shown to be correlated with shading intensity [57–59]. In celeriac, nitrate accumulation is also cultivar-dependent [60]. Therefore, future trials should investigate whether nitrate concentrations in vegetables like celeriac are affected by cultivation underneath AV. Our results show that crude protein content was affected by AV, indicating that protein and N metabolism were altered in some way. However, this effect was not significant and differed between the years. Carbohydrate concentration, as a relevant constituent with respect to the nutritional quality of celeriac bulbs, should also be analyzed in the future. This could offer further evidence on how carbon assimilation is affected by AV. According to the values stated in the literature, total carbohydrate content in celeriac ranges from approximately 2.0% to 3.0% of fresh matter [18,61]. Apart from celeriac, no other studies on the effects of AV on crop production have addressed their impact on the chemical composition of the harvested crops. In wheat, cultivation under AV was found to shift grain size distribution towards smaller classes [9]. Although the chemical composition of the grains was not analyzed, alterations in quality parameters can be assumed to be a consequence of an altered bran/endosperm ratio [9].

4. Conclusions

The production of celeriac was found to be affected in several ways by cultivation underneath an AV system. Under AV, photosynthetic active radiation was reduced by about 30% in both years studied. Both crop height and leaf area index increased in response to shaded conditions, leading to significantly higher aboveground biomass in 2018. Neither bulb yields nor their chemical composition were significantly affected by AV. In 2017, yields tended to be lower under AV, whereas in 2018 they increased slightly. The results were linked to lower soil temperatures and reduced PAR under AV, which may have become advantageous in the hot and dry weather conditions of 2018. We therefore conclude that celeriac can be considered a suitable crop for cultivation under AV. However, as climatic conditions were quite extreme in both years, leading to comparably low yield levels in general, further field trials are necessary to investigate how yields would develop under more optimal conditions and over a longer term. Chemical analysis of C and fiber content provided evidence of an altered carbon metabolism and potentially delayed ripening under AV. Thus, further studies are required to examine whether a prolongation of the celeriac vegetation period can be beneficial for final bulb yields through exploitation of the full potential stored in the increased shoot biomass under AV. Furthermore, quality parameters such as carbohydrate and nitrate content should be assessed. The impact of altered water distribution and increased humidity under AV on infestations with fungal disease should be examined. As a coproductive system, the advantages of AV clearly predominate: increased income through additional energy production, conservation of limited land resources through increased land productivity, and potential benefits for crop production in dry climates. Nevertheless, in view of the land use conflict between energy and food production, these benefits need to be weighed up against potential losses in agricultural productivity. This emphasizes the need to define criteria for assessing the extent to which potential drawbacks in agricultural use can be tolerated in such dual-use systems.

Author Contributions: Conceptualization, A.W., P.H., S.Z. and I.L.; Investigation, A.W. and A.B.; Formal analysis, A.W.; Data curation, A.W.; Visualization, A.W.; Writing—Original Draft, A.W., Writing—Review & Editing, A.W., A.B., S.Z., I.L., P.H.; Supervision, P.H. and I.L, Project administration, P.H.; Funding acquisition, P.H. All authors read and approved the final manuscript.

Funding: This work was carried out within the APV-RESOLA project, which received funding from the German Federal Ministry for Education and Research (BMBF) under grant no. 033L098G. In addition, the corresponding author was supported by a state graduate scholarship received from the Ministry of Science, Research and Arts (MWK) Baden-Württemberg.

Institutional Review Board Statement: Not applicable.

Informed Consent Statement: Not applicable.

Data Availability Statement: Data available on request due to restrictions eg privacy or ethical.

Acknowledgments: The authors are grateful to Florian Reyer and his colleagues from Heggelbachhof for managing the field experiment, and to Moritz Krug for assistance with data loggers and rain gauges. Particular thanks go to Nicole Gaudet for revising the manuscript and to all scientific and practical partners within the APV-RESOLA project: Stephan Schindele and Maximilian Trommsdorff from Fraunhofer Institute for Solar Energy Systems ISE, BayWa r.e., EWS Schönau and the Institute for Technology Assessment and Systems Analysis (ITAS) of Karlsruhe Institute of Technology (KIT).

Conflicts of Interest: The authors declare that they have no conflict of interest.

References

1. Schindele, S.; Trommsdorff, M.; Schlaak, A.; Obergfell, T.; Bopp, G.; Reise, C.; Braun, C.; Weselek, A.; Bauerle, A.; Högy, P.; et al. Implementation of agrophotovoltaics: Techno-economic analysis of the price-performance ratio and its policy implications. *Appl. Energy* **2020**, *265*, 114737. [CrossRef]
2. Weselek, A.; Ehmann, A.; Zikeli, S.; Lewandowski, I.; Schindele, S.; Högy, P. Agrophotovoltaic systems: Applications, challenges, and opportunities. A review. *Agron. Sustain. Dev.* **2019**, *39*, 545. [CrossRef]
3. Dupraz, C.; Marrou, H.; Talbot, G.; Dufour, L.; Nogier, A.; Ferard, Y. Combining solar photovoltaic panels and food crops for optimising land use: Towards new agrivoltaic schemes. *Renew. Energy* **2011**, *36*, 2725–2732. [CrossRef]
4. Dinesh, H.; Pearce, J.M. The potential of agrivoltaic systems. *Renew. Sustain. Energy Rev.* **2016**, *54*, 299–308. [CrossRef]
5. Valle, B.; Simonneau, T.; Sourd, F.; Pechier, P.; Hamard, P.; Frisson, T.; Ryckewaert, M.; Christophe, A. Increasing the total productivity of a land by combining mobile photovoltaic panels and food crops. *Appl. Energy* **2017**, *206*, 1495–1507. [CrossRef]
6. Gorjian, S.; Minaee, S.; M.Mirchegini, L.; Trommsdorff, M.; Shamshiri, R. Applications of solar PV systems in agricultural automation and robotics. In *Photovoltaic Solar Energy Conversion*; Academic Press: Cambridge, MA, USA, 2020; pp. 191–235. ISBN 978-0-12-819610-6.
7. Barron-Gafford, G.A.; Pavao-Zuckerman, M.A.; Minor, R.L.; Sutter, L.F.; Barnett-Moreno, I.; Blackett, D.T.; Thompson, M.; Dimond, K.; Gerlak, A.K.; Nabhan, G.P.; et al. Agrivoltaics provide mutual benefits across the food–energy–water nexus in drylands. *Nat. Sustain.* **2019**, *2*, 848–855. [CrossRef]
8. Marrou, H.; Wery, J.; Dufour, L.; Dupraz, C. Productivity and radiation use efficiency of lettuces grown in the partial shade of photovoltaic panels. *Eur. J. Agron.* **2013**, *44*, 54–66. [CrossRef]
9. Weselek, A.; Bauerle, A.; Hartung, J.; Zikeli, S.; Lewandowski, I.; Högy, P. Microclimate, crop development and harvestable crop yield of grass-clover, potatoes and winter wheat in an agrivoltaic system under organic management in south-west Germany. *Agron. Sustain. Dev.* **2021**. under review.
10. Sekiyama, T.; Nagashima, A. Solar Sharing for Both Food and Clean Energy Production: Performance of Agrivoltaic Systems for Corn, A Typical Shade-Intolerant Crop. *Environments* **2019**, *6*, 65. [CrossRef]
11. Homma, M.; Doi, T.; Yoshida, Y. A Field Experiment and the Simulation on Agrivoltaic-systems regarding to Rice in a Paddy Field. *J. Jpn. Soc. Energy Resour.* **2016**, *37*, 23–31. [CrossRef]
12. Marrou, H.; Dufour, L.; Wery, J. How does a shelter of solar panels influence water flows in a soil–crop system? *Eur. J. Agron.* **2013**, *50*, 38–51. [CrossRef]
13. International Federation of Organic Agriculture Movements. Principles of Organic Agriculture. Available online: https://www.ifoam.bio/sites/default/files/2020-03/poa_english_web.pdf (accessed on 6 February 2021).
14. Trommsdorff, M.; Kang, J.; Reise, C.; Schindele, S.; Bopp, G.; Ehmann, A.; Weselek, A.; Högy, P.; Obergfell, T. Combining food and energy production: Design of an agrivoltaic system applied in arable and vegetable farming in Germany. *Renew. Sustain. Energy Rev.* **2021**, *140*, 110694. [CrossRef]
15. Pascaris, A.S.; Schelly, C.; Pearce, J.M. A First Investigation of Agriculture Sector Perspectives on the Opportunities and Barriers for Agrivoltaics. *Agronomy* **2020**, *10*, 1885. [CrossRef]
16. Eurostat. The Fruit and Vegetable Sector in the EU—A Statistical Overview. Available online: https://ec.europa.eu/eurostat/statistics-explained/index.php/The_fruit_and_vegetable_sector_in_the_EU_-_a_statistical_overview (accessed on 2 April 2021).
17. Lim, T.K. Apium graveolens var. rapaceum. In *Edible Medicinal and Non Medicinal Plants: Volume 9, Modified Stems, Roots, Bulbs*; Lim, T.K., Ed.; Springer: Dordrecht, The Netherlands, 2015; pp. 367–373. ISBN 978-94-017-9511-1.
18. Hadley, P.; Fordham, R. Vegetables of Temperate Climates. Miscellaneous Root Crops. In *Encyclopedia of Food Sciences and Nutrition*, 2nd ed.; Caballero, B., Ed.; Academic Press: Oxford, UK, 2003; pp. 5948–5951. ISBN 978-0-12-227055-0.
19. Destatis. Anbauflächen und Erntemengen von Ökologisch Angebautem Gemüse. Available online: https://www.destatis.de/DE/Themen/Branchen-Unternehmen/Landwirtschaft-Forstwirtschaft-Fischerei/Obst-Gemuese-Gartenbau/Tabellen/oekologisches-gemuese.html (accessed on 11 February 2020).

20. Feller, C.; Bleiholder, H.; Buhr, L.; Hack, H.; Hess, M.; Klose, R. Phänologische Entwicklungsstadien von Gemüsepflanzen: II. Fruchtgemüse und Hülsenfrüchte. In *Compendium of Growth Stage Identification Keys for Mono- and Dicotyledonous Plants, Extended BBCH Scale*; German Federal Biological Research Centre for Agriculture and Forestry: Berlin, Germany, 1995; Volume 47, pp. 109–111.

21. LTZ Augustenberg. Agrarmeterologie Baden-Württemberg. Available online: www.wetter-bw.de (accessed on 7 February 2021).

22. VDLUFA. *10.8.1.2 Mikrowellenbeheizter Druckaufschluss, 8. Ergänzung; Darmstadt, Band III Futtermittel*; VDLUFA: Darmstadt, Germany, 2012.

23. European Committee for Standardization. *Animal Feeding Stuffs—Methods of Sampling and Analysis—Determination of Calcium, Sodium, Phosphorus, Magnesium, Potassium, Sulphur, Iron, Zinc, Copper, Manganese and Cobalt after Pressure Digestion by ICP-AES*; EN 15621:2017-10; European Committee for Standardization: Brussels, Belgium, 2017.

24. VDLUFA. *17.9.1 Bestimmung von Ausgewählten Elementen in Pflanzen Sowie in Grund- und Mischfuttermitteln Mittels Messenspektromie mit Induktiv Gekoppeltem Plasma, 8. Ergänzung*; Darmstadt, Band III Futtermittel; VDLUFA: Darmstadt, Germany, 2012.

25. VDLUFA. *2.2.2.2 Bestimmung von Nitrat in Pflanzlichem Material Mittels Ionenchromatographie*; Darmstadt, Band VII Umweltanalytik; VDLUFA: Darmstadt, Germany, 1983.

26. VDLUFA. *2.2.2.3 Bestimmung des Gehaltes von Extrahierbarem Jod in Futtermitteln Mittels Induktiv Gekoppeltem Plasma und Massenspektrometrie (ICP-MS)*; Darmstadt, Band VII Umweltanalytik; VDLUFA: Darmstadt, Germany, 1983.

27. VDLUFA. *4.1.2 Bestimmung von Rohprotein mittels DUMAS-Verbrennungsmethode, 5. Ergänzung*; Darmstadt, Band III Futtermittel; VDLUFA: Darmstadt, Germany, 2004.

28. Regulation (EC) No 152/2009. *Laying down the Methods of Sampling and Analysis for the Official Control of Feed. Annex III: Methods of Analysis to Control the Composition of Feed Materials and Compounds Feed*; CR (EC) No. 152/2009 III; European Union: Brussels, Belgium, 2009.

29. VDLUFA. *6.5.1 Bestimmung der Neutral-Detergenzien-Faser nach Amylasebehandlung (aNDF) Sowie Nach Amylasebehandlung und Veraschung (aNDFom), 8. Ergänzung*; Darmstadt, Band III Futtermittel; VDLUFA: Darmstadt, Germany, 2012.

30. VDLUFA. *6.5.2 Bestimmung der Säure-Detergenzien-Faser (ADF) und der Säure-Detergenzien-Faser nach Veraschung (ADFom), 8. Ergänzung*; Darmstadt, Band III Futtermittel; VDLUFA: Darmstadt, Germany, 2012.

31. VDLUFA. *6.5.3 Bestimmung des Säure-Detergenzien-Lignins (ADL), 8. Ergänzung*; Darmstadt, Band III Futtermittel; VDLUFA: Darmstadt, Germany, 2012.

32. Marrou, H.; Guilioni, L.; Dufour, L.; Dupraz, C.; Wery, J. Microclimate under agrivoltaic systems: Is crop growth rate affected in the partial shade of solar panels? *Agric. For. Meteorol.* **2013**, *177*, 117–132. [CrossRef]

33. Amaducci, S.; Yin, X.; Colauzzi, M. Agrivoltaic systems to optimise land use for electric energy production. *Appl. Energy* **2018**, *220*, 545–561. [CrossRef]

34. McMaster, G.S.; Morgan, J.A.; Willis, W.O. Effects of Shading on Winter Wheat Yield, Spike Characteristics, and Carbohydrate Allocation1. *Crop Sci.* **1987**, *27*, 967–973. [CrossRef]

35. Li, H.; Jiang, D.; Wollenweber, B.; Dai, T.; Cao, W. Effects of shading on morphology, physiology and grain yield of winter wheat. *Eur. J. Agron.* **2010**, *33*, 267–275. [CrossRef]

36. Kuruppuarachchi, D.S.P. Intercropped potato (*Solanum* spp.) Effect of shade on growth and tuber yield in the northwestern regosol belt of Sri Lanka. *Field Crops Res.* **1990**, *25*, 61–72. [CrossRef]

37. Gommers, C.M.M.; Visser, E.J.W.; Onge, K.R.S.; Voesenek, L.A.C.J.; Pierik, R. Shade tolerance: When growing tall is not an option. *Trends Plant Sci.* **2013**, *18*, 65–71. [CrossRef] [PubMed]

38. Ruberti, I.; Sessa, G.; Ciolfi, A.; Possenti, M.; Carabelli, M.; Morelli, G. Plant adaptation to dynamically changing environment: The shade avoidance response. *Biotechnol. Adv.* **2012**, *30*, 1047–1058. [CrossRef]

39. Valladares, F.; Niinemets, Ü. Shade Tolerance, a Key Plant Feature of Complex Nature and Consequences. *Annu. Rev. Ecol. Evol. Syst.* **2008**, *39*, 237–257. [CrossRef]

40. United Nations. UNECE Standard FFV-59 Concerning the Marketing and Commercial Quality Control of Root and Tubercle Vegetables. Available online: https://www.unece.org/fileadmin/DAM/trade/agr/standard/fresh/FFV-Std/English/59_RootandTubercleVeg.pdf (accessed on 13 December 2020).

41. Hole, C.C.; Thomas, T.H.; Barnes, A.; Scott, P.A.; Rankin, W.E.F. Dry Matter Distribution between Shoot and Storage Root of Carrot, Parsnip, Radish and Red Beet. *Ann. Bot.* **1984**, *53*, 625–631. [CrossRef]

42. Suojala, T. Growth of and partitioning between shoot and storage root of carrot in a northern climate. *AFSci* **2000**, *9*. [CrossRef]

43. Hole, C.C.; Morris, G.E.L.; Cowper, A.S. Distribution of dry matter between shoot and storage root of field-grown carrots. I. Onset of differences between cultivars. *J. Hortic. Sci.* **1987**, *62*, 335–341. [CrossRef]

44. Elamri, Y.; Cheviron, B.; Lopez, J.-M.; Dejean, C.; Belaud, G. Water budget and crop modelling for agrivoltaic systems: Application to irrigated lettuces. *Agric. Water Manag.* **2018**, *208*, 440–453. [CrossRef]

45. Zhang, K.; Wang, J.; Yang, Z.; Xin, G.; Yuan, J.; Xin, J.; Huang, C. Genotype variations in accumulation of cadmium and lead in celery (*Apium graveolens* L.) and screening for low Cd and Pb accumulative cultivars. *Front. Environ. Sci. Eng.* **2013**, *7*, 85–96. [CrossRef]

46. Regulation (EC) 1881/2006. *Setting Maximum Levels for Certain Contaminants in Foodstuffs*; CR (EC) No 1881/2006; European Union: Brussels, Belgium, 2006.

47. Rouphael, Y.; Cardarelli, M.; Schwarz, D.; Franken, P.; Colla, G. Effects of Drought on Nutrient Uptake and Assimilation in Vegetable Crops. In *Plant Responses to Drought Stress: From Morphological to Molecular Features*; Aroca, R., Ed.; Springer: Berlin/Heidelberg, Germany, 2012; pp. 171–195. ISBN 978-3-642-32653-0.
48. Heath-Pagliuso, S.; Matlin, S.A.; Fang, N.; Thompson, R.H.; Rappaport, L. Stimulation of furanocoumarin accumulation in celery and celeriac tissues by Fusarium oxysporum f. sp. Apii. *Phytochemistry* **1992**, *31*, 2683–2688. [CrossRef]
49. Schulzová, V.; Babička, L.; Hajšlová, J. Furanocoumarins in celeriac from different farming systems: A 3-year study. *J. Sci. Food Agric.* **2012**, *92*, 2849–2854. [CrossRef] [PubMed]
50. Zobel, A.M.; Brown, S.A. Influence of low-intensity ultraviolet radiation on extrusion of furanocoumarins to the leaf surface. *J. Chem. Ecol.* **1993**, *19*, 939–952. [CrossRef]
51. Popova, M.; Stoyanova, A.; Valyovska-Popova, N.; Bankova, V.; Peev, D. A new coumarin and total phenolic and flavonoids content of Bulgarian celeriac. *Izvestiya Khimiya Bulgarska Akademiya Naukite* **2014**, *46*, 88–93.
52. Leclerc, J.; Miller, M.L.; Joliet, E.; Rocquelin, G. Vitamin and Mineral Contents of Carrot and Celeriac Grown under Mineral or Organic Fertilization. *Biol. Agric. Hortic.* **1991**, *7*, 339–348. [CrossRef]
53. Kaiser, A.; Hartmann, K.I.; Kammerer, D.R.; Carle, R. Evaluation of the effects of thermal treatments on color, polyphenol stability, enzyme activities and antioxidant capacities of innovative pasty celeriac (Apium graveolens L. var. rapaceum (Mill.) DC.) products. *Eur. Food Res. Technol.* **2013**, *237*, 353–365. [CrossRef]
54. MacLeod, G.; Ames, J.M. Volatile components of celery and celeriac. *Phytochemistry* **1989**, *28*, 1817–1824. [CrossRef]
55. Maynard, D.N.; Barker, A.V.; Minotti, P.L.; Peck, N.H. Nitrate Accumulation in Vegetables. In *Advances in Agronomy*; Brady, N.C., Ed.; Academic Press: Cambridge, MA, USA, 1976; pp. 71–118. ISBN 0065-2113.
56. Bruning-Fann, C.; Kaneene, J. The effects of nitrate, nitrite and N-nitroso compounds on human health: A review. *Vet. Hum. Toxicol.* **1994**, *35*, 521–538.
57. Schuthan, W.; Bengtsson, B.; Bosund, I.; Hylmö, B. Nitrate accumulation in spinach. *Plant Foods Hum. Nutr.* **1967**, *14*, 317–330. [CrossRef]
58. Wright, M.J.; Davison, K.L. Nitrate Accumulation In Crops And Nitrate Poisoning In Animals. In *Advances in Agronomy*; Norman, A.G., Ed.; Academic Press: Cambridge, MA, USA, 1964; pp. 197–247. ISBN 0065-2113.
59. Wojciechowska, R.; Siwek, P. The effect of shading on nitrate metabolism in stalks and blades of celery leaves (Apium graveolens L. var. dulce). *Folia Hortic.* **2006**, *18*, 25–35.
60. Derolez, J.; Vulsteke, G. Accumulation of nitrate: A cultivar-linked property with celeriac (Apium graveolens L.var. rapaceum). *Plant Foods Hum. Nutr.* **1985**, *35*, 375–378. [CrossRef]
61. Dambrauskienė, E.; Maročkienė, N.; Viskelis, P.; Rubinskiene, M. Evaluation of productivity and biochemical composition of edible root celery. *Acta Hortic.* **2009**, *830*, 115–119. [CrossRef]

Article

A First Investigation of Agriculture Sector Perspectives on the Opportunities and Barriers for Agrivoltaics

Alexis S. Pascaris [1,*] , **Chelsea Schelly** [1] **and Joshua M. Pearce** [2,3,4]

[1] Department of Social Sciences, Michigan Technological University, 1400 Townsend Drive, Houghton, MI 49931, USA; cschelly@mtu.edu

[2] Department of Materials Science and Engineering, Michigan Technological University, 1400 Townsend Drive, Houghton, MI 49931, USA; pearce@mtu.edu

[3] Department of Electrical & Computer Engineering, Michigan Technological University, 1400 Townsend Drive, Houghton, MI 49931, USA

[4] School of Electrical Engineering, Aalto University, Aalto, 02150 Espoo, Finland

* Correspondence: aspascar@mtu.edu; Tel.: 906-487-2113

Received: 21 October 2020; Accepted: 24 November 2020; Published: 28 November 2020 

Abstract: Agrivoltaic systems are a strategic and innovative approach to combine solar photovoltaic (PV)-based renewable energy generation with agricultural production. Recognizing the fundamental importance of farmer adoption in the successful diffusion of the agrivoltaic innovation, this study investigates agriculture sector experts' perceptions on the opportunities and barriers to dual land-use systems. Using in-depth, semistructured interviews, this study conducts a first study to identify challenges to farmer adoption of agrivoltaics and address them by responding to societal concerns. Results indicate that participants see potential benefits for themselves in combined solar and agriculture technology. The identified barriers to adoption of agrivoltaics, however, include: (i) desired certainty of long-term land productivity, (ii) market potential, (iii) just compensation and (iv) a need for predesigned system flexibility to accommodate different scales, types of operations, and changing farming practices. The identified concerns in this study can be used to refine the technology to increase adoption among farmers and to translate the potential of agrivoltaics to address the competition for land between solar PV and agriculture into changes in solar siting, farming practice, and land-use decision-making.

Keywords: agrivoltaics; solar energy; agriculture; energy innovation; technology adoption; photovoltaics

---

1. Introduction

The Intergovernmental Panel on Climate Change (IPCC) Carbon and Other Biogeochemical Cycles report [1] reveals the predominant sources of anthropogenic greenhouse gas (GHG) emissions are the use of fossil fuels as sources of energy and land use changes, particularly agriculture. Agrivoltaics, the strategic codevelopment of land for both solar photovoltaic (PV) energy production and agriculture, can meet growing demands for energy and food simultaneously while reducing fossil fuel consumption [2–4]. Integrated energy and food systems have the potential to increase global land productivity by 35–73% [2] and to minimize agricultural displacement for energy production [5–7]. Agrivoltaic systems are a strategic and innovative approach to combine renewable energy with agricultural production, effectively addressing the predominant sources of anthropogenic GHG emissions as identified by the IPCC.

The viability of emerging agrivoltaic innovation has been investigated in various contexts. In conjunction with solar PV, there are emu farms in Australia [8] as well as sheep grazing [6,9,10] and pollinator-friendly sites proliferating in the U.S. (e.g., [11]). There is also the potential to use agrivoltaics with rabbits [12] and aquaponics (aquavoltaics) [13]. Experimental agrivoltaic research is occurring in diverse locations and climates. Examples include cultivation of corn and maize [14,15], lettuce [16,17], aloe vera [18], grapes [19], and wheat [20]. Mow [6] describes agrivoltaics as low-impact solar development that can alleviate agricultural displacement and assume varied designs: a solar-centric design that prioritizes solar output while growing low-lying vegetation; a vegetation-centric design that prioritizes crop production but incorporates solar panels and a colocation design that integrates both solar and agriculture for equal maximum dual output. Colocation designs have produced an estimated 3–8% per watt reduction in overall installation cost during site preparation due to cost reductions in land clearing and grubbing, soil stripping and compaction, grading and foundation for vertical supports, when compared to conventional solar industry development practices [6]. Further, Mavani et al. [4] found over a 30% increase in economic value for farms deploying such systems. Previous studies demonstrated that the dual-use of land for both PV and agriculture generates a mutually beneficial partnership that provides unique market opportunities to farmers and reduced operation and maintenance fees to solar developers, particularly in the case of grazing livestock [3,6,21–23].

The growing land footprint of solar PV presents social and spatial challenges, which are exacerbating the competition for land between agriculture versus energy production [5,23–25]. The U.S. Department of Energy Sunshot Vision Study forecasts that solar energy capacity will be nearly 329GW by 2030, which will necessitate approximately 1.8 million acres of land for ground-mounted systems [26]. Guerin [23] posits that the colocation of energy and agriculture will be stunted if there is absence of support from farmers and rural landowners, as the potential of agrivoltaic systems to address land-use competition will be contingent on farmer acceptance of agrivoltaics as a sociotechnological innovation. Brudermann et al. [27] found that PV adoption by farmers is primarily driven by environmental and economic considerations, which suggests factors that will be critical in agriculture sector decision-making concerning agrivoltaics.

Diffusion is a spatial and temporal phenomenon by which an innovation disseminates amongst adopters through a gradual process of filtering, tailoring and acceptance [28–30]. Rogers' [28] diffusion of innovations theory explains how and why some technological innovations are widely accepted while some are not, specifically referring to the adoption of an innovation by farmers over time in a rural diffusion model. The diffusion of innovations theory has been used to study diffusion of an innovation among physicians [31], among industrialized firms [32] and in terms of policy diffusion [33], among many other applications. Wilson & Grübler [34] applied the theory distinctly to energy innovations and described four phases of diffusion in which agrivoltaics can be categorized as existing in the first stage of an extended period of experimentation, learning, diversity of designs and small unit and industry-scale technologies. Grübler [30] warns that the existence of an innovation in itself does not promise proper diffusion, and while innovations have the capacity to induce change, it is the process of diffusion that realizes this potential as changes in social practice. By applying the diffusion of i theory to the agrivoltaic innovation, this study seeks to offer insight into potential refinements to the innovation of agrivoltaics in terms of its social acceptance to enable continued diffusion. This study uses Rogers' theory [28] as a practical framework for informing the diffusion of agrivoltaic innovation to discern the future potential and challenges for this technology to diffuse sufficiently to address energy and agricultural demands sustainably. While the technical viability of colocating solar PV and agriculture has been demonstrated [2,3,16,17], research in this field is incomplete with regard to placing the innovation within a social context to determine barriers to diffusion as perceived by industry experts.

Recognizing the fundamental importance of farmer adoption in the successful diffusion of agrivoltaics, this study investigates agriculture sector experts' perceptions on the opportunities and barriers to dual land-use agrivoltaic systems. Using in-depth, semistructured interviews, this study seeks to further the potential of agrivoltaics by identifying challenges to farmer adoption in an effort to address them by responding to societal concerns. In the following sections, the results are discussed, and conclusions are drawn on barriers to be overcome for agrivoltaic diffusion as identified by industry experts. The organization of the results and discussion are based on concepts from the diffusion of innovations theory [28], with a focus on relevant innovation characteristics (observability, relative

advantage and compatibility), stages of the adoption process and categories of adopters. Finally, the implications of these findings for the future development of agrivoltaics and farmer adoption are considered.

2. Materials and Methods

This study investigates agriculture sector experts' perceptions of the opportunities and barriers to agrivoltaics using in-depth, semistructured interviews. Interview methodology is exploratory by nature and, most appropriately, collects and analyzes data about perceptions, opinions and attitudes of people [35]. Aimed at providing an inclusive and nuanced perspective of the phenomenon under study, interviews were employed to directly engage relevant informants related to agriculture and agrivoltaics.

Prior to commencement, this research obtained approval from Michigan Technological University's Institutional Human Subjects Review Board (code: 1524021-1) to ensure compliance with institutional ethics in human subjects research. The initial interview protocol can be found in Appendix A. Email was used to introduce the agrivoltaic concept and the study while inviting prospective participants to video conferencing discussions, which resulted in 10 online interviews lasting between 30 to 90 min. All participants provided informed consent for the recording of conversations, which were anonymized for the protection of their privacy. Data collection occurred between February and July 2020 until saturation was attained, known as the point when no new additional insight is derived from conversations with participants and stabilization of data patterns occur [36,37].

A total of 10 interviews were conducted with 11 agriculture sector professionals (one interview engaged two individuals simultaneously), including livestock and crop farmers, solar grazers (individuals who graze their livestock underneath solar panels) and an agriculture policy expert. Sampling for logical representativeness, variance, diversity, and relevance to agriculture, participants were pursued based on their potential to provide insight into the opportunities and barriers to agrivoltaics because they have direct experience in the agricultural sector. Both theoretical and snowball sampling methods are nonprobability techniques that were employed to construct a sample capable of representing a wide range of perceptions. Theoretical sampling intentionally captures individuals with certain characteristics [38,39], whereas snowball sampling progressively follows a chain of referrals from study participants to other potential contributors [40,41]. Table 1 details the sample of participants that was generated using these sampling methods, ranging in profession, geographic location and gender. While credible and valuable, samples constructed through nonprobability sampling do not lend themselves to generalization [42], nor are the findings generated through interview methodology suitable for statistical generalization or analysis. However, all of the themes discussed as findings were raised by the majority of participants and identify the primary opportunities and barriers to agrivoltaics according to this sample but cannot be quantified or suggested to represent a broader population. Therefore, the findings are not discussed quantitatively to steer clear from suggesting these results are statistically generalizable to the entire agriculture sector.

Table 1. Interview Participant Characteristics.

Profession	Geographic Region (United States)	Gender
Livestock farmer: 5	North East: 4	
Crop farmer: 1	South East: 1	Male: 5
Solar grazer: 4	Midwest: 5	Female: 6
Policy: 1	South West: 1	

Drawing from grounded theory methodology [41,43], data collection and data analysis occurred in parallel to strategically shape subsequent inquiry. Responses that emerged in initial interviews instructed the development of ensuing questions, allowing for gradual pursuit and refinement of relevant issues. Interview themes were generally organized around: (1) the participants' experience in agriculture and details of their current operation; (2) experience with and perceptions of agrivoltaics (e.g., attitudes, opinions, perceived opportunities and barriers); (3) willingness to engage in an agrivoltaic project (e.g., perceived benefits and challenges). Interview protocol matured over time to explicate what agriculture sector professionals perceived as relevant opportunities and barriers to agrivoltaic development.

All interviews were recorded, manually transcribed and analyzed using the qualitative data analysis program NVivo 12 Pro (QSR International, Melbourne, Australia) [44]. Data were studied on a line-by-line basis using a series of coding and analytic induction to explore relationships, patterns and processes. Line-by-line coding is the fundamental step in interview analysis that moves beyond concrete statements to make analytic interpretations [41]. Coding in grounded theory methodology helps anchor analysis to participants' perspectives, explore nuances of meaning, identify implicit and explicit issues, as well as cluster similarities and observe differences among responses [41]. As outlined by Znaniecki [45] and Robinson [46], analytic induction involves identifying patterns, themes and categories in qualitative data in preparation for comparison amongst the varied findings. Employing rigorous, iterative and comparative grounded theory techniques, analysis of these data has captured and condensed the most relevant opportunities and barriers to agrivoltaics according to this sample of agriculture sector professionals.

3. Results

This section organizes findings based on frequency and expressed magnitude of the barriers and opportunities to agrivoltaics as defined by study participants. Both direct quotations (italicized) and analysis of results are presented jointly. Sections 3.2 and 3.3 are aligned with three of the five innovation characteristics defined by Rogers' diffusion of innovations theory [28] (observability, relative advantage and compatibility), which were identified by participants as the most critical when considering the adoption of agrivoltaic technology. These results offer insights into the main challenges to farmer adoption of agrivoltaics and suggest opportunities for interested stakeholders to further diffuse this innovation. A discussion considering the implications of these results is followed in Sections 4 and 5.

3.1. Long-Term Land Productivity and Planning

The underlying fundamental challenge of agrivoltaic systems, as perceived by participants, concerns long-term land viability. Land viability is intrinsically proportionate to the livelihood of agriculturalists, as farmers explained that the quality of their land is of critical importance and cannot be compromised. Interviews with farmers revealed their temporal approach to decision-making as they prioritize the protection of long-term land viability above all. One farmer expressed this concern when considering the use of an agrivoltaic system:

> *I'm concerned too, if you're pouring a bunch of concrete and putting in permanent structures, what does this look like in the end of 20 or 30 years?*

Encompassed within concerns of long-term land viability are more nuanced challenges related to land productivity in the presence of permanent solar panel structures. Participants explained that in order to maintain their agricultural land status and thrive in their farming venture, land must stay actively agricultural. The challenge that permanent solar structures could potentially impose on land productivity was unsettling:

> *Given the permanency of all of the solar panels and the permanency of the size of the plot, maintaining it to be continually productive for the animals would be a challenge. One of the challenges that I foresee is learning how to get the production that you want navigating around all of those structures.*

When considering an agrivoltaic system, participants' concerns were largely technical and economic in nature, reflecting their dependence on land productivity. Considerations about long-term land use and farmland preservation constituted the basis for decision-making, suggesting that anything that jeopardizes land viability will not be tolerated by farmers. Thinking beyond protecting the soil itself, various participants expressed potential opportunities that agrivoltaic systems could bring to agriculturalists:

> *When we talk about farmland preservation, it's not just about preserving the physical ground, it's also about preserving the viability of the farm. If a farmer is going to go under because of lack of revenue, why wouldn't you want them to open up an additional revenue stream to be able to actually preserve that land?*

> *There's going to be ground that goes into the solar panels and I think the idea that here you can integrate mixed-use with this makes a lot of sense. I think you have to have the right farmers and the right producers that are committed to making some of these things work.*

Participants explained that long-term land viability and productivity implies required long-term planning. When discussing the prospect of engaging in an agrivoltaic project, participants proposed that incorporating some type of land-use agreement or long-term plan would relieve concerns around the future of their farm. Providing certainty of farmland preservation surfaced as a recurring consideration of agrivoltaic adoption, as articulated by one participant:

> *Restoring the land back to what it was having the right land agreements to where when that lease is up, they have to return it to prelease form.*

To address the need for long-term planning and prioritization of agricultural interests, agrivoltaic project contracts are widely used by current stakeholders. As described by interviewees who identify as solar grazers, agrivoltaic contracts provide certainty and prevent against loss for both parties involved. The temporal concerns of agriculturalists with regards to long-term land viability can be reassured by agreement and engagement on both sides, as a solar grazer explains:

> *You can't have any business planning when you have that degree of uncertainty. So, it was getting people to have contracts. What the contract did is give certainty to both sides. It meant the farmers could plan their businesses, because there is a whole bunch of this remote targeted grazing, there's tons of mechanics, tons of money, staffing, and planning around breeding schedules, you name it. And then on the other side you got people wanting to make sure that the insurance is okay, and that their wiring is going to be okay, and how they'll interface with all their service work, the whole picture. I just knew the contract was the first key to the puzzle.*

> *If you don't have a real contract and if you don't have someone really interested engaging in a 10-year kind of way on both sides, the whole thing is not going to work.*

The majority of participants communicated that to the extent that the solar infrastructure of an agrivoltaic project does not threaten long-term land productivity, there are opportunities for increased revenue to farmers and mutually beneficial land-use agreements. These interviews reveal that addressing concerns about the viability of land after project decommissioning and protecting the livelihoods of farmers will involve long-term planning and partnership between agriculture and solar industries. The establishment of agrivoltaic contracts has proven valuable to current solar grazers and provides a direct way to alleviate uncertainties in land-use planning.

3.2. Market (Un)certainty and Observability of Benefits

When considering barriers to farmer adoption of agrivoltaics, economic concerns were raised by participants only second to concerns described above regarding long-term planning for technical considerations. At a basic level, farming is a business, and is thus accompanied by a set of risks, uncertainties and investments. Participants explained that risk is especially unwelcome in the business of farming and that certainty in productivity and security in investment are vital. One participant articulated that the market unknowns are potentially more critical than the technical unknowns of agrivoltaics:

There's a lot of unknowns for the producer in this as well. Having established markets, alleviating some of the unknowns and the risks are probably as much of a piece of this as anything. So, sketching out the long-term financial return of like, "Here's what these markets look like for livestock production." And what the guaranteed revenue is for solar panels, for instance. In terms of just making it happen out there in the field, there's some requirements to make that happen, but they aren't insurmountable, I wouldn't imagine.

Others stressed the need for a secure market for an agrivoltaic system to be successful:

You would probably want to package it more as, "Do we have a food and farm system in place that allows somebody to have solar and grow these crops that are tolerant to that condition?" And then importantly, "Do we have a market to send that stuff to?" Because then all of a sudden it becomes this closed loop, kind of circular economy feel to it. But without that end market side of it, I think people would say, "That's great if you want to grow that stuff."

As long as the market is there, I would think a lot of these things could work.

As business owners, considerations of financial return and security in the marketplace are at the forefront of decision-making for farmers. For the majority of participants, the agrivoltaic innovation is unfamiliar and imposes constraints on business planning borne of unknowns and uncertainties. Building flexibility into the system to accommodate for changes in market conditions and farming practice could potentially alleviate some of the concern of uncertainty, as explained:

If we're looking at a 25-year kind of investment with the solar panels and when you're talking about integrating them within the livestock species too, that market for livestock might look totally different within 10 years. So, implementing some flexibility there that if we're not going to run rabbits, maybe we're running something else in there in 20 years. But having some flexibility in the system that you could respond to the livestock markets in there as well, I think is important.

Flexibility and adaptation to changing market conditions emerged as key elements to be incorporated into planning for an agrivoltaic system, highlighting again the temporal component to farmer decision-making and identifying concerns to be addressed for successful adoption. While the future unknowns of market acceptance of a product are difficult to ascertain, participants suggested that integrating flexibility into system design would reduce financial unease.

Coupled with concerns of a stable and reliable market for their product, were expectations for just compensation and tangible benefits from participation in an agrivoltaic project. When considering the adoption of the agrivoltaic innovation, participants also questioned if such an endeavor would be justified in terms of monetary gains. Participants perceived the adoption of such technology as an increased labor commitment and thus expected to reasonably gain from it. When asked if they would engage in an agrivoltaic project, one participant answered:

Essentially, they would have to pay me if they wanted me to be there because it's so much work to remediate soil and bring it up to a productive level, especially if this has been formally row cropped conventionally. So, it would really depend on what it had been earlier, how much I trusted the people who were starting this operation, and how much I felt that there would be ease of incorporating it into my schedule. I also think that it's not free pasture, you know what I mean? Even if they didn't charge me a single thing, there would be a lot of investment. So, I'd be going for like- I don't even know- I almost want to see like co-ownership, we own this land together, you get the profits from the solar and I get whatever everything else is. Or putting the solar panels on my own farm and then I get the revenue from the solar panels.

When judging the adoption of agrivoltaic innovation, participants expressed critical valuations of its worth and asserted that observable and substantial benefits would have to be derived in order for them to commit. Of the 10 farmers interviewed, four were already engaging with the technology and five others said they would get involved if they would derive more benefit than cost from it. Thus, the vast majority (nine of 10) of the farmers interviewed were open to using or already using agrivoltaics. Improving the agrivoltaic innovation to increase diffusion to these interested farmers will require establishment of just compensation for farmers, as explained by two solar grazers:

> The biggest misconception to clear up immediately when people start thinking about this is that it can be anything like free grass. Because there's so much commitment on my end, and the cost of setting up all that equipment is very high. The time and labor of going there and servicing the sheep is a big commitment.

> I'm really trying to get out of is the idea that the farmer should be doing all this work for free. The solar firms are making—maybe not tons of money—but reasonable amounts of money off these investments. For them, they need to know that the performance guarantee is there, the sun has to shine on their panels, there shouldn't be interference with that. They need that steady assurance. And the farmers need to get paid for recognizing that there is a performance guarantee to meet.

Participants explained that their willingness to be involved with the agrivoltaic innovation would be contingent on the near-term observability of direct benefits to them and the long-term certainty and security in the marketplace for their product. Observability is an innovation characteristic explained by Rogers (1962) that concerns the degree to which the results of an innovation are visible to potential adopters. When assessing their potential adoption of agrivoltaics, agriculture sector experts framed their considerations in terms of direct and tangible benefits, suggesting that observability of benefits is a characteristic of the agrivoltaic innovation that is of decisive importance to adopters. As discussed by participants in Section 3.1, agrivoltaic contracts are currently recognizing the rights and duties of involved parties, and provide opportunity to establish legitimate, mutually beneficial partnerships. With nine of 10 farmers inclined to partake in an agrivoltaic partnership, the above concerns about economic uncertainty and gains are active considerations for all involved stakeholders in project development.

Relative Advantage

The degree to which agrivoltaics are perceived by participants to be advantageous to current practice was identified as important when considering adoption. While participants expressed that financial compensation for farmers is both necessary and attractive, they also spoke of other benefits they anticipate as a result of engaging with the agrivoltaic technology. Participants discussed potential marketing advantages:

> It's got a great story; it's got a wonderful marketing edge from that perspective. So, your advantage is a great story to tell from a marketing standpoint.

> I think that's where you have a very unfair advantage for whoever would be doing this rabbit production, you might be getting paid for land maintenance and then have rabbits for free. So, your profitability could be way up or your price could be way lower because you wouldn't have land expenses. There's a lot of opportunity to create some advantage from a production standpoint. From that perspective they may sell better or have an [edge] in the marketplace because of that aspect.

Another participant expressed other technical synergies when grazing animals underneath solar panels:

> I think it sounds like a great idea. It sounds like a great way to maintain, and not have to mow. I can see the panels providing shade and protection from the rain in a way that seems very valuable.

Perceiving a multitude of potential benefits, participants speculated how the adoption of the agrivoltaic innovation could provide them benefits and competitive advantages in the marketplace. Foreseeing a unique opportunity to derive a revenue stream from land maintenance, some participants postulated that there were economic gains associated with combined solar and agriculture systems. Rogers' (1962) innovation characteristic, relative advantage, explains that innovations that are perceived to be superior to business as usual have higher potential for adoption. Participants described the relative advantage of agrivoltaics worthwhile, and thus identified this innovation characteristic as critical when considering the adoption of the innovation, suggesting that if an agrivoltaic system could provide an advantage to a farmer, the likelihood of adoption would be greater.

3.3. Compatibility with Current Practice

A considerable opportunity for farmers in agrivoltaic projects is the potential for integration of the innovation into their current practice. Participants expressed disinterest in increased complications in their business, and rather actively seek ways to reduce labor through harnessing the synergies of innovative practices. The ease of integration and compatibility of solar with current production was frequently considered amongst participants, highlighting the opportunity to plan overlapping operations to increase farmer acceptance. The attractiveness of agrivoltaic integration was explained by two participants:

Most of my exposure to this is from sheep, and I think that it's a great idea. For my own particular system, it would definitely reduce the amount of labor for one aspect of the system, which is moving the fencing. So, I'm all for it. I think it'd be a really nice mesh.

Alternative energy is expensive to people like us. But it's something that I guess, if it could be integrated into something I'm already doing and could potentially help protect the animals, or do whatever, and then also run the homestead, it's just another perk of having something like that. It's another reason to have it besides just having the electricity.

As elucidated by participants, compatibility of the agrivoltaic innovation with current practice could reduce labor and create an incentive to engage in the technology. When considering the value of agrivoltaics to them personally, farmers offered calculated and context-dependent perspectives, making judgments on the benefits in terms of their own operation rather than speaking generally about dual-use solar systems. Speaking from a place of personal considerations and interests, participants revealed that there is a context-dependent nature of success for agrivoltaic projects. Reflecting their own practices, one participant stated:

I've also heard them say in meetings the fact that we're going to farm soybeans underneath solar panels, which is just asinine. Like, it's not going to happen. The size of our equipment doesn't permit that kind of thing. Putting livestock under, kind of a grazing operation, seems to make sense.

Compatibility with current practice not only includes size of equipment, but also scale of the farming operation, as explained by one participant:

The work that would be involved with that, I think, or potentially having to hire someone to manage them, it would decrease our profit so much that it wouldn't make sense. I could see how that would be to someone's benefit though, but not at our scale.

To justify the labor involved in engaging in an agrivoltaic project, farmers evaluated their own enterprise by mentally applying the innovation and determining the potential compatibilities. As suggested by participants, the benefits of agrivoltaics are noteworthy, but will only be fully realized if there is ease of integration into their current farming practice. Compatibility is an innovation characteristic defined by Rogers (1962) that explains the degree to which an innovation is perceived to be consistent with needs, norms and sociocultural values is decisive to potential adopters. The theme of compatibility among most participants was viewed as an opportunity rather than a barrier for agrivoltaics, suggesting that the innovation's context-dependent nature provides flexibility and potential to leverage the solar system to derive synergistic benefits to compliment current farming practices.

4. Discussion: The Opportunities & Barriers for Agrivoltaic Diffusion

This research provides insight from the agricultural sector into the challenges and opportunities for farmer adoption of the agrivoltaic innovation. Results indicate that participants see potential benefits for themselves in combined solar and agriculture technology and identify barriers to adoption including desired certainty of long-term land productivity, market potential and just compensation, as well as the need for predesigned system flexibility to accommodate different scales of operation and adjustment to changing farming practice. The findings suggest that these barriers to adoption are not insurmountable and can be sufficiently addressed through prudent planning and mutually beneficial land agreements between solar and agriculture sector actors. Table 2 below organizes the identified barriers and opportunities to address them. All of the participants of this study assented to agrivoltaics as a synergistic and innovative approach to combined land-uses, while nine of the 10 participants who are currently active farmers stated they would engage in the use of a dual-use system given the discussed concerns are considered (four of the nine already are). Interviews with industry professionals informed the current state of diffusion of the agrivoltaic innovation and identified opportunities to further stimulate farmer adoption of the technology. These findings may be used to translate the potential of agrivoltaics to address the competition for land between solar PV and agriculture into changes in solar siting, farming practice and land-use decision-making.

Table 2. Barriers, opportunities, and directions for future work regarding the diffusion of agrivoltaics.

Barrier	Opportunity	Future Work
End-of-life impacts from solar infrastructure	-Driven piles (constructed of galvanized steel I-beams, channel-shaped steel or posts), helical piles (galvanized steel posts with split discs welded to the bottom at an angle) and ground screws (galvanized steel posts with welded or machined threads) can be removed and recycled [47,48]. -Photovoltaic (PV) racking can be put on removeable ballasted foundations or skids of precast or poured-in-place concrete ballasts to minimize land disturbances [47]. -Impacts from modules such as leaching of trace metals [49–51] and compromised future agricultural productivity [52] have been proven highly unlikely. -Contracted agreements that establish plans to return land back to prelease form after decommissioning of solar system.	-Empirical research investigating the magnitude of long-term impacts of solar infrastructure on land (e.g., [53]), soil, and pasture-grass productivity.
Permanent structures interfering with agricultural production and future farming practice	-A variety of plants have proven to maintain higher soil moisture, greater water efficiency, and experience increase in late season biomass underneath PV panels [54]. -Improvements in water productivity and additional shading are projected to increase crop production in arid regions experiencing climate change [55]. -Semitransparent PV [56] (Thompson et al., 2020) or vertical bifacial PV [57]. -Raised racking systems provide clearance for agricultural equipment, which could allow for nearly any crop to be used in agrivoltaic production [58]. -Design flexible open source racking systems [59,60] that have adjustable panel height, tilt angle and spacing [61], as well as a combination of permanent and portable fencing. -East-west tracking array configurations allow optimal conditions for plant growth when compared to conventional south-facing designs [62].	-Empirical research aimed at understanding the implications of solar PV infrastructure on perennial pasture grass maintenance. -Optimized agrivoltaic PV -Cost-benefit analysis of open source PV racking systems designed with adjustable panel height, tilt angle and spacing. -Cost-benefit analysis of permanent and portable fencing for animal grazing agrivoltaics.

Table 2. *Cont.*

Barrier	Opportunity	Future Work
Uncertainties in operation and business planning	-Legitimate partnerships and contracts that establish up-front costs and compensation for both parties -Local government policy aimed at supporting development of solar PV [63,64] -Education and outreach from PV industry to farming industry to reduce barriers to knowledge and increase trust.	-Policy research focused on market mechanisms to incentivize agrivoltaic systems for both solar and agriculture sector. -Increased efforts from university extension programs to increase information sharing and partnership between energy and agriculture.

4.1. Diffusing the Agrivoltaic Innovation—Where Are We Now?

The diffusion of innovations theory [28] identifies five stages in the process of technology adoption. Participants of this study predominantly fell into the decision or evaluation stage of adoption, which is understood as the stage in which an individual mentally applies an innovation to their present and perceived future circumstances to arrive at a decision to try it or not. Beyond the initial knowledge or interest stages of Rogers' adoption model [28], the majority of participants (six of 11) considered their potential adoption of agrivoltaics beneficial but dependent on factors related to context. Speaking from a place of receptivity, these participants saw value in the innovation and felt inclined to engage with it, while voicing a few concerns about compatibility with their practice and uncertainties about long-term land productivity. Four of the 11 participants were already functioning in the confirmation or adoption stage of the adoption process, making full use of the innovation. Based on these findings, it is observed that the current state of the diffusion of agrivoltaics is advancing towards wider implementation and has surpassed initial phases of information gathering and persuasion. Participants in the decision or evaluation stage of adoption identified barriers to their engagement with agrivoltaics, giving interested stakeholders the ability to directly respond to these concerns by improving the technology to enable further diffusion.

Further, most participants of this study were early majority adopters, characterized by wanting proven and reliable applications, reference from trusted peers and being prudent in financial risk and uncertainty. Rogers [28] asserts that an innovation must meet the needs of all categories of adopters, making clear in the context of agrivoltaic adoption where efforts should be focused to successfully move early majority adopters into acceptance of the innovation. Technological diffusion is a process of filtering, tailoring and accepting [30], and the identified concerns of the agriculture sector professionals in this study can be used to tailor or refine the technology to increase adoption among farmers. The following section will elaborate upon the critical characteristics of agrivoltaic systems as identified by participants and suggest recommendations for improvement with the intention of facilitating accelerated diffusion.

4.2. Diffusing the Agrivoltaic Innovation—What Needs to Happen?

Rogers [28] posited that there are five distinct innovation characteristics that help explain why some innovations are widely accepted and some are not. Understanding the characteristics of the agrivoltaic innovation is valuable for interested stakeholders when assessing areas for improvement and pursuing further acceptance of the technology. The results of this study identify the most critical characteristics of agrivoltaics and point to opportunities to directly respond to farmers concerns.

Of these five characteristics, observability of benefits, relative advantage and compatibility with current practice were identified by participants as the most critical when considering their personal adoption of the agrivoltaic technology. What this means for further diffusion is that the solar industry actors involved in the development of agrivoltaic systems must devise mutually beneficial land agreements with farmers that establish compensation for their labor, articulate plans for land restoration after the decommissioning of the system and be sensitive to contextual differences among agriculturalists by designing a system that is flexible enough to meet the needs of the current and future users. Participants in this study saw immediate value in personal adoption of the technology but sought long-term security in terms of farmland preservation and financial return.

There are a handful of practical actions to be taken to enable further diffusion of agrivoltaics. Table 2 presents a summary of the identified barriers, existing opportunities to overcome them and directions for future work. First, the establishment of agrivoltaic contracts has proven valuable to current solar grazers. Robust and forward-thinking land use agreements will provide a direct way to alleviate uncertainties in land-use planning and secure compensation for farmer's labor. Second, system designers need to integrate flexibility in design by accommodating current land practices and allowing for future changes. Concerns about market uncertainty and rigid systems can be addressed by crafting a combined solar and agricultural project that is adaptable to changing market and farming conditions. Third, agrivoltaics systems should be designed with compatibility in mind. By strategically harnessing the synergy of compatibility with current practice, these results suggest that farmers would be more inclined to engage with a project if it generated advantages in their operation. Being sensible in scaling a system to current practice, rather than creating increased labor burden on farmers, will increase the likelihood of their participation with the technology.

The potential for increased utilization of the agrivoltaic technology is ripe. While previous research has demonstrated its technical viability, this study recognizes that technology innovations exist within a social context and thus depend upon social acceptance and adoption. It is concluded that continued farmer adoption of agrivoltaics is likely, yet contingent on observable benefits in farming practice and assurance of financial gain. Future research should investigate how perceptions vary across geographic regions and agriculture professions (i.e., animal versus crop farming) to study the unique opportunities and barriers for agrivoltaics in the context of local climate and agricultural practice. Increased education and outreach concerning the end-of-life impacts, negligible effects of solar PV on agricultural productivity and potential for agrivoltaic systems to protect crop production during climate change, is necessary to inform and stimulate further farmer adoption. Empirical experimental research should investigate the long-term impacts of solar PV infrastructure on perennial pasture grasses to better understand the possible effects of agrivoltaic systems on future grazing productivity. Economic cost-benefit analysis will be valuable for quantifying the potential cost disadvantages of designing flexible PV arrays that can be adjusted to accommodate different panel heights and spacing requirements. Future policy research can investigate the role of market mechanisms, such as incentives, in prompting further development of agrivoltaics. Based on these findings, policy makers should consider implementing financial instruments that stimulate both solar and agriculture sector adoption of the technology, while building flexibility into such policies to allow diverse, innovative and contextually appropriate system designs. To do this, agrivoltaic proponents can model their efforts on the successful diffusion of wind farm/solar farm integration that focuses on local support [65,66]. Previous research examining diffusion of solar as an innovation among residential adopters highlighted the role of communities of information sharing for promoting adoption [67]. The study presented here is unique in examining the diffusion of agrivoltaic solar innovation as a community level consideration, but also demonstrates how diffusion of innovation can occur within a social context. Moving forward, placing the agrivoltaic technology in a social context will be essential to identify the barriers to its diffusion and will offer relevant solutions to increase its adoption.

5. Conclusions

Agrivoltaic systems are a strategic and innovative approach to combine renewable energy with agricultural production. Recognizing the fundamental importance of farmer adoption in the successful diffusion of agrivoltaics, this study investigates agriculture sector experts' perceptions on the opportunities and barriers to dual land-use systems. Results indicate that participants saw potential benefits for themselves in combined solar and agriculture technology and identified barriers to adoption including desired certainty of long-term land productivity, market potential and just compensation, as well as the need for predesigned system flexibility to accommodate different scales and types of operations and adjustment to changing farming practice. The identified concerns of the agriculture sector professionals in this study can be used to refine the technology to increase adoption among farmers and to translate the potential of agrivoltaics to address the competition for land between solar PV and agriculture into changes in solar siting, farming practice and land-use decision-making. Ultimately, building integrated energy and food systems can increase global land productivity, minimize agricultural displacement and reduce greenhouse gas emissions from fossil fuels. Informed and concerted efforts at enabling further diffusion of this innovation are imperative for meeting growing demands for energy and food simultaneously.

Author Contributions: Conceptualization, C.S. and J.M.P.; methodology, A.S.P. and C.S.; validation, A.S.P. and C.S.; formal analysis, A.S.P. and C.S.; investigation, A.S.P.; resources, C.S. and J.M.P.; data curation, A.S.P.; writing—original draft preparation, A.S.P.; writing—review and editing, A.S.P., C.S. and J.M.P.; supervision, C.S. and J.M.P.; project administration, J.M.P.; funding acquisition, C.S. and J.M.P. All authors have read and agreed to the published version of the manuscript.

Funding: This material is based upon work supported by the U.S. Department of Energy's Office of Energy Efficiency and Renewable Energy (EERE) under the Solar Energy Technology Office Award Number DE-EE0008990 and the Witte Endowment.

Conflicts of Interest: The authors declare no conflict of interest. The funders had no role in the design of the study; in the collection, analyses, or interpretation of data; in the writing of the manuscript, or in the decision to publish the results.

Appendix A

Initial interview protocol as approved by IRB

1. Please tell me about your experience as a farmer.

    a. What is your geographic location?
    b. How long have you been doing it?

2. Who [markets, restaurants] are your biggest customers?

    a. How do you go about opening new accounts with potential customers?
    b. What is your greatest barrier to gaining access to new markets/customers?

3. How large is your operation? Would you consider it small-medium-large?
4. Are you familiar with both crop and animal farmers that incorporate solar panels on their land?

    a. If so, what are your thoughts on this?

5. Would you ever consider embracing the mixed-use of solar on your farm to harness co-benefits of solar energy generation and agricultural production?

    a. If so, why?

        i. What is your minimum acceptable rate of return?
    b. It not, why?

        i. What type of barriers are there?

6. Would you consider renting land on a prefenced solar-farm meant for agricultural production?

    a. If so, why?

        i. What is your minimum acceptable rate of return?

    b. It not, why?

        i. What type of barriers are there?

7. What is needed to make a mixed-use solar farm more attractive to you?
8. A new study that is sponsored by the D.O.E. has shown an opportunity to incorporate rabbit farming with solar photovoltaic farms that make electricity. This study has shown substantial economic opportunity from this mixed-use scheme: upwards of 24% increase in site revenue. Now I would like to ask you specifically about mixed-use solar involving farmed meat rabbits.

    a. What do you think are the biggest opportunities for this kind of mixed-use solar development?

    b. What do you think are the biggest barriers for this kind of mixed -use solar development?

9. Do you anticipate solar farm pasture-raised livestock selling for a premium or increasing sales?
10. Is there anything else you would like to tell me about your perspectives of mixed-use solar PV development?
11. Do you have suggestions of other experienced farmers I should speak with?

References

1. Ciais, P.; Sabine, C.; Bala, G.; Bopp, L.; Brovkin, V.; Canadell, J.; Chhabra, A.; DeFries, R.; Galloway, J.; Heimann, M.; et al. Carbon and Other Biogeochemical Cycles. In *Climate Change 2013: The Physical Science Basis. Contribution of Working Group I to the Fifth Assessment Report of the Intergovernmental Panel on Climate Change*; Stocker, T.F., Qin, D., Plattner, G.K., Tignor, M., Allen, S.K., Boschung, J., Nauels, A., Xia, Y., Bex, V., Midgley, P.M., Eds.; Cambridge University Press: Cambridge, UK; New York, NY, USA, 2013.
2. Dupraz, C.; Marrou, H.; Talbot, G.; Dufour, L.; Nogier, A.; Ferard, Y. Combining solar photovoltaic panels and food crops for optimising land use: Towards new agrivoltaic schemes. *Renew. Energy* **2011**, *36*, 2725–2732. [CrossRef]
3. Dinesh, H.; Pearce, J.M. The potential of agrivoltaic systems. *Renew. Sustain. Energy Rev.* **2016**, *54*, 299–308. [CrossRef]
4. Mavani, D.D.; Chauhan, P.M.; Joshi, V. Beauty of Agrivoltaic System regarding double utilizati on of same piece of land for Generation of Electricity & Food Production. *Int. J. Sci. Eng. Res.* **2019**, *10*.
5. Calvert, K.; Mabee, W. More solar farms or more bioenergy crops? Mapping and assessing potential land-use conflicts among renewable energy technologies in eastern Ontario, Canada. *Appl. Geogr.* **2015**, *56*, 209–221. [CrossRef]
6. Mow, B. Solar Sheep and Voltaic Veggies: Uniting Solar Power and Agriculture|State, Local, and Tribal Governments|NREL [WWW Document], n.d. URL. 2018. Available online: https://www.nrel.gov/state-local-tribal/blog/posts/solar-sheep-and-voltaic-veggies-uniting-solar-power-and-agriculture.html (accessed on 2 July 2020).
7. Adeh, E.H.; Good, S.P.; Calaf, M.; Higgins, C.W. Solar PV Power Potential is Greatest Over Croplands. *Sci. Rep.* **2019**, *9*, 11442. [CrossRef]
8. Renewable Energy World (REW). Getting Out of the Weeds: How to Control Vegetative Growth under Solar Arrays. 2014. Available online: https://www.renewableenergyworld.com/articles/2014/07/weed-control-at-solar-installations-what-works-best.html (accessed on 7 February 2020).
9. Ouzts, E. Farmers, Experts: Solar and Agriculture 'Complementary, not Competing' in North Carolina [WWW Document]. Energy News Network. 2017. Available online: https://energynews.us/2017/08/28/southeast/farmers-experts-solar-and-agriculture-complementary-not-competing-in-north-carolina/ (accessed on 2 July 2020).
10. Andrew, A.C. *Lamb Growth and Pasture Production in Agrivoltaic Production System*; Oregon State University: Corvallis, OR, USA, 2020.

11. Dunbar, E. Pollinator-Friendly Solar Energy Becomes the Norm in Minnesota. 2019. Available online: https://www.mprnews.org/story/2019/06/20/pollinatorfriendly-solar-energy-becomes-the-norm-in-minnesota (accessed on 23 September 2020).

12. Lytle, W.; Meyer, T.K.; Tanikella, N.G.; Burnham, L.; Engel, J.; Schelly, C.; Pearce, J.M. Conceptual Design and Rationale for a New Agrivoltaics Concept: Pastured-Raised Rabbits and Solar Farming. *J. Clean. Prod.* **2020**, 124476. [CrossRef]

13. Pringle, A.M.; Handler, R.M.; Pearce, J.M. Aquavoltaics: Synergies for dual use of water area for solar photovoltaic electricity generation and aquaculture. *Renew. Sustain. Energy Rev.* **2017**, *80*, 572–584. [CrossRef]

14. Amaducci, S.; Yin, X.; Colauzzi, M. Agrivoltaic systems to optimize land use for electric energy production. *Appl. Energy* **2018**, *220*, 545–561. [CrossRef]

15. Sekiyama, T.; Nagashima, A. Solar Sharing for Both Food and Clean Energy Production: Performance of Agrivoltaic Systems for Corn, A Typical Shade-Intolerant Crop. *Environments* **2019**, *6*, 65. [CrossRef]

16. Marrou, H.; Wery, J.; Dufour, L.; Dupraz, C. Productivity and radiation use efficiency of lettuces grown in the partial shade of photovoltaic panels. *Eur. J. Agron.* **2013**, *44*, 54–66. [CrossRef]

17. Elamri, Y.; Cheviron, B.; Lopez, J.M.; Dejean, C.; Belaud, G. Water budget and crop modelling for agrivoltaic systems: Application to irrigated lettuces. *Agric. Water Manag.* **2018**, *208*, 440–453. [CrossRef]

18. Ravi, S.; Macknick, J.; Lobell, D.; Field, C.; Ganesan, K.; Jain, R.; Elchinger, M.; Stoltenberg, B. Colocation opportunities for large solar infrastructures and agriculture in drylands. *Appl. Energy* **2016**, *165*, 383–392. [CrossRef]

19. Malu, P.R.; Sharma, U.S.; Pearce, J.M. Agrivoltaic potential on grape farms in India. *Sustain. Energy Technol. Assess.* **2017**, *23*, 104–110. [CrossRef]

20. Marrou, H.; Guilioni, L.; Dufour, L.; Dupraz, C.; Wéry, J. Microclimate under agrivoltaic systems: Is crop growth rate affected in the partial shade of solar panels? *Agric. For. Meteorol.* **2013**, *177*, 117–132. [CrossRef]

21. Santra, P.; Pande, P.C.; Kumar, S.; Mishra, D.; Singh, R.K. Agri-voltaics or Solar farming-the Concept of Integrating Solar PV Based Electricity Generation and Crop Production in a Single Land use System. *Int. J. Renew. Energy Res.* **2017**, *7*, 694–699.

22. Barron-Gafford, G.A.; Pavao-Zuckerman, M.A.; Minor, R.L.; Sutter, L.F.; Barnett-Moreno, I.; Blackett, D.T.; Thompson, M.; Dimond, K.; Gerlak, A.K.; Nabhan, G.P.; et al. Agrivoltaics provide mutual benefits across the food–energy–water nexus in drylands. *Nat. Sustain.* **2019**, *2*, 848–855. [CrossRef]

23. Guerin, T.F. Impacts and opportunities from large-scale solar photovoltaic (PV) electricity generation on agricultural production. *Environ. Qual. Manag.* **2019**. [CrossRef]

24. Nonhebel, S. Renewable energy and food supply: Will there be enough land? *Renew. Sustain. Energy Rev.* **2005**, *9*, 191–201. [CrossRef]

25. Marcheggiani, E.; Gulinck, H.; Galli, A. Detection of Fast Landscape Changes- The Case of Solar Modules on Agricultural Land. In *International Conference on Computational Science and Its Applications*; Springer: Berlin/Heidelberg, Germany, 2013; pp. 315–327.

26. U.S. Department of Energy (DOE). *SunShot Vision Study (Rep.)*. Available online: https://www.energy.gov/sites/prod/files/2014/01/f7/47927.pdf (accessed on 24 February 2020).

27. Brudermann, T.; Reinsberger, K.; Orthofer, A.; Kislinger, M.; Posch, A. Photovoltaics in agriculture: A case study on decision making of farmers. *Energy Policy* **2013**, *61*, 96–103. [CrossRef]

28. Rogers, E. *Diffusion of Innovations*, 1st ed.; Free Press: New York, NY, USA, 1962.

29. Robertson, T.S. The process of innovation and the diffusion of innovation. *J. Mark.* **1967**, *31*, 14–19. [CrossRef]

30. Grübler, A. Time for a change: On the patterns of diffusion of innovation. *Daedalus* **1996**, *125*, 19–42.

31. Coleman, J.; Katz, E.; Menzel, H. The diffusion of an innovation among physicians. *Sociometry* **1957**, *20*, 253–270.

32. Mansfield, E. Technical change and the rate of imitation. *Econom. J. Econom. Soc.* **1961**, *29*, 741–766. [CrossRef]

33. Shipan, C.R.; Volden, C. The mechanisms of policy diffusion. *Am. J. Political Sci.* **2008**, *52*, 840–857. [CrossRef]

34. Wilson, C.; Grübler, A. Lessons from the History of Technological Change for Clean Energy Scenarios and Policies. In *Natural Resources Forum*; Blackwell Publishing Ltd.: Oxford, UK, 2011; Volume 35, pp. 165–184.

35. Goetz, J.P.; LeCompte, M.D. *Ethnography and Qualitative Design in Educational Research*; Academic Press: New York, NY, USA, 1984.

36. Strauss, A.; Corbin, J. *Basics of Qualitative Research-Grounded Theory Procedures and Techniques*; Sage Publications: Newbury Park, CA, USA, 1990.

37. Fusch, P.I.; Ness, L.R. Are we there yet? Data saturation in qualitative research. *Qual. Rep.* **2015**, *20*, 1408.
38. Corbin, J.; Strauss, A. Theoretical sampling. In *Basics of Qualitative Research*; SAGE Publications: Thousand Oaks, CA, USA, 2008.
39. Lindlof, T.R.; Taylor, B.C. Sensemaking: Qualitative data analysis and interpretation. *Qual. Commun. Res. Methods* **2011**, *3*, 241–281.
40. Biernacki, P.; Waldorf, D. Snowball sampling: Problems and techniques of chain referral sampling. *Sociol. Methods Res.* **1981**, *10*, 141–163. [CrossRef]
41. Charmaz, K. *Constructing Grounded Theory*; SAGE Publications: Thousand Oaks, CA, USA, 2014.
42. Bhattacherjee, A. *Social Science Research: Principles, Methods, and Practices*; Open University Press: Tampa Bay, FL, USA, 2012.
43. Charmaz, K.; Belgrave, L.L. Grounded Theory. In *The Blackwell Encyclopedia of Sociology*; Blackwell Publications: Malden, MA, USA, 2007.
44. Unlock Insights in Your Data with Powerful Analysis. QSR International. NVivo (Version 12.0 Pro) [Computer Software]. Available online: https://www.qsrinternational.com/nvivo-qualitative-data-analysis-software/home (accessed on 11 May 2020).
45. Znaniecki, F. *The Method of Sociology*; SAGE Publications: Thousand Oaks, CA, USA, 1934.
46. Robinson, W.S. The logical structure of analytic induction. *Am. Sociol. Rev.* **1951**, *16*, 812–818. [CrossRef]
47. Lorenz, E. Types of PV Racking Ground Mounts. 2016. Available online: https://www.cedgreentech.com/article/types-pv-racking-ground-mounts (accessed on 14 October 2020).
48. Turney, D.; Fthenakis, V. Environmental impacts from the installation and operation of large-scale solar power plants. *Renew. Sustain. Energy Rev.* **2011**, *15*, 3261–3270. [CrossRef]
49. Fthenakis, V. *Could CdTe PV Modules Pollute the Environment?* National Photovoltaic Environmental Health and Safety Assistance Center, Brookhaven National Laboratory: Upton, NY, USA, 2002.
50. *Potential Health and Environmental Impacts Associated with the Manufacture and Use of Photovoltaic Cells*; EPRI: Palo Alto, CA, USA; California Energy Commission: Sacramento, CA, USA, 2003.
51. NC Clean Energy Technology Center. Balancing Agricultural Productivity with Ground-Based Solar Photovoltaic (PV) Development. 2017. Available online: https://energizeohio.osu.edu/sites/energizeohio/files/imce/Handout_Balancing-Ag-and-Solar-final-version-update.pdf (accessed on 14 October 2020).
52. Armstrong, A.; Ostle, N.; Whitaker, J. Solar Park Microclimate and Vegetation Management Effects on Grassland Carbon Cycling. *Environ. Res. Lett.* **2016**, *11*, 074016. [CrossRef]
53. Hernandez, R.R.; Hoffacker, M.K.; Murphy-Mariscal, M.L.; Wu, G.C.; Allen, M.F. Solar energy development impacts on land cover change and protected areas. *Proc. Natl. Acad. Sci. USA* **2015**, *112*, 13579–13584. [CrossRef] [PubMed]
54. Hassanpour Adeh, E.; Selker, J.S.; Higgins, C.W. Remarkable agrivoltaic influence on soil moisture, micrometeorology and water-use efficiency. *PLoS ONE* **2018**, *13*, e0203256. [CrossRef] [PubMed]
55. Weselek, A.; Ehmann, A.; Zikeli, S.; Lewandowski, I.; Schindele, S.; Högy, P. Agrophotovoltaic systems: Applications, challenges, and opportunities. A review. *Agron. Sustain. Dev.* **2019**, *39*, 25. [CrossRef]
56. Thompson, E.; Bombelli, E.L.; Simon, S.; Watson, H.; Everard, A.; Schievano, A.; Bocchi, S.; Zand Fard, N.; Howe, C.J.; Bombelli, P. Tinted semi-transparent solar panels for agrivoltaic installation. *Adv. Energy Mater.* **2020**, *10*, 1614–6840. [CrossRef]
57. Riaz, M.H.; Younas, R.; Imran, H.; Alam, M.A.; Butt, N.Z. Module Technology for Agrivoltaics: Vertical Bifacial vs. Tilted Monofacial Farms. *arXiv* **2019**, arXiv:1910.01076.
58. REM TEC (Revolution Energy Maker TEC). 2017. Available online: http://www.remtec.energy/en/#agrovoltaico; https://www.youtube.com/watch?v=gmbfb4vZOuQ (accessed on 14 October 2020).
59. Buitenhuis, A.J.; Pearce, J.M. Open-source development of solar photovoltaic technology. *Energy Sustain. Dev.* **2012**, *16*, 379–388. [CrossRef]
60. Wittbrodt, B.; Pearce, J.M. 3-D printing solar photovoltaic racking in developing world. *Energy Sustain. Dev.* **2017**, *36*, 1–5. [CrossRef]
61. Denholm, P.; Margolis, R.M. *Impacts of Array Configuration on Land-Use Requirements for Large-Scale Photovoltaic Deployment in the United States (No. NREL/CP-670-42971)*; National Renewable Energy Lab (NREL): Golden, CO, USA, 2008.

62. Perna, A.; Grubbs, E.K.; Agrawal, R.; Bermel, P. Design Considerations for Agrophotovoltaic Systems: Maintaining PV Area with Increased Crop Yield. In Proceedings of the 2019 IEEE 46th Photovoltaic Specialists Conference (PVSC), Chicago, IL, USA, 19 June 2019; pp. 0668–0672.

63. Prehoda, E.; Pearce, J.M.; Schelly, C. Policies to overcome barriers for renewable energy distributed generation: A Case study of utility structure and regulatory regimes in Michigan. *Energies* **2019**, *12*, 674. [CrossRef]

64. Corwin, S.; Johnson, T.L. The role of local governments in the development of China's solar photovoltaic industry. *Energy Policy* **2019**, *130*, 283–293. [CrossRef]

65. Mulvaney, K.K.; Woodson, P.; Prokopy, L.S. A tale of three counties: Understanding wind development in the rural Midwestern United States. *Energy Policy* **2013**, *56*, 322–330. [CrossRef]

66. Groth, T.M.; Vogt, C.A. Rural wind farm development: Social, environmental and economic features important to local residents. *Renew. Energy* **2014**, *63*, 1–8. [CrossRef]

67. Schelly, C. Residential solar electricity adoption: What motivates, and what matters? A case study of early adopters. *Energy Res. Soc. Sci.* **2014**, *2*, 183–191. [CrossRef]

 *agronomy*

Article

Modeling of Stochastic Temperature and Heat Stress Directly Underneath Agrivoltaic Conditions with *Orthosiphon Stamineus* Crop Cultivation

Noor Fadzlinda Othman [1,2], Mohammad Effendy Yaacob [2,3,4,*], Ahmad Suhaizi Mat Su [1], Juju Nakasha Jaafar [1], Hashim Hizam [4,5], Mohd Fairuz Shahidan [6], Ahmad Hakiim Jamaluddin [2], Guangnan Chen [7] and Adam Jalaludin [8]

[1]   Department of Agriculture Technology, Faculty of Agriculture, Universiti Putra Malaysia, Serdang, Selangor 43400, Malaysia; fadzlin013@gmail.com (N.F.O.); asuhaizi@upm.edu.my (A.S.M.S.); jujunakasha@upm.edu.my (J.N.J.)

[2]   Hybrid Agrivoltaic Systems Showcase (HAVs) eDU-PARK, Universiti Putra Malaysia, Serdang, Selangor 43400, Malaysia; ahmadhakiimjamaluddin@gmail.com

[3]   Department of Process & Food Engineering, Faculty of Engineering, Universiti Putra Malaysia, Serdang, Selangor 43400, Malaysia

[4]   Centre for Advanced Lightning, Power and Energy Research (ALPER), Universiti Putra Malaysia, Serdang, Selangor 43400, Malaysia; hhizam@upm.edu.my

[5]   Department of Electrical & Electronics Engineering, Faculty of Engineering, Universiti Putra Malaysia, Serdang, Selangor 43400, Malaysia

[6]   Department of Landscape Architecture, Faculty of Design and Architecture, Universiti Putra Malaysia, Serdang, Selangor 43400, Malaysia; mohdfairuz@upm.edu.my

[7]   Faculty of Health, Engineering and Sciences, University of Southern Queensland, Toowoomba, QLD 4350, Australia; Guangnan.Chen@usq.edu.au

[8]   Department of Agriculture and Fisheries, Agri-Science Queensland, Leslie Research Facility, 13 Holberton Street, Toowoomba, QLD 4350, Australia; adam.jalaludin@gmail.com

*   Correspondence: fendyupm@gmail.com

Received: 28 July 2020; Accepted: 9 September 2020; Published: 25 September 2020

Abstract: This paper presents the field measured data of the ambient temperature profile and the heat stress occurrences directly underneath ground-mounted solar photovoltaic (PV) arrays (monocrystalline-based), focusing on different temperature levels. A previous study has shown that a 1 °C increase in PV cell temperature results in a reduction of 0.5% in energy conversion efficiency; thus, the temperature factor is critical, especially to solar farm operators. The transpiration process also plays an important role in the cooling of green plants where, on average, it could dissipate a significant amount of the total solar energy absorbed by the leaves, making it a good natural cooling mechanism. It was found from this work that the PV system's bottom surface temperature was the main source of dissipated heat, as shown in the thermal images recorded at 5-min intervals at three sampling times. A statistical analysis further showed that the thermal correlation for the transpiration process and heat stress occurrences between the PV system's bottom surface and plant height will be an important factor for large scale plant cultivation in agrivoltaic farms.

Keywords: transpiration; PV heat conversion; plant heat stress; agrivoltaic system; sustainable integration; thermal analysis

---

1. Introduction

Dramatic changes and increasing public interest in solar photovoltaic (PV) landscapes show that the dual beneficial use of land may have better impacts on energy production and future agriculture transdisciplinary design. Some highlights and recent research in solar PV projects by higher education institutions show that the solar industry has broadened its stakeholders and interest in the future, reflecting a significant shift in the dynamics of the market [1,2]. The PV industry for large scale solar projects is dominated by energy companies but, based on the effort above, it is shown that experts in higher education within the research environment have the capabilities to compete with energy companies in the solar PV industry. This trend has been transferred to ecological efficiency and positive effects, consequently upscaling the number and size of PV systems installed on the land. Rapidly decreasing price of PV modules in the world market in line with the increasing demand of fresh produce promotes the idea of agro-PV integration, commonly known as an agrivoltaic system.

This type of solar power system is a power generation system that incorporates several parts, namely PV modules, solar inverters, mounting, cabling and other electrical components, which are integrated in the balance of systems (BOS) [3,4]. This PV device absorbs rays from sunlight and translates them into a direct current (DC) via semiconductor materials. Malaysia, a tropical country in Southeast Asia, has given years of commitment to culturing green initiatives, especially PV systems and applications. This statement is evidenced by the increasing quota specifically for large scale solar (LSS) PV systems and the commitment by the Ministry of Energy, Science, Technology, Environment and Climate Change (MESTECC) [5] to persistently aim for a 20% energy mix by the year 2025 with multiple initiatives [6].

Generally, based on PV projects in University Putra Malaysia, where the size and ground conditions are put into a factor that generates empty areas under the panels, 1 kWp solar PV arrays may occupy roughly 8 to 12 square meters of land [7,8]. Based on their high demand, solar PV models in the market nowadays are ground-mounted arrays and require a fixed PV panel arrangement. There is a call for futuristic features from the market, with application in large-scale areas by enhancing their design while maintaining cost-effective deployment [9]. Temperature plays an important role in DC generation via PV modules. Park et al. [10], in their research on building-integrated PV (BIPV), defined such significant effects of the PV module's thermal characteristics, where approximately a 0.5% reduction in energy is generated based on a 1 °C increase of the module temperature. This statement is supported by Kim et al. [11], with additional information on the energy efficiency from a common PV module that can be increased due to a drop in surface temperature, especially on the highest heated portions of PV cells and ribbons.

The concept of agrivoltaics, or solar farming, aspired to creatively convert agriculture to photovoltaics, applied on the same land to maximize the yield [12]. The agrivoltaic system, as shown in Figure 1, contemplates specific plant attributes: height, productivity, water consumption and shading resistance. The figure demonstrates the idea of the agrivoltaic method employed in several countries by plotting vacant land with various types of crops. This method of farming under the solar panel is an innovation of incorporating green energy into agriculture and it is a part of introducing modern aspects to the agricultural community [13]. Some of the published results in [9,12–14] relating to agrivoltaic projects summarized the importance and successful integration of the systems by assessing whether:

- The AV system improved environmental efficiency.
- The AV system promoted effective usage of light and space for concurrent energy and food output.
- The AV system boosted the technological capacity for PV and agricultural production conjointly by implementing a hybrid simulation model.
- The AV system yielded more crop as compared to the period before the deployment.

This integrated system will maximize crop production, enhancing the system's performance while addressing land management and sustainability issues. The integration of these two resources would optimize the yield, improve clean system efficiency and solve the issue of land resource sustainability. The issue of the agrivoltaic concept implemented in ground-mounted PV systems and the shading effect of the PV arrays on crop canopy have been discussed by [15] recently.

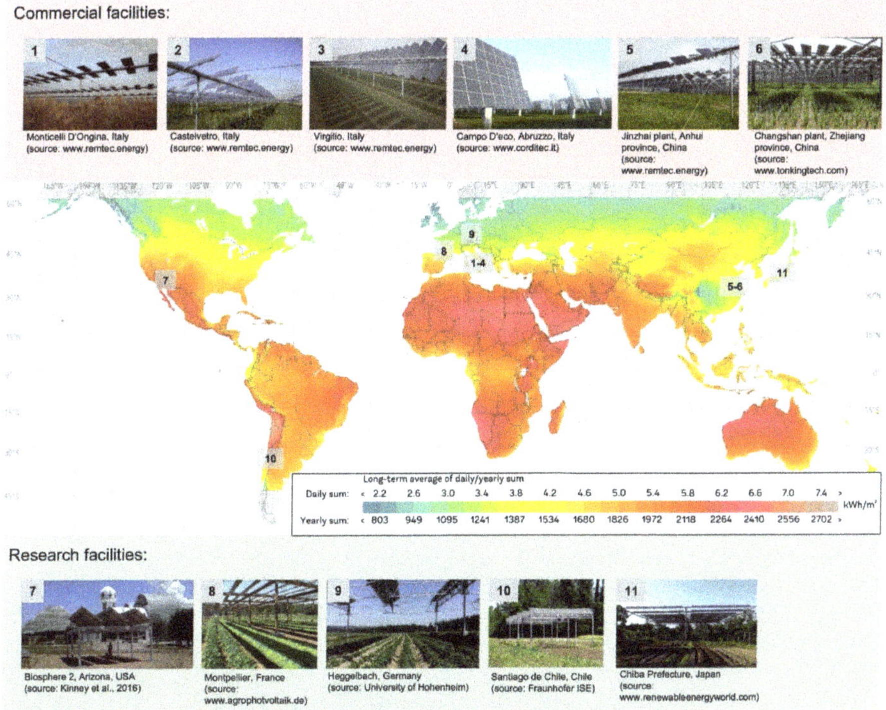

Figure 1. Typical agrivoltaic research facilities worldwide. [15].

The group suggested that the density of the PV arrays should be reduced adequately to enable ample amounts of light penetration while also maintaining a respectable production of DC electricity. The concept of agrivoltaics is in line with the Kyoto Protocol [16] and the United Nations Sustainable Development Goals (UN-SDG) [17,18], which promote the usage of clean and affordable energy towards sustainable urban infrastructure and further reducing the usage of fossil fuels.

In Malaysia, most planned and retrofitted agrivoltaic facilities are based on existing ground-mounted solar PV farm infrastructures where the primary activity is to sell the electricity generated to the National Grid. The issue of ground-mounted photovoltaic systems can be explained based on several factors, namely:

- The fact that existing solar PV farms do not allow any intervention or disturbance to any wiring, operation, structure or subsurface of the PV systems.
- The difficulties and hazards for farmers working under PV arrays result in lower production yield.
- Semi-confined working spaces, as workers have to bend down and inspect plants under PV array structures for growth monitoring and harvesting activities.
- The need for some tools to ease the process of planting, harvesting and post-harvest under agrivoltaic farming (most crop yields four cycle harvest per annum).

Heat stress normally occurs when temperatures rise above a certain level for a certain period and bear deleterious and permanent effects on a crop cycle, thus affecting yield [19,20]. Generally, heat stress is set to occur when a transient temperature rises over the average temperature of 10–15 °C [20–25]. The degree to which it happens in a particular climate zone relies on the frequency and amount of extreme temperatures happening during the day and/or the night. Some general definitions by [20] have also discussed the tendency of plants to grow with good economic yield under high temperature conditions. The extent to which this occurs in specific climatic zones depends on the probability and period of high temperatures occurring during the day and/or the night.

The transpiration process plays an important role in the cooling of green plants where, on average, it could dissipate around 32.9% of the total solar energy absorbed by the leaves, making it a good natural cooling mechanism [26–28]. However, the magnitude of its impact varies from species to species. Increased transpiration levels do have an impact on water stress because the increase in ambient temperature increases the water evaporation from ground soil, thus, some plants have a tendency to grow slowly or even die at an early stage. *Orthosiphon stamineus* was chosen as the herbal plant for a project where, based on field evaluation (40 days under tropical climate), remarkably, the crop proved growth sustainability [29]. Compared to the four other types of herbal plants in the assessment, *Orthosiphon stamineus* showed healthy growth and its morphological aspects were enhanced compared to the normal conditions. The roots and fresh branches showed aggressive growth, mostly due to the soil's moisture content, thus, it could be harvested on time. The method of cultivation underneath solar PV arrays used a drip fertigation system (DFS) directly to polybags, to maintain the soil's moisture level and to prevent any disturbances to the electrical cablings and trenches. This method also eased the process of harvesting and replanting under such restricted conditions.

Herbal plants tend to possess valuable bioactive chemical compound reserves with an abundance of possible applications in pharmaceutical and agrochemical industries. [30] explained the basic concept of microclimate conditions as a set of climate parameters assessed in a specified area near the surface of the planet, including a variation of temperature, light, wind intensity and relative humidity (RH), which are significant measures for habitat selection and other ecological practices. One of the critical elements calculated based on these parameters was the vapor pressure deficit (VPD), which is defined as the discrepancy between the volume of moisture in normal settings with saturated condition (VPD in a greenhouse range of 0.45 kPa to 1.25 kPa with an idle of 0.84 kPa) [30]. Leonardi, Guichard and Berlin, in [31], explained that during daylight hours, where the high VPD condition was enhanced, the transpiration rates were better for plants to grow because the VPD exerted a substantial rise of soluble solids but lowered the fruits' fresh weight and internal fluid levels. A plant's transpiration, and the correct VPD under a controlled environment, can effectively help to optimize the plant's ideal growth and plant health [32,33]. Hot and dry surrounding air under shade can produce high VPD and causes stress to the plant.

In agrivoltaic systems, plants, or crops, are one of the crucial elements that need to be considered. The transpiration process in plant growth takes place when water is biologically released from the aerial parts of the plants in the form of water vapor. During the process of transpiration, as illustrated in Figure 2, water molecules are transmitted from roots to stomata, the small pores underneath the leaves, where vaporization takes place, and the molecules are transpired through the surrounding air. The effect of vaporization increases with the number of plants being deposited under the PV panels, which results in an increased RH value.

Crawford et al., in [28], explained that extreme temperatures multiply the risk of plant damage due to the heat and, simultaneously, water shortage, which enhances the plant cooling capability, as shown in Figure 3. The increase in transpiration rate is directly correlated with the increased in stomata opening thus, this increases photosynthesis activities.

The transpiration characteristics of plants in different surrounding temperatures and relative humidities portray a significant heat dissipation value (transpirative heat transfer through leaves). In relation to this, a study by [27] in Wuxi, China, during the summer and winter seasons reflected a 55.8% and 24.3% transpiratory heat flux for each season, respectively, accounting for the total heat dissipation of the cinnamon. Temperature difference, ΔT, is a crucial factor to be analyzed in agrivoltaic conditions, especially the effect of plant height for each growth cycle. Mittler, in [35], explained that heat is one of the prominent elements in the abiotic stress effect on plant growth where, during heat stress, plants open their stomata to cool their leaves by transpiration. If the condition is prolonged or under an increasing rate, this will eventually create a greater detrimental effect on the plant's growth and productivity. Therefore, this study aims to measure the ambient temperature profile and the impacts of heat stress occurrences directly underneath ground-mounted solar PV arrays, focusing on different temperature levels.

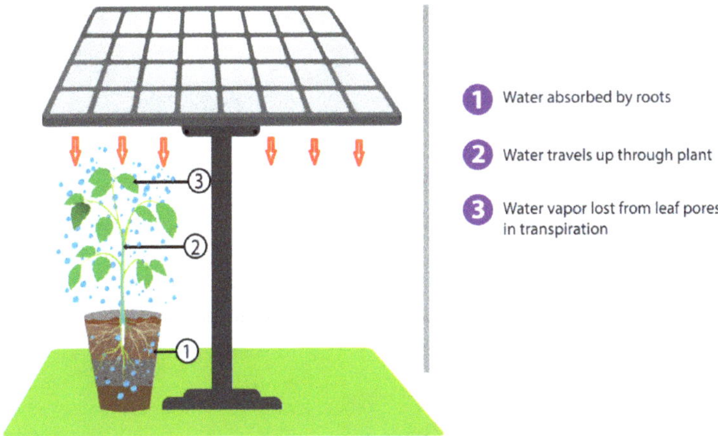

Figure 2. A simple analogy of the plant transpiration process directly underneath a photovoltaic (PV) module (heat source). Original source from [34].

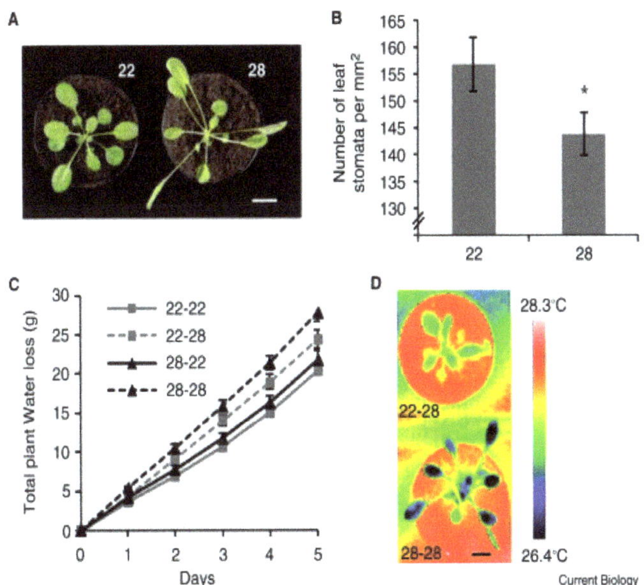

Figure 3. Crop responses at high temperatures indicate an increase in transpiration and enhanced leaf cooling capacity. (**A**) shows the plant at two different temperature levels and the thermal image of this condition is shown in (**D**), (**B**) proves the increasing number of leaves at lower temperature with respect to the lower value of water loss as shown in (**C**).

2. Methodology

This work was carried out based on a straightforward process so as to study the actual effects of temperature on planting cultivations under agrivoltaic conditions, comprising site setup, installation of sensors, data loggers, weather stations and thermal imagers, with an emphasis on the statistical analysis of the field temperature parameters.

2.1. Site Setup

The site setup was located at the Hybrid Agrivoltaic System Showcase (HAVs), Faculty of Engineering, University Putra Malaysia. A weather station was installed on site to measure the environmental factors. The location of the station was near the PV array at a 2 m height to negate any ground disturbances, whilst the PV structure height ranged from 1 m to 1.5 m. The Arduino-based data acquisition (DAQ) compartment, type-K thermo sensor and wind sensor are shown in Figure 4a.

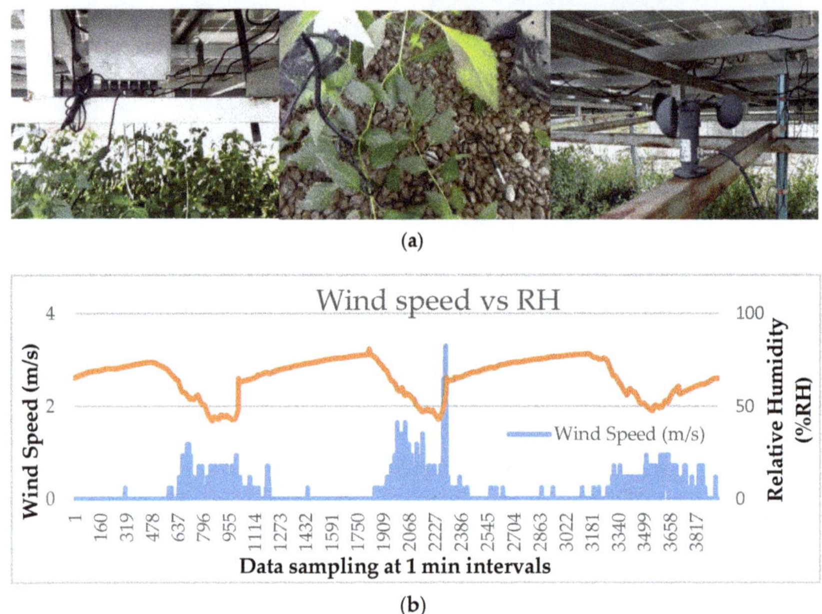

(a)

(b)

Figure 4. (a) Installation of the data acquisition (DAQ) compartment, thermo sensor and other environmental sensors; (b) Data plots for relative humidity (RH) and wind speed for agrivoltaic plots.

Based on 24 h data monitoring, shown in Figure 4b, a total of 3956 data samples were recorded for temperature value (°C), wind speed (m/s) and RH. It was observed that the average wind speed was only 0.098 m/s, due to the stagnant condition most of the time and the location of the wind sensor under the PV array (approx. 4 feet from ground level). The maximum recorded wind speed was 3.3 m/s. The maximum value for RH was 80.71%, with an average reading of 65.67% throughout the three-day duration.

The ambient temperature surrounding the plant leaves was the main component to be recorded and analyzed in this project. A Fluke thermal imager was used to record videos and images of surrounding temperatures and it was located at a 2 feet distance from the edge of the PV array, with an infrared lens focusing on the leaves (middle angle), as shown in Figure 5.

Figure 5. Agrivoltaic system with a Fluke thermal imager on a tripod for video recording.

2.2. Calculation for Vapor Pressure Density

Vapor pressure density (VPD) in kilopascals (kPa) can be measured by subtracting the actual vapor pressure of the air with the saturated vapor pressure (VP$_{sat}$ − VP$_{air}$), as shown in Equation (1).

$$\text{VPD} = \text{VP}_{sat} - \text{VP}_{air} \tag{1}$$

where:

$\text{VP}_{sat} = T_a/1000$
$\text{VP}_{air} = \text{VP}_{sat} \times \text{RH}/100$

The value for VPD was also summarized and simplified by the University of Arizona's College of Agriculture and Life Sciences [36] using their online VPD calculator, where the user only inserts the values for air temperature (T_a) and relative humidity. The information related to the microclimate for a specified location reflects the ecological processes and wildlife behavior, covering some elements of plant regeneration and growth which depict their unique spatial and temporal responses to change [37,38]. It is also a crucial measure to identify permutations in the local environment for tracking and evaluating the results of various management regimes.

Extreme high-temperature events affect the demand for atmospheric water vapor, which could be represented by the energy balance of a leaf, shown in Equation (2).

$$S_t(1 - a_\iota) + L_d - \varepsilon\sigma T_\iota^4 = \frac{pC_p\,(T_\iota - T_a)}{r_a} + \frac{pC_p(e^* - e_a)}{r(r_s - r_a)} \tag{2}$$

where:

S_t is the incoming solar radiation,
a_ι is the albedo of the leaf or canopy,
L_d is the incoming longwave radiation,
ε is the emissivity of the leaf or canopy,
σ is the Stefan–Boltzmann constant,

T_l is the leaf canopy temperature,

T_a is the ambient temperature,

p is the density of dry air,

C_p is the volumetric heat capacity of dry air,

r_a is the aerodynamic conductance,

r_s is the canopy conductance,

e^* is the saturation vapor pressure,

e_a is the saturation ambient pressure.

Saturation vapor pressure (e^*) is exponentially relative to air temperature, thus, the changing of the e^* value would affect the energy balance. Based on this correlation, an increase in VPD causes more water to be transpired by a leaf, leading to a reduction in photosynthesis [39].

Thermal images using the Fluke device are shown in Figure 6, where all the thermal images were taken using the same device and the same PV panel arrangements at different times of shooting (Figure 6 shows the thermal conditions at 11 a.m. and 3 p.m.). The images show a much higher temperature below the PV panels, which was reflected in the surrounding temperature condition and in the scope directly underneath the PV panels. A sample video clip of the thermal conditions underneath the PV array is enclosed with the document.

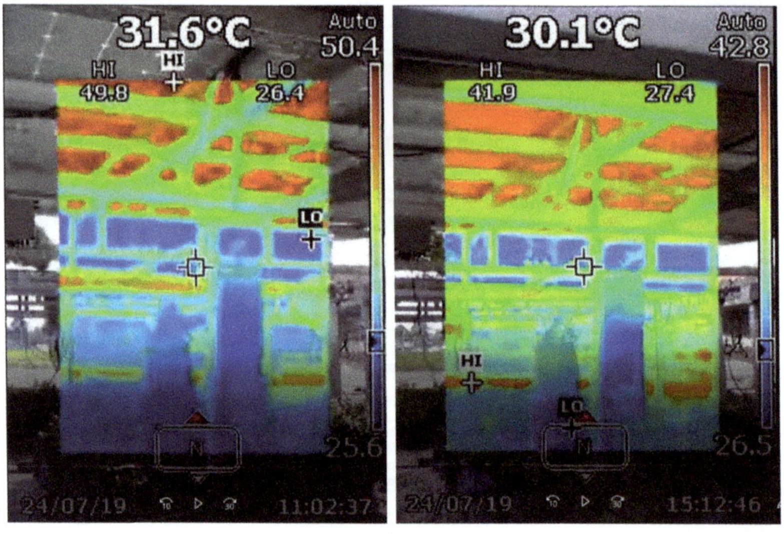

Figure 6. Thermal images of the agrivoltaic conditions for hourly sampling at the University Putra Malaysia (UPM) site.

The thermal imager provided some insight into the temperature under agrivoltaic conditions, although the readings might not be too precise because they only showed one spot value at a time. Figure 6 and Video S1 show the temperature values at different locations, i.e., below the PV panel, the surrounding air underneath PV, the surrounding air at plant level, around the leaves and the ground surface temperature taken randomly at different times (5-min intervals). Assumptions were made for the temperature values at each location and level based on the color indicator on the right side.

3. Results and Discussion

The contribution from this work can be shown in the temperature elements plotted in Figure 7, where the actual temperature pattern for six different heights under agrivoltaic conditions is portrayed, using 3600 data samples for five consecutive days from 7 a.m. to 7 p.m., daily. Each temperature value came from a thermal sensor (Type K: DS18B20, Maxim Integrated, San Jose, CA, US), starting from T_g, which was the ground surface temperature, up to the bottom of the PV array, ($T_{b,pv}$) which was directly glued to the PV array's bottom surface. The other four temperature locations ($T_{1ft,2ft,3ft,4ft}$) were based on readings from a hanging sensor to measure the surrounding air temperature.

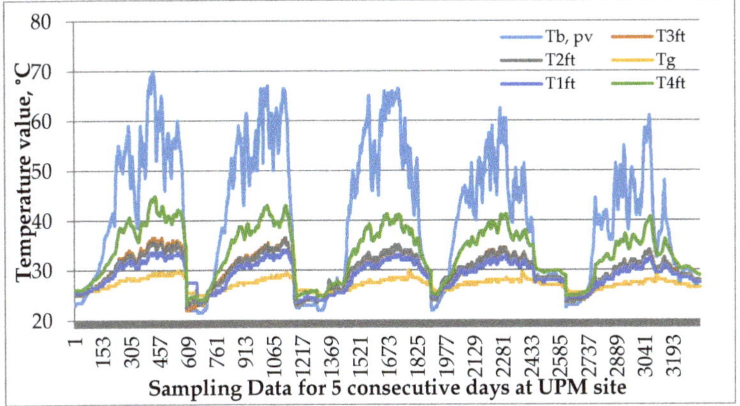

Figure 7. Temperature trends under agrivoltaic conditions at 1 min intervals (12 h daily). Abbreviations: T_g: Ground temperature; $T_{1ft,2ft,3ft,4ft}$: Temperature at 1 foot intervals; $T_{b,pv}$: PV panel's bottom surface temperature.

Based on the temperature values in Table 1, the maximum recorded temperature for T_{1ft}, T_{2ft} and T_{3ft} was 34 °C, 36.5 °C and 36.5 °C, respectively, where, at this height, the plant started growing under agrivoltaic conditions. The value for ΔT_{max} was increasing with the plant height–temperature difference (1–2 feet) ranging below 3 °C. The ground temperature (T_g) was considered as the reference value based on its effect on plant seedlings, and T_b (the bottom surface of PV module) as the maximum plant height. Hatfield and Prueger [39] explained that the rate of plant growth and development is heavily dependent on the surrounding temperature (min, max and optimum temperature values) and the annual temperature increment due to global warming over the next 50 years is likely to reach 1.5 °C between 2030 and 2052 [40].

Table 1. Values of temperature difference, ΔT, (in °C) based on a 1 foot height distribution.

	T_g	T_{1ft}	T_{2ft}	T_{3ft}	T_{4ft}	$T_{b,pv}$
Average	27.14	29.06	29.78	29.83	33.47	40.97
Max	30	34	36.5	36.5	44.88	70
Min	25	23.5	22.5	22	23.33	21.5
ΔT_{ave}	1.92	2.64	2.69	6.33	13.83	
ΔT_{max}	4	6.5	6.5	14.88	40	
ΔT_{min}	−1.5	−2.5	−3	−1.67	−3.5	

Abbreviations: T_g: Ground temperature; $T_{1ft,2ft,3ft,4ft}$: Temperature at 1 foot intervals; $T_{b,pv}$: PV panel's bottom surface temperature.

Based on Equation (1) and an online calculator software, the values for VPD are summarized in Table 2. The value for T_{1ft} was used to represent the designated surrounding air temperature (T_a) because the location was at par with the plant at a 1 foot height and touching the polybags and soil.

Table 2. Vapor pressure density (VPD) calculations based on 1 foot height under agrivoltaic conditions.

Reading	Ta (°C)	% RH	SVP (kPa)	VP (kPa)	VPD (kPa)
Average Value	27.24	70.36	3.618	2.546	1.072
Max Value	34	89.7	5.324	4.776	2.005
Min Value	23.5	30.77	2.897	0.891	0.548

Abbreviations: T_a: Ambient temperature; % RH: Relative humidity; SVP: Saturated vapor pressure; VP: Vapor pressure; VPD: Vapor pressure density.

The optimum value for VPD under a greenhouse condition ranges from 0.45 kPa to 1.25 kPa, ideally sitting at around 0.85 kPa [31]. For agrivoltaic conditions, the VPD value ranged between 2.005 kPa (max) to 0.548 kPa (min), with an average value of 1.072 kPa.

For the temperature analysis, the field data measured were segregated into five sampling hours (daily) with different temperature levels, as shown in Table 3.

Table 3. Analysis of temperature distributions based on sampling hours.

		Measure	Early Sun	Moderate Sun (Morning)	Peak Sun	Moderate Sun (Afternoon)	Mild Sun (Evening)
Time			7:00–8:59	9:00–10:59	11:00–14:59	15:00–16:59	17:00–18:59
	Tg (°C)	Average	25.5409	26.5125	27.9233	28.2828	26.3924
		Min	25.0000	25.5000	26.5000	25.5000	25.5000
		Max	26.5000	27.5000	29.5000	30.0000	27.5000
	T1ft (°C)	Average	25.1210	27.8650	31.4333	30.7819	26.9340
		Min	23.5000	25.0000	28.0000	24.0000	23.5000
		Max	27.5000	30.5000	33.5000	34.0000	29.0000
	T2ft (°C)	Average	25.2473	28.7383	32.6763	31.6755	26.4902
		Min	23.0000	25.0000	28.5000	23.0000	22.5000
		Max	29.5000	31.5000	35.5000	36.5000	29.5000
Temperature	T3ft (°C)	Average	25.0249	28.6808	32.8779	31.8989	26.3888
		Min	23.0000	25.0000	28.5000	23.0000	22.0000
		Max	28.0000	31.5000	36.5000	36.5000	29.5000
	T4ft (°C)	Average	26.0758	32.0005	38.4923	35.7665	27.9299
		Min	24.0000	26.1000	31.9500	26.9100	23.3300
		Max	30.1600	37.6400	44.8800	43.0300	31.1700
	Tb, pv (°C)	Average	24.6806	39.9392	53.6071	42.3635	26.0416
		Min	21.5000	26.0000	35.0000	23.0000	22.0000
		Max	35.5000	55.0000	70.0000	66.5000	30.5000

Abbreviations: T_g: Ground temperature; $T_{1ft,2ft,3ft,4ft}$: Temperature at 1 foot intervals; $T_{b,\,pv}$: PV panel's bottom surface temperature.

Based on Table 3 and R programming, the heat stress contour throughout the five sampling hours was plotted as shown in Figure 8.

An illustration of heat stress occurrences in % value with respect to the 1 foot height–temperature level under agrivoltaic conditions is shown in Figure 8. These field data were further analyzed as shown in Figure 9, where dependencies on the bottom of the PV panel and at a 4 foot height can be observed.

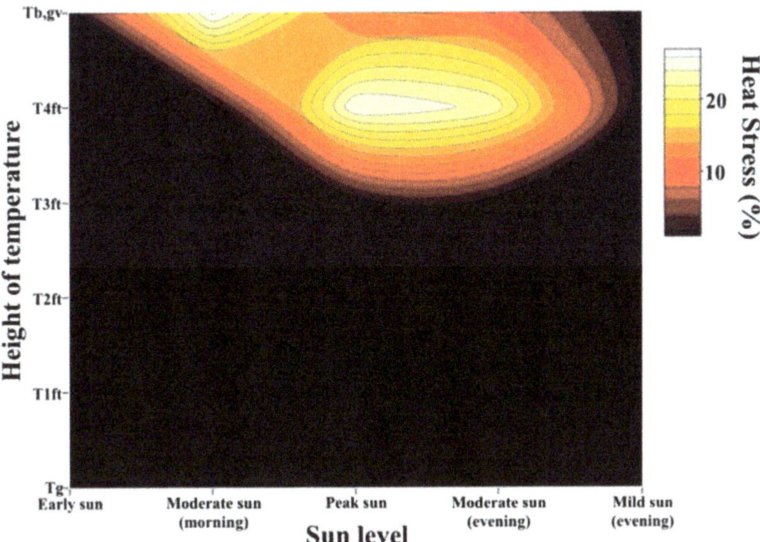

Figure 8. Heat stress occurrences (%) at five sampling hours.

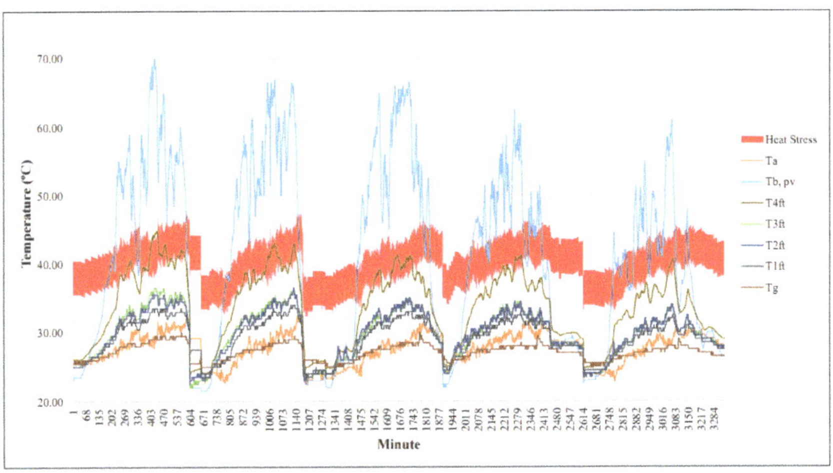

Figure 9. Field observation for heat stress directly underneath the PV arrays. Abbreviations: T_a: Ambient temperature; T_g: Ground temperature; $T_{1ft,2ft,3ft,4ft}$: Temperature at 1 foot intervals; $T_{b,pv}$: PV panel's bottom surface temperature.

Based on Figure 8, the percentage of heat stress occurrences shows at what specific time in the day the plant will possibly experience a high surrounding temperature, above the normal ambient temperature. Based on the data sample, the highest heat stress occurred at a 4 foot height during peak sun and moderate sun (afternoon), with more than 23% heat stress points, as shown in Table 4. This is due to the bottom of the PV panel producing a much higher temperature after the photonic conversion and heat dissipation process. The ground heat's effect in this agrivoltaic condition was relatively low due to the PV array shading, as per temperature values for T_g until T_{2ft}, thus, it can be assumed that no heat stress was caused by this.

Table 4. Percentages of heat stress (T_h) occurrence across sun level and height.

	Temperature Level	Early Sun	Moderate Sun (Morning)	Peak Sun	Moderate Sun (Afternoon)	Mild Sun (Evening)
Time		7:00–8:59	9:00–10:59	11:00–14:59	15:00–16:59	17:00–18:59
Percentage of Occurrence (%)	Tg	0	0	0	0	0
	T1ft	0	0	0	0	0
	T2ft	0	0	0	0	0
	T3ft	0	0	0	0	0
	T4ft	0	0	25.9167	23.2270	0
	Tb, pv	0	26.5000	10.3333	9.3972	0

A two-sample proportion test and a Chi-square test were used as the statistical approaches as shown in Tables 5 and 6, respectively.

Table 5. Count of heat stress (T_h) cases across temperature–height levels during peak sun.

		Heat Stress Status		Total
		Heat Stress	Non Heat Stress	
Temperature–Height and Sun Level	$T_{4ft_PeakSun}$	311	889	1200
	$T_{bpv_PeakSun}$	124	1076	1200

Table 6. Chi-square test for difference in proportions of heat stress (T_h) occurrence during peak sun.

	Value	df	Asymptotic Significance (2-Sided)	Asymptotic Significance (1-Sided)
Pearson Chi-Square	98.184	1	0.000	0.000

Based on the Chi-square test, T_{4ft} had a higher percentage of heat stress occurrence than $T_{b,pv}$ during peak sun at 99% confidence level ($p < 0.00001$). The same test was conducted for height level during moderate sun (afternoon) and these results also proved that T_{4ft} had a higher percentage of heat stress occurrence than $T_{b,pv}$ during moderate sun (afternoon) at 99% confidence level ($p < 0.00001$).

Based on the correlations of $T_{b,pv}$ and T_{4ft} towards heat stress (T_h) under agrivoltaic conditions, a summary of the findings of both the minimum and maximum values of heat stress, $T_{h,min}$ and $T_{h,max}$, is modelled as shown in Table 7. Some preliminary assessments were conducted to assess the fitness of data for regression modelling and the findings are displayed in Figures S1a–d, S2a–d, S3a–b, S4a–b and Table S7. Since all assumptions were fulfilled, regression models were developed and detailed findings are presented in Table S1–S6 which were simplified into Tables 7 and 8. The coefficient of determination (R squared) was 0.739, which indicates that 73.9% of the variation in $T_{h,min}$ and $T_{h,max}$ could be explained by the variation in both $T_{b,pv}$ and T_{4ft}, and both the $T_{h,min}$ and $T_{h,max}$ models were significantly fit at a 99% confidence level ($F = 4724.462$, p-value < 0.001).

Table 7. Regression statistics and analysis of variance (ANOVA).

	df	SS	MS	F	Significance F
Regression	2	13,722.334	6861.167	4724.462	0.000
Residual	3332	4838.944	1.452		
Total	3334	18,561.278			

Multiple R = 0.860; R Square = 0.739; Adjusted R Square = 0.739; Standard Error = 1.205; Observation Counts = 3335.

Table 8. Individual t-test on independent variables.

	Coefficients	Standard Error	t Stat	*p*-Value	Lower 95%	Upper 95%
$T_{h,max}$ Intercept	21.553	0.232	93.023	0.000	21.098	22.007
$T_{h,min}$ Intercept	16.553	0.232	71.442	0.000	16.098	17.007
$T_{b,pv}$	−0.293	0.005	−57.141	0.000	−0.303	−0.283
T_{4ft}	0.987	0.013	78.155	0.000	0.962	1.011

A t-test on independent variables, as shown in Table 8, confirmed that both $T_{b,pv}$ and T_{4ft} significantly affected the $T_{h,min}$ and $T_{h,max}$ at 99% confidence level ($t_{Tb,pv} = -57.141$, $t_{T4ft} = 78.155$; *p*-value < 0.001). Hence, both were significant predictors of $T_{h,min}$ and $T_{h,max}$. Meanwhile, a unit increase of $T_{b,pv}$, $T_{h,min}$ and $T_{h,max}$ would decrease by 0.293 °C, and a unit increase in T_{4ft} would increase $T_{h,min}$ and $T_{h,max}$ by 0.987 °C.

$T_{h,min}$ and $T_{h,max}$ could be expressed by the following new equations:

$$T_{h,min} = 16.553 - 0.293T_{b,pv} + 0.987T_{4ft} \tag{3}$$

$$T_{h,max} = 21.553 - 0.293T_{b,pv} + 0.987T_{4ft} \tag{4}$$

Or both equations could be simplified into a heat stress temperature model:

$$T_h \text{ (Heat stress temperature)} = [16.553, 21.553] - 0.293T_{b,pv} + 0.987T_{4ft} \tag{5}$$

4. Conclusions

As a major source of renewable energy, many photovoltaic farms have now been constructed in the world. The agrivoltaic system is a further concept that aims to combine commercial agriculture and photovoltaic electricity generation in the same space, in order to maximize crop production while addressing land management and sustainability issues.

This paper has presented the field measured data of ambient temperature profile and the heat stress occurring directly underneath solar photovoltaic (PV) arrays (monocrystalline-based) in a tropical climate condition (in Malaysia). With reference to the plant heat stress at 10 °C to 15 °C above the ambient temperature, the percentage of heat stress occurrences was the highest at a 4 foot height during peak sun and moderate sun (afternoon), with more than 23% heat stress points. It has also been found that the ground heat effect in this agrivoltaic condition was relatively low due to the PV array shading. A heat stress model for ground-mounted agrivoltaic conditions has been developed. It has been found that the coefficient of determination (R squared) for the model is 0.739, indicating that 73.9% of variation in $T_{h,min}$ and $T_{h,max}$ could be explained by the variations in both T_b, pv and T_{4ft}. Both $T_{h,min}$ and $T_{h,max}$ models were significantly fit at 99% confidence level. This paper has contributed to the understanding of plant physiological processes in response to environmental conversion factors. The model developed could also be used for further exploring the integration of crop cultivation and PV energy generation for optimum land use.

Supplementary Materials: The following are available online at http://www.mdpi.com/2073-4395/10/10/1472/s1, Table S1: Regression statistics of $T_{h,min}$ Model, Table S2: Analysis of variance (ANOVA) $T_{h,min}$ Model, Table S3: Individual t-test on independent variable $T_{h,min}$ Model, Table S4: Regression statistics $T_{h,max}$ Model, Table S5: Analysis of variance (ANOVA) $T_{h,max}$ Model, Table S6: Individual t-test on independent variable $T_{h,max}$ Model, Table S7: Variance inflation factor (VIF) for all independent variables, Figure S1: Boxplots for outlier detection. (a) Tb,pv; (b) T4ft; (c) Th,min; (d) Th,max, Figure S2: Scatter plots between dependent and independent variables for linearity. (a) Tb,pv against Th,min; (b) T4ft against Th,min; (c) Tb,pv against Th,max; (d) T4ft against Th,max, Figure S3: Normal QQ plot for residuals for normality; (a) Th,min model; (b) Th,max model, Figure S4: Residuals against fitted values plots for homoscedasticity; (a) Th,min model; (b) Th,max model.

Author Contributions: Conceptualization, N.F.O., M.E.Y., and A.S.M.S.; Methodology, N.F.O.; Software, A.H.J., N.F.O.; Validation, formal analysis, and investigation, N.F.O., M.E.Y., A.S.M.S., and A.H.J.; Resources, M.E.Y. and A.S.M.S.; Data curation, N.F.O. and A.H.J.; Writing—original draft preparation, N.F.O. and M.E.Y.; Writing—review and editing, A.S.M.S., J.N.J., H.H., M.F.S., G.C., and A.J.; Visualization, N.F.O.; Supervision,

M.E.Y. and A.S.M.S.; Project administration, N.F.O.; Funding acquisition, M.E.Y. and A.S.M.S. All authors have read and agreed to the published version of the manuscript.

Funding: This research was funded by the Ministry of Energy, Science, Technology, Environment and Climate Change (MESTECC) under the MESITA (Malaysia Energy Supply Industry Trust Account) Research Fund (Vote no. 6300921), and the Research Management Center (RMC), University Putra Malaysia, for the approval of research funding under the IPS Putra Grants Scheme (Vote no. 9667400).

Acknowledgments: The authors delegate our thanks to the Ministry of Energy, Science, Technology, Environment and Climate Change (MESTECC) under the MESITA (Malaysia Energy Supply Industry Trust Account) Research Fund (Vote no. 6300921) and the Research Management Center (RMC), University Putra Malaysia, for the approval of research funding under the IPS Putra Grants Scheme (Vote no. 9667400).

Conflicts of Interest: The authors declare no conflict of interest.

References

1. Australia's Evolving Energy Future. Available online: https://aibe.uq.edu.au/files/1811/AIBE_Industry_Research_Series_Energy_Final.pdf (accessed on 15 October 2019).
2. Available online: sustainability.uq.edu.au (accessed on 3 November 2019).
3. Dhere, N.G. Reliability of PV modules and balance-of-system components. In *Conference Record of the Thirty-First IEEE Photovoltaic Specialists Conference*; IEEE: Lake Buena Vista, FL, USA, 2005. [CrossRef]
4. Mason, J.E.; Fthenakis, V.M.; Hansen, T.; Kim, H.C. Energy payback and life-cycle CO_2 emissions of the BOS in an optimized 3.5MW PV installation. *Prog. Photovolt. Res. Appl.* **2006**. [CrossRef]
5. Available online: https://www.mestecc.gov.my/web/ (accessed on 1 November 2019).
6. Available online: http://inisiatif.mestecc.gov.my/core/submenu1_ms.html (accessed on 12 November 2019).
7. Othman, N.F.; Ya'Acob, M.E.; Abdul-Rahim, A.S.; Othman, M.S.; Radzi, M.A.M.; Hizam, H.; Wang, Y.D.; Ya'acob, A.M.; Jaafar, H.Z.E. Embracing new agriculture commodity through integration of Java Tea as high Value Herbal crops in solar PV farms. *J. Clean. Prod.* **2015**, *91*, 71–77. [CrossRef]
8. Ya'acob, M.E.; Hizam, H.; Htay, M.T.; Radzi, M.A.M.; Khatib, T. Calculating electrical and thermal characteristics of multiple PV array configurations installed in the tropics. *Energy Convers. Manag.* **2013**, *75*, 418–424. [CrossRef]
9. Scognamiglio, A. Photovoltaic landscapes: Design and assessment. A critical review for a new transdisciplinary design vision. *Renew. Sustain. Energy Rev.* **2016**, *55*, 629–661. [CrossRef]
10. Park, K.E.; Kang, G.H.; Kim, H.I.; Yu, G.J.; Kim, J.T. Analysis of thermal and electrical performance of semi-transparent photovoltaic (PV) module. *Energy* **2010**, *35*, 2681–2687. [CrossRef]
11. Kim, J.P.; Lim, H.; Song, J.H.; Chang, Y.J.; Jeon, C.H. Numerical analysis on the thermal characteristics of photovoltaic module with ambient temperature variation. *Sol. Energy Mater. Sol. Cells* **2011**, *95*, 404–407. [CrossRef]
12. Dupraz, C.; Marrou, H.; Talbot, G.; Dufour, L.; Nogier, A.; Ferard, Y. Combining solar photovoltaic panels and food crops for optimising land use: Towards new agrivoltaic schemes. *Renew. Energy* **2011**, *36*, 2725–2732. [CrossRef]
13. Dinesh, H.; Pearce, J.M. The potential of agrivoltaic systems. *Renew. Sustain. Energy Rev.* **2016**, *54*, 299–308. [CrossRef]
14. Malu, P.R.; Sharma, U.S.; Pearce, J.M. Agrivoltaic potential on grape farms in India. *Sustain. Energy Technol. Assess.* **2017**, *23*, 104–110. [CrossRef]
15. Weselek, A.; Ehmann, A.; Zikeli, S.; Lewandowski, I.; Schindele, S.; Högy, P. Agrophotovoltaic systems: Applications, challenges, and opportunities. A review. *Agron. Sustain. Dev.* **2019**, *39*, 35. [CrossRef]
16. Ki-moon, B. Kyoto Protocol Reference Manual. *USA Framew. Conv. Clim. Chang.* **2008**. [CrossRef]
17. Franco, I.B.; Power, C.; Whereat, J. SDG 7 Affordable and Clean Energy. In *Actioning the Global Goals for Local Impact 2020*; Springer: Singapore, 2020; pp. 105–116.
18. United Nations. SDG GOALS 11: Make cities inclusive, safe, resilient and sustainable. *Sustain. Dev. Goals* **2018**.
19. Hall, A.E. *Crop Responses to Environment*; CRC Press LLC: Boca Raton, FL, USA, 2001.
20. Wahid, A.; Gelani, S.; Ashraf, M.; Foolad, M.R. Heat tolerance in plants: An overview. *Environ. Exp. Bot.* **2007**, *61*, 199–223. [CrossRef]
21. Bajguz, A. Brassinosteroid enhanced the level of abscisic acid in Chlorella vulgaris subjected to short-term heat stress. *J. Plant Physiol.* **2009**, *166*, 882–886. [CrossRef] [PubMed]

22. Goswami, A.; Banerjee, R.; Raha, S. Mechanisms of plant adaptation/memory in rice seedlings under arsenic and heat stress: Expression of heat-shock protein gene HSP70. *AoB Plants* **2010**, *2010*, 1–9. [CrossRef]
23. Maya, M.A.; Matsubara, Y. Influence of arbuscular mycorrhiza on the growth and antioxidative activity in cyclamen under heat stress. *Mycorrhiza* **2013**, *23*, 381–390. [CrossRef]
24. Walter, M.H. The induction of phenylpropanoid biosynthetic enzymes by ultraviolet light or fungal elicitor in cultured parsley cells is overriden by a heat-shock treatment. *Planta* **1989**, *177*, 1–8. [CrossRef]
25. Carvalho, L.C.; Coito, J.L.; Colaço, S.; Sangiogo, M.; Amâncio, S. Heat stress in grapevine: The pros and cons of acclimation. *Plant. Cell Environ.* **2015**, *38*, 777–789. [CrossRef]
26. Gates, D.M. Transpiration and Leaf Temperature. *Annu. Rev. Plant Physiol.* **1968**. [CrossRef]
27. Xu, K.; Zheng, C.; Ye, H. The transpiration characteristics and heat dissipation analysis of natural leaves grown in different climatic environments. *Heat Mass Transf.* **2020**. [CrossRef]
28. Crawford, A.J.; McLachlan, D.H.; Hetherington, A.M.; Franklin, K.A. High temperature exposure increases plant cooling capacity. *Curr. Biol.* **2012**. [CrossRef] [PubMed]
29. Othman, N.F.; Ya'acob, M.E.; Abdul-Rahim, A.S.; Othman, M.S.; Ramlan, M.F.; Stanslas, J. Morphological Analysis and Sustainability of Four Types of Herbal Plants Under Fix Solar PV Panel Structure in Malaysia. *Acad. J. Sci.* **2015**, 183–190.
30. Chen, J.; Saunders, S.C.; Crow, T.R.; Naiman, R.J.; Brosofske, K.D.; Mroz, G.D.; Brookshire, B.L.; Franklin, J.F. In Forest Microclimate and Ecosystem Ecology Landscape Variations in local climate can be used to monitor and compare the effects of different management regimes. *Bioscience* **1999**, 49288–49297. [CrossRef]
31. Leonardi, C.; Guichard, S.; Bertin, N. High vapour pressure deficit influences growth, transpiration and quality of tomato fruits. *Sci. Hortic. Amst.* **2000**, *84*, 285–296. [CrossRef]
32. Gholipoor, M.; Prasad, P.V.V.; Mutava, R.N.; Sinclair, T.R. Genetic variability of transpiration response to vapor pressure deficit among sorghum genotypes. *Field. Crop. Res.* **2010**, *119*, 85–90. [CrossRef]
33. Ryan, A.C.; Dodd, I.C.; Rothwell, S.A.; Jones, R.; Tardieu, F.; Draye, X.; Davies, W.J. Gravimetric phenotyping of whole plant transpiration responses to atmospheric vapour pressure deficit identifies genotypic variation in water use efficiency. *Plant Sci.* **2016**, *251*, 101–109. [CrossRef]
34. Available online: https://byjus.com/ (accessed on 14 January 2020).
35. Mittler, R. Abiotic stress, the field environment and stress combination. *Trends Plant Sci.* **2006**, *11*, 15–19. [CrossRef]
36. Available online: https://cals.arizona.edu/vpdcalc/ (accessed on 13 January 2020).
37. Rosenberg, N.J.; Blad, B.L.; Verma, S.B. *Microclimate: The Biological Environment*; Wiley: New York, NY, USA, 1983.
38. Sutherst, R.W.; Maywald, G.F. A computerised system for matching climates in ecology. *Agric. Ecosyst. Environ.* **1985**, *13*, 281–299. [CrossRef]
39. Hatfield, J.L.; Prueger, J.H. Temperature extremes: Effect on plant growth and development. *Weather Clim. Extrem.* **2015**, *10*, 4–10. [CrossRef]
40. IPCC. Global Warming of 1.5 °C. 2018. Available online: https://www.ipcc.ch/sr15/ (accessed on 2 February 2020).

MDPI
St. Alban-Anlage 66
4052 Basel
Switzerland
Tel. +41 61 683 77 34
Fax +41 61 302 89 18
www.mdpi.com

Agronomy Editorial Office
E-mail: agronomy@mdpi.com
www.mdpi.com/journal/agronomy